A
Critical Appraisal
of
Viral Taxonomy

Editor

R. E. F. Matthews, Ph.D.
Professor of Microbiology
University of Auckland
Auckland, New Zealand

CRC Press
Taylor & Francis Group
Boca Raton London New York

CRC Press is an imprint of the
Taylor & Francis Group, an **informa** business

First published 1983 by CRC Press
Taylor & Francis Group
6000 Broken Sound Parkway NW, Suite 300
Boca Raton, FL 33487-2742

Reissued 2018 by CRC Press

© 1983 by CRC Press, Inc.
CRC Press is an imprint of Taylor & Francis Group, an Informa business

No claim to original U.S. Government works

Library of Congress in Publication Data

Main entry under title:

A Critical appraisal of viral taxonomy.

   Bibliography: p.
   Includes index.
   1. Virology—Classification. I. Matthews,
R. E. F. (Richard Ellis Ford), 1921-
[DNLM: 1. Viruses—Classification. QW. 15 C934]
QR394.C74   1983   576'.64'012   82-17795
ISBN 0-8493-5648-2.

A Library of Congress record exists under LC control number: 82017795

Publisher's Note
The publisher has gone to great lengths to ensure the quality of this reprint but points out that some imperfections in the original copies may be apparent.

Disclaimer
The publisher has made every effort to trace copyright holders and welcomes correspondence from those they have been unable to contact.

ISBN 13: 978-1-315-89212-2 (hbk)
ISBN 13: 978-1-351-07122-2 (ebk)

Visit the Taylor & Francis Web site at http://www.taylorandfrancis.com and the
CRC Press Web site at http://www.crcpress.com

PREFACE

By the early 1960s many hundreds of viruses had been found infecting vertebrates, invertebrates, plants, and bacteria. The classification and nomenclature of these agents were in a chaotic state. At a Microbiology Congress in Moscow in 1966 the International Committee for Nomenclature of Viruses was launched. Many problems were encountered, particularly in the early years, but steady progress was made. By 1981, 54 virus families and groups, with a total of 1372 member viruses, had been delineated and had received formal international approval. In spite of this progress many problems remain for the future.

The overall aim of this volume is to review critically the current state of, and future prospects for developments in viral taxonomy. Most of the contributors have recently had substantial periods of service on the Executive Committee and subcommittees of the International Committee on Taxonomy of Viruses (ICTV) as follows:

H. -W. Ackermann (3 years as vice-chairman and 6 years as chairman of the Bacterial Virus Subcommittee); J. G. Atherton (6 years as chairman of the Code and Data subcommittee); K. W. Buck (6 Years as a member of the Fungal Virus subcommittee and 1 year as chairman); R. I. B. Francki (3 years as a member of the Plant Virus subcommittee and 6 years as chairman); J. F. Longworth (3 years as a member of Executive Committee and the Invertebrate Virus subcommittee); R. E. F. Matthews (6 years as President of the ICTV); F. A. Murphy (3 years as vice-chairman and 6 years as chairman of the Vertebrate Virus subcommittee). Chairmen of subcommittees were also *ex officio* members of the Executive Committee.

All these workers except K. W. Buck and J. F. Longworth retired from the ICTV Executive Committee in August 1981. It should be stressed that the views express in each chapter are those of the individual author. They are not necessarily shared either by the editor, or by the ICTV.

In the first chapter I have outlined the historical development of the subject, emphasizing the difficulties along the way and the extent to which these have been overcome. The chapter concludes with a summary of the state of viral taxonomy at August 1981. Chapters 2 to 6 consider past, present, and future problems that are particularly relevant for the taxonomy of viruses infecting vertebrates, plants, bacteria, invertebrates, and fungi. In Chapter 7, J. A. Dodds gives a brief account of the structure and biology of the new viruses and virus-like particles being found in eukaryotic algae and in protozoa. The taxonomy of these agents poses significant problems. Chapter 8 summarizes the progress that has been made in developing a universal system for the collection, and computer-based storage and retrieval of virus data. Last, Chapter 9 outlines my personal views on the future prospects for viral taxonomy, with particular emphasis on the difficult problems associated with the delineation and naming of virus species.

I wish to thank all the contributors for their ready cooperation in the production of this volume. I thank particularly the following colleagues who read and commented upon parts or all of the manuscripts for my own contributions: H. -W. Ackermann, A. R. Bellamy, D. W. Dye, F. Fenner, R. I. B. Francki, A. J. Gibbs, B. D. Harrison, D. W. Kingsbury, J. F. Longworth, J. Maurin, F. A. Murphy. H. G. Pereira, and P. Wildy.

THE EDITOR

Richard E. F. Matthews, Ph.D., Sc.D., is Professor Microbiology and Head of the Department of Cell Biology in the University of Auckland, Auckland, New Zealand. He received an M.Sc. degree in Botany from the University of New Zealand in 1941 and a Ph.D. from the University of Cambridge, England in 1948, followed by the Sc.D. in 1964.

He has published 130 research papers mainly in the field of plant virology, together with two books in this field. From 1974 to 1981 he was President of the International Committee for Taxonomy of Viruses and was editor of two reports for the committee published in 1979 and 1982. He was made a life member of the organization in 1981. He is a member of the Society for General Microbiology and the Biochemical Society of London. He is a Fellow of the New Zealand Institute of Chemists and the Royal Society of New Zealand. In 1974 he was elected a Fellow of the Royal Society of London.

CONTRIBUTORS

Hans-Wolfgang Ackermann, M. D.
Professor
Department of Microbiology
Faculty of Medicine
Laval University
Quebec, Canada

J. G. Atherton, Ph.D.
Reader in Virology
University of Queensland
St. Lucia, Queensland
Australia

K. W. Buck, Ph.D.
Reader in Fungal and Plant Virology
Department of Pure and Applied Biology
Imperial College of Science and
 Technology
London, England

J. Allan Dodds, Ph.D.
Associate Professor
Department of Plant Pathology
University of California
Riverside, California

R. I. B. Francki, Ph.D.
Reader in Plant Pathology
Waite Agricultural Research Institute
University of Adelaide
Glen Osmond, South Australia

I. R. Holmes
Department of Microbiology
University of Queensland
St. Lucia, Queensland
Australia

E. H. Jobbins
Department of Microbiology
University of Queensland
St. Lucia, Queensland
Australia

John F. Longworth, M. S.
Director
Entomology Division
Department of Scientific and Industrial
 Research
Auckland, New Zealand

R. E. F. Matthews, Ph.D.
Professor of Microbiology
University of Auckland
Auckland, New Zealand

F. A. Murphy, D.V.M., Ph.D.
Professor of Microbiology
College of Veterinary Medicine and
 Biomedical Sciences
Colorado State University
Ft. Collins, Colorado

TABLE OF CONTENTS

Chapter 1

THE HISTORY OF VIRAL TAXONOMY

R. E. F. Matthews

TABLE OF CONTENTS

I. INTRODUCTION

Viruses have been found in almost all the major groups of prokaryotic and eukaryotic organisms. The number of distinct viruses that have been described cannot be ascertained with any precision, but the total must be several thousand. There are over 2000 descriptions of viruses infecting bacteria alone. Man has an innate desire to classify and name all of the natural objects which he studies, and viruses are no exception. However, there are many different possible ways of classifying and naming biological objects, and judgments must be made that do not have a strictly scientific basis. Thus there is ample room for differences of opinion, emotional attachment to a particular point of view, and unwillingness to compromise. For these reasons the development of viral taxonomy has been marked by some stormy interludes. In this introductory chapter the development of viral taxonomy over the past half century or so will be outlined and the state the subject had reached at the end of the Fifth International Congress for Virology in Strasbourg, 1981, will be summarized.

The historical development can usefully be divided into four periods as follows:

1.   The period up to 1961. This period was characterized by premature efforts on the part of individual virologists to establish their particular scheme, accompanied by slow-moving and largely ineffective international cooperation.
2.   1962-1966. In this period important events took place which led up to the meetings held during the International Congress for Microbiology in Moscow in 1966, and to the establishment of the International Committee on Nomenclature of Viruses (I.C.N.V.).
3.   1966-1970. The first period of I.C.N.V., under the 4-year presidency of Wildy, was a critical one. The policy decisions made and the taxonomic work carried out laid the foundations for further development of a universal taxonomy of viruses based on international and interdisciplinary cooperation.
4.   1971-1981. This period was mainly one of development and consolidation.

II. THE PERIOD UP TO 1961

From the very beginnings of virology, names have been given to viruses as they were discovered and described. For viruses infecting animals and plants most of these names were trivial or vernacular, and commonly included an important host or an important disease symptom as part of the name. Bacterial viruses were commonly given code symbols such as T1 or C16 sometimes in a quite haphazard fashion. However, some workers in this period provided their viruses with more high-sounding names. For example d'Herelle[1] applied the name *Bacteriophagum intestinale* to a culture of a bacterial virus (or viruses) which he studied, but names such as this were seldom of any lasting significance.

As far as virus classification was concerned in this period, the custom developed of grouping viruses according to the kind of host—animal, plant, or bacterial. The reasons for this were mainly historical and practical. Most virologists worked with viruses replicating (or thought to replicate) in only one kind of host. Doubts concerning this system arose in the 1940s when it became apparent that some plant viruses could replicate in their insect vectors. During the period up to 1961 attempts were made both by individuals and by organizations to introduce more detailed systems of classification and nomenclature. I will deal first with efforts made by individual virologists. I can find no record of any phage worker attempting a comprehensive classification of bacterial viruses in this period.

A. Individual Plant Virologists

Johnson[2] made the first approach to the problem of classifying plant viruses. He was well aware of the problem. He wrote: "A system of nomenclature for plant viruses is greatly needed. The present system of applying names on the basis of host attacked or symptoms exhibited is quite inadequate for present needs." He confined his attention to a set of 11 viruses affecting tobacco and other solanaceous hosts. The separation and classification of the viruses described was based on the symptoms (if any) produced on 10 or more different host species, their longevity in vitro, thermal death points, lethal effects of chemicals, and such properties as relative infectivity and length of incubation period. He suggested that properties of the virus in vitro would be more reliable criteria than disease symptoms. He named the viruses with the English name of the host followed by a number, e.g., Tobacco virus 1, Tobacco virus 2, etc.

Using the same naming system, Johnson and Hoggan[3] compiled a descriptive key based on five characters — modes of transmission, natural or differential hosts, longevity in vitro, thermal death point, and distinctive or specific symptoms. About 50 viruses were identified and placed in groups.

Smith[4] outlined a scheme in which the known viruses or virus diseases were divided into 51 groups. Viruses were named and grouped according to the generic name of the host in which they were first found. Successive members in a group were given a number. For example, tobacco mosaic virus was *Nicotiana virus* 1, and there were 15 viruses in the *Nicotiana* virus group. Viruses that were quite unrelated in their basic properties were given names that might have been thought to imply that they were related. Although Smith's list served for a time as a useful catalogue of the known plant viruses, it was not a classification.

Holmes[5] published a classification of plant viruses based primarily on host reactions and methods of transmission. He used a Latin binomial-trinomial system of naming. For example tobacco mosaic virus became *Marmor tabaci*, Holmes. His classification was based on diseases rather than the viruses. The viruses were split into 10 families. The family *Marmoraceae* (the Mosaic group) had a single genus *Marmor* which included 53 of the 89 viruses considered as species by Holmes. This genus contained viruses known even at the time to differ widely in their properties (e.g., tobacco mosaic and tomato bushy stunt viruses). Nevertheless, the Latin binomial system appealed to a number of workers and various modified schemes were put forward.[6-11] The Holmes system received the blessing of the Nomenclature Committee of the American Phytopathological Society.[12] Subsequently Holmes[13] put forward a revised scheme which was published as a supplement in the sixth edition of *Bergey's Manual of Determinative Bacteriology*, which covered all viruses. This version still contained many groupings that were even then obviously inappropriate. These are discussed further below in relation to animal viruses. Hansen[14-16] was the most recent and probably the last individual plant virologist to propose a system for nomenclature and classification. His system for deriving generic names was based on characters of the virus particle, the diseases caused, and the methods of transmission. His system of pseudo-Latinized names would have provided the generic names for a Latinized binomial system.

While the Holmes and similar systems received some support from plant virologists particularly in the U.S., in the main they were ignored. Bawden[17] was a leading critic of the use of disease symptoms in classification. He made an important contribution by suggesting that any permanent classification must be based on the peculiarities of the viruses themselves and not on those of the host plants. He emphasized the importance of serological relationships. He pointed out the widespread existence of virus strains which although closely related on both the physical and chemical properties of the particle and serologically, may cause markedly different diseases. He stated, "it would seem as reasonable to expect to classify flowering plants because of their reactions to a number of viruses as it is to classify viruses by the symptoms they produce in a given number of hosts."

Bawden[17] and Valleau[21] made the useful, but unheeded suggestion that while well-studied viruses might be usefully classified, little-studied entities should be placed aside in a category equivalent to the Fungi Imperfecti. Nevertheless Bawden's published comments and criticisms undoubtedly helped to slow the rush of many plant virologists in the 1940s towards premature classification and nomenclature.

One of the few useful specific contributions made to plant virus taxonomy in this period was that of Brandes and Wetter.[22] They classified elongated rod-shaped plant viruses into 12 groups on the basis of particle morphology, and in particular on the "modal" length of the rods. While some of their subdivisions turned out to be too fine, seven of the groups have survived in the present taxonomy of plant viruses.

B. Individual Animal Virologists

Over the period up to 1961 animal virologists appear not to have felt the necessity to erect comprehensive systems of orderly classification and nomenclature. Would-be taxonomists were warned by Andrewes[23] who wrote, "judgment must be suspended...in the case of the invisible viruses or so-called 'filter-passing' organisms. Here our ignorance is almost complete; they are possibly a heterogeneous group but in the case of creatures which we cannot see and whose very existence is, in many cases, a matter of inference only, it is idle to talk of classification in the usual sense."

However some attempts were made and those classifications that were devised used tissue affinities as the main criterion. Thus Levaditi and Lépine[24] classified some viruses infecting vertebrates into the following eight groups, (1) Ectodermoses (foot-and-mouth disease virus, vesicular stomatitis, ectromelia, vaccinia); (2) Ecto-endodermoses (canary-pox, laryngotracheitis); (3) Mesodermo-endodermoses (fowl plague, and Newcastle disease); (4) Ectodermoses neurotropes (a number of purely neurotropic virus infections); (5) Endodermoses neurotropes (yellow fever); (6) Mesodermoses (lymphogranuloma venereum); (7) Septicémies (influenza, Rift valley fever); (8) Ultravirus neoformatif (Rous sarcoma). This and similar classifications of the period met with little favor.

As far as nomenclature is concerned Goodpasture[25] proposed a generic name *Borreliota* for the Pox group and three specific names within that genus. Apart from such occasional small-scale attempts there was little activity in nomenclature until Holmes[13] proposed his comprehensive system for all viruses. This was an expanded version of his earlier scheme for plant viruses and bacteriophages.[5]

Although the generic and specific names proposed by Holmes were often quite suitable, the scheme was almost completely ignored and rejected by vertebrate virologists. This was because the classification was so bad. Five of the six families in the suborder *Zoophagineae* were based almost entirely on tissue tropisms and symptomatology. These families grouped together viruses with widely differing properties. For example the family *Borreliotaceae* was characterized as follows, "viruses of the Pox group, inducing diseases characterized in general by discrete primary and secondary lesions of the nature of macules, papules, vesicles, or pustules." It included poxviruses, herpes viruses, and viruses of the foot-and-mouth disease virus group.

Andrewes[26] in a discussion on virus classification wrote, "Dr. Holmes is the credit for stimulating great interest in virus taxonomy and nomenclature; his action has also been invaluable — and here I am still trying to be polite — as a glaring example of how not to classify viruses."

In spite of such severe criticisms of Holmes' classification by many leading virologists, others published similar schemes or formal revisions of Holmes.[13] For example, Zhdanov[27] published a classification scheme similar to that of Holmes but omitted the names he used. Breed and Petraitis[28] gave a summary translation in English. Van Rooyen[29] published a formal revision and extension of Holmes' classification of animal viruses. These publications were almost totally ignored by other virologists.

In the period leading up to 1961 Cooper[30] was the last to suggest a basis for the classification of animal viruses. He made an important contribution by suggesting that properties of the virus particle that are inherently genetically stable should be used to make the primary groupings of viruses. His primary division was on the basis of genome nucleic acid type — DNA or RNA. The DNA viruses were further subdivided on the basis of size and then on ether sensitivity (presence of a lipoprotein membrane). The RNA viruses were further sub-divided on the basis of ether sensitivity. This use of properties of the virus particle fore-shadowed the classification proposed by Lwoff et al.[31] which is discussed in a later section.

C. Attempts at International Cooperation

I will now turn to the attempts that were made to achieve international cooperation concerning viral taxonomy in the period up to 1961. The earliest attempts were made in conjunction with International Congresses of Botany in 1930 and Microbiology in 1939.

In 1930 an appeal was made before the Fifth International Botanical Congress meeting in Cambridge, England for an internationally agreed-upon viral classification and nomen-clature.[32] A committee of five members was formed under the chairmanship of Johnson to propose a standard system of nomenclature with the objective of preventing further wide proliferation and use of synonyms.

A tentative system of nomenclature based on the name of the host with appropriate numbers to indicate the virus species and letters for strain designation was presented to phytopath-ologists at the Sixth Botanical Congress in Amsterdam in 1935. The report was adopted "in principle" by the Congress, but it was never officially published. The committee was empowered to continue working and to put forward a more complete proposal at the next Congress due to be held in Stockholm in 1940. Six additional virologists were appointed to the committee and by 1939 a 32-page mimeographed prospectus had been circulated to many virus workers for comments and suggestions.

The committee met with the International Microbiological Congress in New York City in September 1939, in the hope that some joint system of virus nomenclature could be agreed upon by animal and plant virologists attending the Congress. On the day of the meeting war was declared in Europe, bringing to an end any attempt at international cooperation con-cerning viral nomenclature.

The Seventh International Botanical Congress, planned for Stockholm in 1940, was never held. Bawden[18] criticized the apparent inactivity of Johnson's committee. Johnson[33] defended its work. He pointed out that at the time World War II began, bringing the activity of the committee to a halt, substantial progress had been made in securing world-wide adoption of its plan for the nomenclature of plant viruses. He stated, "had the committee anticipated more closely the oncoming chaos in both virus nomenclature and world politics, it might have proceeded with more speed and dictatorial methods, but perhaps with less caution and diplomacy."

In December 1939 an attempt to rescue and develop the work of Johnson's committee was made by the American Phytopathological Society. It appointed a Standing Committee on Virus Nomenclature. As Johnson[33] describes, this committee was composed mainly of individualists who cherished, and published, their own particular schemes leading to further chaos.

At the First International Congress for Microbiology held in Paris in 1930, an International Committee on Bacteriological Nomenclature was set up. At the Fourth International Congress in Copenhagen in 1947, it was decided that viruses came within the field of jurisdiction of the Microbiological Code. The following wording was authorized: "Bacteriological no-menclature considers bacteria, related organisms and the viruses."

Some important and influential recommendations were made at the Fifth International Congress of Microbiology held at Rio de Janeiro in August 1950. Before this meeting Andrewes had been asked to convene an unofficial committee which was given official status at Rio de Janeiro under the International Nomenclature Committee.[34] The committee was no doubt spurred on by the unilateral publication of Holmes[13] classification as a supplement to *Bergey's Manual of Determinative Bacteriology*. Before the meeting opinions were sought by correspondence from 120 virologists in 22 countries.[34] Over 80 replies were received from plant and animal virologists in about equal numbers. Plant virologists were evenly divided as to whether they wanted binomial nomenclature, while about two out of three animal virologists believed that a binomial system would be appropriate. However the verdict was almost unanimously against Holmes' system.

Of the eight members of Andrewes' virus subcommittee meeting at Rio de Janeiro, only one was a plant virologist. For this reason the committee was unwilling to make far-reaching recommendations concerning plant viruses. As a result of submissions made by the committee, the following resolutions were endorsed unanimously by the Plenary session of the Congress.

1. That consideration of the starting date for scientific nomenclature of viruses be deferred to the next Congress,
2. That until studies on viruses now in progress have advanced further, the use of any comprehensive system of scientific nomenclature for them is unwise. The cooperation of all virologists is requested in assisting the subcommittee now studying the nomenclature and classification of viruses.

The Virus Subcommittee considered and agreed upon a set of principles upon which the classification of animal viruses should be based. These were

1. Morphology and methods of reproduction
2. Chemical composition and physical properties
3. Immunological properties
4. Susceptibility to physical and chemical agents
5. Natural methods of transmission
6. Host, tissue, and cell tropisms
7. Pathology, including inclusion-body formation
8. Symptomatology (symptomatology is purposely placed last, as of minor importance)

The Plant Diseases section of the Congress asked the Virus Subcommittee to divide into sections to consider separately the viruses attacking animals, plants, and bacteria, these sections to deliberate together at a later date.

The Virus Subcommittee considered that a start could be made on the taxonomy of certain animal virus groups. These were

1. Psittacosis-lymphogranuloma group (Chlamydozoaceae)
2. Insect-pathogenic viruses
3. Pox group
4. Influenza group (influenza, mumps, fowl-plague)
5. Insect-borne encephalitides

Convenors for each of these five groups were appointed, and suggestions were made for small groups of working specialists. This working separation of animal, plant, and bacterial virologists and the setting up of small international groups of specialists foreshadowed the organization adopted by the International Committee for Nomenclature of Viruses some 16 years later. In spite of this activity in 1950, a proposal at the International Botanical Congress at Stockholm in 1951 that the nomenclature of plant viruses become the responsibility of microbiologists was not approved. This decision appears to have had little impact on subsequent events.

In January 1952 the New York Academy of Sciences through its section on Biology provided for an International Conference on Virus and Rickettsial Classification and Nomenclature under the chairmanship of Sir McFarlane Burnet.[35]

Most speakers agreed that a system of classification that relied heavily on symptomatology would be unsatisfactory, and that it would be wise to proceed slowly and deal at first only with those groups of viruses about which there was sufficient knowledge. This plan which became known as the 'Rio Approach' received general support. Some individual speakers foreshadowed later trends.

Mayr, a zoological taxonomist, considered that, "virology would avoid many of the nomenclatural difficulties of the zoologists if they would assign the authorship of their scientific names to national or international committees and charge them with the responsibility of issuing a list of official names."

At the meeting Black described several leafhopper-borne viruses that had been found to multiply in their insect vectors. Their existence served "to remind us that the separation of viruses into bacterial, plant and animal virus groups is arbitrary and that there exist forms that bridge the gap between two of these three major classes."

At this meeting two further study groups were set up for viruses related to poliomyelitis and for herpes simplex.

At the Sixth International Congress of Microbiology held in Rome in 1953 the Virus Subcommittee of the International Nomenclature Committee had before it the reports of its study groups. Altogether nine animal virologists and six plant virologists attended the two meetings of the subcommittee.[36-37] They endorsed the conclusion from the 1950 Congress that at present "the use of systems of classification for, and the application of binomials to viruses as a whole are undesirable and should be discouraged." They also recommended that the starting date for valid Linnaean nomenclature for viruses should not yet be determined, and that until this was done names of viruses already proposed should have no standing in bacteriological nomenclature.

However the animal virologists considered that some provisional alternatives to systems such as that of Holmes[13] should be put forward, but that such names should not prejudice future uniformity between animal and plant viruses. They suggested "non-Linnaean" binomial names for 20 well-known viruses infecting vertebrates and for 16 viruses infecting insects. The first name was to be a "group" name rather than a genus, and end in the suffix "virus". The second name would not be a species but a "group member" name. For example smallpox virus was named *Poxvirus variolae*. The laws of priority would not be applied and names would not be ascribed to authors. The main International Nomenclature Committee approved publication of the proposals.

Some of the group or generic names such as Poxvirus, Poliovirus, and Myxovirus became widely used but the species names were almost totally ignored.[38] Although the proposed "non-Linnaean" binomials never came into general use the idea of using the suffix "-virus" for generic names was later adopted by the International Committee for Nomenclature of Viruses.

An important decision at the 1953 Rome meeting concerned the Chlamydozoaceae (Chlamydiae). Up until this time these agents had been classed with the viruses, or were considered as being in a borderland between viruses and bacteria. The Chlamydozoaceae study group held that the agents related to psittacosis were closely related to the Rickettsiae and should be considered with them. This implied that these agents came under the bacterial code, and would no longer be given consideration by virologists. The report was accepted. This decision was important as it made much more evident the gap between bacteria and viruses. Lwoff[39] was able to end his Marjorie Stephenson memorial lecture on "The concept of virus" with the now well-known statement, "viruses should be considered as viruses because viruses are viruses."

Following the period of activity in the years 1950 to 1953, there appears to have been a slackening of interest in viral taxonomy and nomenclature until 1962. Some papers were published by individuals. For example Andrewes and Sneath[38] discussed the species concept among viruses, while Andrewes et al.[40] and Andrewes[41,42] surveyed the taxonomy of vertebrate viruses and proposed two new names. However there appears to have been very little progress made by the Virus Subcommittee of the International Committee on Bacteriological Nomenclature following the Rome meeting. The reasons for this are not clear. The 1950s and early 1960s were years in which the quality and availability of electron microscopes improved greatly, as did techniques of specimen preparation, and in particular the development of negative staining for studying virus structure.[43] The potential of the negative staining technique for providing basic data for taxonomic purposes was emphasized by the work of Horne and Wildy[44] and Wildy and Watson[45] who studied a series of viruses infecting vertebrates. The history of the application of this technique is given in detail by Horne and Wildy.[46] The impact of this technique for virus classification was immediate: (1) particles could be characterized with respect to size, shape, surface structure, and sometimes symmetry; (2) the method could be applied to viruses infecting all kinds of hosts; and (3) virus particles could be characterized in unpurified material, a considerable advantage for diagnostic purposes. Negative staining and other techniques gave rise to a rapid accumulation of data about the physical and chemical properties of the virions of many viruses. This may have reinforced the idea, already widespread, that to develop a sound taxonomy and nomenclature for viruses, it would be best to make haste slowly.

III. THE PERIOD LEADING UP TO THE MOSCOW CONFERENCE
(1962-1966)

This period was dominated by the activities of Lwoff. To many he appeared as a strong willed and charismatic figure. At the Cold Spring Harbor Symposium of 1962 Lwoff, Horne, and Tournier[31] proposed "a system of viruses" that could embrace all known viruses. This was a further development of the proposals put forward by Cooper.[30] It was an hierarchical system based on arbitrarily chosen characters, as follows: the first subdivision was according to type of nucleic acid — DNA or RNA; the second according to capsid symmetry — helical, cubic, or both; the third — presence or absence of an envelope; and fourth, for helical capsids diameter of the particle, and for cubic capsids the number of capsomeres. This gave rise to 16 groups of viruses. The proposals led to considerable discussion and debate right up to and after the Moscow conference.

Also in 1962, the Virus Subcommittee of the International Nomenclature Committee met in Montreal, with Andrewes as chairman. The subcommittee decided that it would be most useful to define and name major groups of viruses and not to propose any further specific names for individual viruses.[47] Thus they proposed no immediate move towards a binomial system. The major groups would be defined on the basis of (1) nucleic acid composition, (2) sensitivity to ether, (3) presence of a limiting membrane, (4) symmetry whether cubic or helical, and (5) number of capsomers. Thus the criteria used were virtually the same as those used by Lwoff et al.[31] but the objective was much more limited. It was to delineate a few well-described virus groups, and not to erect an all-embracing virological hierarchy. The groups they named and described were *Picornavirus* (replacing the name *Enterovirus)*, *Myxovirus, Herpesvirus, Adenovirus, Reovirus,* and *Papovavirus.*

There were only a few plant virologists at the 1962 Montreal meeting and no proposals concerning plant viruses were put forward. Bacterial virologists were not represented.

In 1963 the chairman of the Virus Subcommittee, after consultation with members, submitted a formal proposal to the President of the International Association of Microbiological

Societies (IAMS) requesting that the committee should be dissolved and that a new Virus Committee should be appointed by the Executive Committee of IAMS.

Because of lack of agreement, very few specific nomenclatural or taxonomic proposals had been sent forward to the Judicial Commission in the period since 1950. Most of the proposals had been merely tentative. However, the main reason for this requested change was that many members of the Virus Subcommittee considered that viruses differ so much from bacteria that they needed separate taxonomic treatment. The implication of the request was that the proposed Virus Committee would rank equally with the Committee on Bacteriological Nomenclature and would report directly to the Executive Committee of IAMS.

The Executive Committee of IAMS met in Paris in July 1963 and approved the establishment of an International Committee on Nomenclature of Viruses (ICNV).[48]

It was decided that the members of ICNV should be nominated by the National Societies, and that the first official meeting would be held in Moscow during the Ninth International Microbiological Congress in 1966. However preliminary work before the Congress was needed. For this purpose a Provisional Committee on Nomenclature of Viruses (PCNV) was set up. This committee discussed the problems extensively by correspondence and then met once only, in Paris in June 1965 under the chairmanship of Andrewes. Lwoff was a major contributor to the setting up and subsequent work of the PCNV. In spite of the fact that committee members came from widely differing fields of virology, and had diverging views, substantial agreement was reached on many matters. The main proposals published by the PCNV[49] can be summarized as follows.

With respect to *Principles* the following points were made:

1. An International Nomenclature of Viruses is necessary. Viruses cannot be left in a taxonomic vacuum.
2. The only way to achieve an International Nomenclature of Viruses is by a binomial system.
3. The code of nomenclature of bacteria cannot be applied to viruses.
4. Virologists will therefore have to build their own code.

With respect to naming of taxa the committee made 16 recommendations, the more significant of which were as follows:

1. No taxon should be named from a person
2. Anagrams, siglas, hybrids of names, nonsense names, should be prohibited
3. Names should preferentially be Latin or Latinized Greek names
4. A species shall be selected to typify each genus
5. Specific names can be names or letters or numerals
6. The names of all viral genera end in "virus", e.g., *Poliovirus*
7. A genus shall be selected to typify each family
8. A family is named from its type genus, e.g., Poxvirus, Poxviridae
9. The suffix for the family is *idae*. Therefore all virus families end in *viridae*
10. The name of a taxon whatever its rank is not followed by the name of the author who proposed it

The PCNV proposed a list of names which should be conserved together with a list of new genera. There were 21 families and 39 genera with type species given latinized names. For example genera in the Poxviridae were as follows:

Genus	Type species	Common name
1. *Poxvirus* (type genus)	*variolae*	Variola

2. *Dermovirus*	*orfi*	Contagious pustular dermatitis
3. *Pustulovirus*	*ovis*	Sheep pox
4. *Avipoxvirus*	*galli*	Fowl pox
5. *Fibromavirus*	*myxomatosis*	Rabbit myxoma
6. *Molluscovirus*	*hominis*	Molluscum contagiosum

Finally the PCNV boldly proposed that a scheme for classification of viruses into subphyla, classes, orders, suborders, and families should be adopted. The scheme was the one proposed by Lwoff et al.[31]

The PCNV proposals for Latin binomial names were influenced by the past proposals of the defunct Virus Subcommittee. The overall scheme of classification was clearly influenced by Lwoff's ideas.

The proposals generated a substantial controversy both before and during the Moscow meetings. Lwoff and Tournier[50] gave an expanded account of, and justification of their all-embracing hierarchical scheme. Gibbs, Harrison, Watson, and Wildy[51] opposed the arbitrary decisions on the relative importance of different virus characters that were essential in any hierarchical classification. They considered it essential to follow the principles of Adanson[52] who suggested that all available information should be used in classification and that all characters should be given equal weight. At least 60 characters are needed for each virus to give a satisfactory classification of this sort. They pointed out that while computers make the task relatively easy, not enough information was available for most viruses. One of the main points made by Gibbs et al.[51] was that any temporary measures should not prejudice the later development of a more satisfactory classification.

As far as nomenclature is concerned, Gibbs et al.[51] introduced the idea of the cryptogram. They suggested that virus names should consist of two parts. The first part would be the vernacular name now in use and would be invariant. The second part would consist of a cryptogram which was essentially a coded summary of eight characters. These were considered to be the minimum set of characters that distinguished between most of the obvious groups of viruses known at that time. They were set out in pairs as follows:

Type of nucleic acid	MW of nucleic acid	Outline of particle	Hosts
Strandedness	Percent of nucleic acid	Shape of nucleocapsid	Vector

Letters and numbers were used to code the information, with an asterisk indicating data not available. For example the cryptogram for influenza virus A was given as:

$$\frac{R}{1} : \frac{2}{1} : \frac{S}{E} : \frac{V}{O}$$

meaning

RNA	2×10^6 MW	Spherical particle	Vertebrate hosts
Single stranded	1% of particle weight	Elongated nucleocapsid	No vector known

For many viruses much of the information was missing in 1966. For example the cryptogram for beet curly top virus was

$$\frac{*}{*} : \frac{*}{*} : \frac{*}{*} : \frac{S}{Au}$$

S = seed plant hosts; Au = hopper vector. The cryptogram was further discussed by Gibbs and Harrison[53] and Gibbs.[54]

Lwoff[55] criticized the idea of the cryptogram and supported the use of a binomial system for naming virus species.

At the Ninth International Congress of Microbiology in Moscow the first and critical meeting of the new International Committee for Nomenclature of Viruses (ICNV) was held on 22 July 1966 attended by 43 members. The meeting was opened by Lwoff who outlined the circumstances leading to the formation of ICNV. Pereira (U.K.) was elected chairman.

It was unanimously decided that the committee was effective. It was soon decided (by acclamation) that:

1. The code of bacterial nomenclature should not be applied to viruses.
2. Nomenclature should be international.
3. Nomenclature should be universally applied to all viruses.

A substantial debate ensued concerning what sort of nomenclature system should be applied to viruses. Lwoff (France) was the prime supporter of a Latin binomial system and the other proposals from the PCNV.

Gibbs (U.K.) was among those against Latinized binomials and a hierarchical classification. He promoted the use of vernacular names and the cryptogram idea. Gibbs had polled all the known virologists in the U.K. on their views on the PCNV and the cryptogram proposals. He had found that the great majority disapproved of a move to introduce Latinized binomials at that time, and would prefer to try other schemes, or none at all. Melnick stated that of 200 virologists (mainly vertebrate) polled in the U.S. 90% had favored no change yet. Best reported that a survey of plant virologists in Australia had shown that a great majority were opposed to Latinized binomial names. Wildy was firmly against the PCNV proposals. Pereira took a position some distance away from either extreme. Best (Australia) considered that plant viruses should be named and regarded in the same terms as enzymes. Hansen (Denmark) made a plea for consideration of his system of a Latinized code for generating generic names.[15]

Finally three proposals were supported:

1. The committee considers that an international nomenclature for viruses is desirable (carried unanimously)
2. An effort should be made towards a Latinized binomial nomenclature (41 for, 2 against)
3. If and when Latinized binomials are introduced, the existing names should be retained wherever feasible (42 for, 1 against)

During the Ninth Congress, the International Association of Microbiological Societies formally established a permanent International Committee on Nomenclature of Viruses. This committee was constituted so that as far as possible each country would be represented by one member, with no country having more than five. Nominations for membership were to be made by the Microbiological Society in each country.

A second meeting of the ICNV was held on 23 July when Wildy (U.K.) was elected President, Ginsberg (U.S.) Vice-President, and Maurin (France) and Brandes (Federal Republic of Germany) as secretaries. These officers together with eight elected members made the first executive committee.

IV. THE PRESIDENCY OF P. WILDY (1966-1970)

In this period many important decisions were taken that set the framework for the future mode of operation of the committee and for the development of an effective taxonomy and nomenclature for viruses. Besides the President, three other people played an important role in this period. Behind the scenes Pereira worked in a steadfast and unpretentious way to aid progress while maintaining a nondogmatic approach. Gibbs and Harrison did a great deal of work. They promoted the interests and views of plant virologists in the face of the more numerous group working with viruses of vertebrates. They were also an important counterweight to those who supported the views of Lwoff.

At the first meeting of the Executive on 25 July four subcommittees were set up to study invertebrate viruses, plant viruses, vertebrate viruses, and bacteriophages. It was expected that the chairman of each subcommittee would circulate proposals of his group for comment to a sizeable representation of relevant virologists. A working subcommittee was set up to consider the practicability, form, and general utility of the cryptogram.

There was substantial interest in viral taxonomy among virologists attending the Congress. On July 26th a symposium on virus classification held at Moscow University was attended by 650 persons in the hall with 200 in an overflow room. There were six presentations. Maramorosch, for example, explained how the PCNV proposals would apply to plant viruses, while Harrison[56] criticized the PCNV proposals and developed the theme of the cryptogram. A lively discussion ensued. The third and final Moscow meeting of the ICNV was held following the symposium. As an outcome of these meetings, statutes and a code of nomenclature were elaborated. The following set of rules were approved:

1.   The code of bacterial nomenclature shall not be applied to viruses.
2.   Nomenclature shall be international.
3.   Nomenclature shall be universally applied to all viruses.
4.   An effort will be made towards a Latinized binomial nomenclature.
5.   Existing Latinized names shall be retained whenever feasible.
6.   The law of priority shall not be observed.
7.   New sigla* shall not be introduced.
8.   No person's names shall be used.
9.   No nonsense names shall be used.
10. For pragmatic purposes the species is considered to be a collection of viruses with like characters.
11. The genus is a group of species sharing certain common characters.
12. The rules of orthography of names and epithets are listed in the proposed code of nomenclature.

Many of these rules reflected the principles and recommendations made by the PCNV in 1965. However no approvals were given for any of the 21 families and 39 genera proposed by the PCNV. Likewise the all-encompassing hierarchical classification proposed by the PCNV was ignored. The idea of the cryptogram was not formally adopted. Thus by the end of the Moscow meetings Wildy's committee had managed to avoid the extremes of either the Lwoff or the Gibbs and Harrison viewpoints. Taxonomically speaking they could begin work with a clean slate. The question was whether this International Committee would be any more effective than those that preceded it. Besides the personality of the first President, the committee had important factors in its favor. It was the first committee to be constituted on a truly international basis; and with each year that passed taxonomically useful information about the viruses themselves was accumulating rapidly.

---

*   Names made up of letters from abbreviations (especially first letters) of a series of words.

The Executive Committee, however, had one weakness. Like earlier committees there was a preponderance of virologists working with viruses of vertebrates. In fact there was only one plant virologist elected (Gibbs). Wildy et al.[57] outlined the background to the formation of the ICNV, the difficulties facing the committee, and the way ahead as they saw it.

Apart from the opposing viewpoints of those who did or did not want a Latinized binomial system of nomenclature and hierarchical classification there were the geographical and linguistic barriers inherent in a truly international organization, and the fact that different branches of virology had developed taxonomically speaking to very different extents.

It was hoped that the four host-oriented subcommittees would ensure cross-fertilization by frequent communication. The Executive Committee would maintain a postal debate and meet once before the next International Congress.

This second meeting of the Executive Committee was held in London in April 1968. Some progress was reported by all committees but no specific taxonomic proposals were approved. The plant virus subcommittee under the chairmanship of Harrison submitted a report in which 16 groups of plant viruses were defined. The report had been circulated to plant virologists world-wide. Replies from 41/46 laboratories supported the groupings. Thirty replies supported the proposed names, some with reservations.

The proposals were published in 1971 under the names of seven out of nine members of the Plant Virus Subcommittee.[58] Two members, who had contributed little to the work of the subcommittee, did not share the views expressed in the paper. In retrospect it can be seen that the 16 proposed groups have stood the test of time. These groups were in fact the first substantial contribution towards a rational taxonomy for plant viruses.

However the proposals did not conform with the ICNV rules in two respects. The names proposed for 12 of the groups ended in virus which would correspond to the ending approved for a genus name by the ICNV in 1970 (see below). However these authors preferred the term group to genus. Furthermore the proposed names were all siglas which were banned under rule 7 from Moscow. Thus in the period 1966 to 1970 there was considerable controversy regarding some of the rules, which developed into a serious rift between most of the plant virologists, and some animal virologists.

There were other undercurrents. For example, vertebrate virologists had to make choices between various names that had been proposed for some genera. There was little interest in taxonomy among bacterial virus workers. Bradley could find only three virologists willing to serve on the first Bacterial Virus Subcommittee.

A major problem for the Invertebrate Virus Subcommittee under Vago was the lack of information about many of the viruses. Nevertheless proposals for two genera and one group of viruses infecting invertebrates were approved at Mexico in 1970.

The Cryptogram Subcommittee under Gibbs prepared a list of cryptograms for the plant viruses. These were incorporated in the list of plant virus names published by the Commonwealth Mycological Institute.[59]

An attempt to get leading virology journals to try out the cryptogram for an experimental period was only partly successful. Little interest was shown in the idea by vertebrate, invertebrate, and bacterial virologists.

Adding to the various undercurrents and differences of opinion between and within Wildy's subcommittees, efforts were being made within the ICNV to revive the idea of a full hierarchical classification that had been dropped in Moscow.

A few quotes from correspondence may help to recapture the flavor of the period:

1.   "I dislike — on both linguistic and aesthetic grounds — the legitimization of such bastard terms as picornavirus and papovavirus."
2.   "Why should they be given *funny names*? Are we not exposing ourselves to the laughter

of the general public? Do we want to join the ranks of old-fashioned botanists and zoologists so soon?''

3.   ''I can see no value in publishing such an indefinite article. There is no point in giving the ICNV seal of approval to a list of generic names when the viruses in the genera are not specified — it will generate chaos.''

4.   ''I fully agree that X was wagging the dog. He is an extremely determined and obstinate young man.''

5.   ''This means that the chairman of ICNV considers our rules are valueless. Why then make rules? And why an International Committee on Nomenclature of Viruses? And why a chairman?''

6.   ''I wouldn't mind if the comment was fair, and presented in a decent way, but it is a mixture of misquotations, misrepresentations etc. which make me seethe, as it is quite obvious that they are done quite deliberately.''

7.   ''I vote against the proposal for allocating genera to families. The existing practice (herpesvirus *group*, poxvirus *group*) is perfectly adequate for talking about a number of different species of herpesviruses etc. Once families have been approved the next step will doubtless be a request for acceptance of the complete fantasy (i.e., a full hierarchical classification).''

In 1968 a congress was held which was to have substantial importance for the future development of viral taxonomy. The First International Congress for Virology organized by virologists was held in Helsinki. Many virologists at the Helsinki congress who had attended the huge and diverse International Congresses for Microbiology decided that it would be desirable to have a separate Virology Section within the International Association of Microbiological Societies, which would then allow for regular international virology meetings. Appropriate submissions were made to the International Association of Microbiological Societies (IAMS). IAMS moved a year or so later to set up a section of virology and the Executive Committee of the ICNV made a request that the ICNV should be an integral part of the new section. This was approved.

The new organizational arrangements greatly facilitated the work of the ICNV and its committees because various meetings of the ICNV, its Executive Committee and subcommittees, could be held concurrently with each international virology congress.

In spite of the marked differences between the ideas of various virologists, substantial progress was made when it came to the meetings of the ICNV held in Mexico City in 1970 during the Tenth International Congress for Microbiology. Substantial credit must go to the first President, Wildy, for the skill and patience with which he managed to keep the organization intact to the stage where the first set of taxonomic decisions were approved.

Various attempts were made to change the most controversial rules approved in Moscow. Such changes were resisted and at the ICNV meetings in Mexico City the following additional rules were approved:

13.   The ending of the name of a virus genus is *-virus*.

14.   To avoid changing accepted usage, numbers, letters, or combinations may be accepted as the names of species.

15.   These symbols may be preceded by an agreed abbreviation of the Latinized name of a selected host genus or, if necessary, by the full name.

16.   Should families be required, a specific termination to the name of the family will be recommended.

17.   Any family name will end in *-idae*.

18.   A family is a group of genera with common characters.

Table 1
SUMMARY OF THE FIRST TAXONOMIC GROUPS APPROVED BY THE ICNV
(1970, IN MEXICO CITY)

A. Viruses infecting vertebrates

Taxonomic status	Vernacular name	Approved name	Type species
Family	Papovavirus group	*Papovaviridae*	
Type genus	Papillomavirus	*Papillomavirus*	Rabbit papilloma virus
Genus		*Polyomavirus*	Polyomavirus
Family	Picornavirus group	*Picornaviridae*	
Type genus	Enterovirus	*Enterovirus*	Poliovirus type 1
Genus	Vesicular exanthema	*Calicivirus*	Vesicular exanthema type A
	Rhinovirus	*Rhinovirus*	Rhinovirus 1A
Genus	Poxvirus group	*Poxvirus*	Vaccinia virus
Genus	Herpesvirus group	*Herpesvirus*	Herpes simplex virus
Genus	Adenovirus group	*Adenovirus*	Adenovirus type 1
Genus	Parvovirus group	*Parvovirus*	Latent rat virus (Kilham)
Genus	Reovirus group	*Reovirus*	Reovirus type 1
Genus	Leukosis virus complex	*Leukovirus*	Rous sarcoma virus
Genus	Paramyxovirus	*Paramyxovirus*	Newcastle disease virus
Genus	Myxovirus (influenza group)	*Orthomyxovirus*	Influenza virus
Genus	Vesicular Stomatitis group	*Rhabdovirus*	Vesicular stomatitis virus
Genus	Arbovirus group A	*Alphavirus*	Sindbis virus
Genus	Arbovirus group B	—	Yellow fever virus
Genus	Infectious bronchitis group	*Coronavirus*	Avian infectious bronchitis virus
Genus	LCM group	*Arenavirus*	Lymphocytic choriomeningitis virus

B. Viruses infecting invertebrates

Genus	Iridescent virus group	*Iridovirus*	*Tipula* iridescent virus
Genus	Nuclear polyhedrosis virus group and granulosis viruses	*Baculovirus*	*Bombyx mori* nuclear polyhedrosis virus
Genus	Cytoplasmic polyhedrosis virus	—	*Bombyx mori* cytoplasmic polyhedrosis virus

C. Viruses infecting bacteria

Genus	T-even phages	—	Coliphage T4
Genus	λ phage	—	Coliphage λ
Genus	Lipid phage PM2	—	Lipid phage PM2
Genus	ØX group	—	Coliphage ØX174
Genus	Filamentous phages	—	Coliphage fd
Genus	Ribophage group	—	Coliphage f2

D. Viruses infecting plants

Group	Tobacco ringspot virus group	*Nepovirus*	Tobacco ringspot virus
Group	Brome mosaic virus group	*Bromovirus*	Brome mosaic virus
Group	Cucumber mosaic virus group	*Cucumovirus*	Cucumber mosaic virus (S isolate)
Group	Cauliflower mosaic virus group	—	Cauliflower mosaic virus (cabbage B isolate)
Group	Cowpea mosaic virus group	—	Cowpea mosaic virus (SB isolate)
Group	Alfalfa mosaic virus group	—	Alfalfa mosaic virus
Group	Tobacco rattle virus group	—	Tobacco rattle virus (PRN isolate)
Group	Tobacco mosaic virus group	—	Tobacco mosaic virus (U1 strain)
Group	Turnip yellow mosaic virus group	—	Turnip yellow mosaic virus (Cambridge isolate)

Table 1 (continued)
SUMMARY OF THE FIRST TAXONOMIC GROUPS APPROVED BY THE ICNV
(1970, IN MEXICO CITY)

Taxonomic status	Vernacular name	Approved name	Type species
Group	Tomato bushy stunt virus group	—	Tomato bushy stunt virus
Group	Tobacco necrosis virus group	—	Tobacco necrosis virus
Group	Pea enation mosaic virus group	—	Pea enation mosaic virus
Group	Potato virus Y group	—	Potato virus Y
Group	Carnation latent virus group	—	Carnation latent virus
Group	Potato virus X group	—	Potato virus X
Group	Tomato spotted wilt virus group	—	Tomato spotted wilt virus

The specific taxonomic proposals that had been worked out by the four host-oriented subcommittees in the period up to 1970 received a mixed reception by the ICNV at Mexico City. Changes, deletions, and alterations were made by the Executive Committee and by the ICNV. The main proposals that were approved are summarized in Table 1.

For viruses infecting vertebrates, two families were approved, one with two genera and one with three. Names for all these taxa were also approved. The other 12 vertebrate virus taxa were given generic status, with approved names, even though some of the groups such as *Poxvirus*, *Herpesvirus*, and *Leukovirus* clearly required further subdivision even at this stage. It was anticipated that such genera would soon be given family status.

For all the approved genera, type species were given specific names by the ICNV. However these names had not been approved by the relevant subcommittees. They did not survive and do not appear in the second report of ICNV.[60]

A close and continuous correspondence based on 26 circulars from Dr. Vago resulted at Mexico City in the approval of two genera and one group of viruses infecting invertebrates.

The bacterial virus subcommittee developed two alternative sets of genera for bacterial viruses. The first based solely on nucleic acid gave three genera (ssDNA, dsDNA, and ssRNA). The second was based on morphology of the particle and gave rise to six genera. Names were proposed for all the genera, and also for the type species in the nucleic acid scheme, which the subcommittee preferred. In the event the ICNV adopted the six morphologically based genera, but did not approve any of the proposed names.

The substantial effort that the Plant Virus Subcommittee had made under Harrison to delineate groups of plant viruses, based on as many properties as possible, bore fruit at Mexico City. All 16 groups were approved by the ICNV. The plant virologists did not consider either a family or genus designation suitable. The ICNV deferred to this view and the plant groups remained groups.

The Plant Virus Subcommittee suggested names for 12 of their groups based on sigla. Nine of these were rejected as they contravened rule 7. One, which had established usage (*Nepovirus*) was allowed, as were two others which could be regarded as Latinized (*Bromovirus* and *Cucumovirus*). It was recognized that some plant viruses probably belonged in the *Rhabdovirus* and *Reovirus* genera established by the vertebrate virologists, and a number of plant viruses were listed as possible members of these genera.

At Mexico City it was further decided to broaden the scope for the activities of the Cryptogram Subcommittee so that it could become involved in the development of a data acquisition, storage, and retrieval system that would be applicable to all viruses. In recognition of this wider role it was renamed the "Code and Data Subcommittee".

Last, at Mexico City, the statutes of IAMS were changed restricting the membership of international bodies like the ICNV to one representative from each member society (plus life members).

Although there were still marked differences of opinion at Mexico City as how best to proceed in developing a taxonomy for viruses, there is little doubt that all concerned had the same long-term objective in view, as the following anecdote indicates. Lwoff, in conversation with Harrison was overheard to remark to this effect, "It occurs to me that if you have your way or I have my way we shall arrive at the same conclusion by 1984; it is only our methods that are different."

The publication of the first report of the ICNV[61] was a landmark in virus taxonomy and for virology in general. It set the basic pattern for further developments.

V. 1970 TO 1981

Fenner was elected President in 1970. His term ran for 5 years because the venue for the ICNV meetings was changed from the Microbiology to the Virology Congresses. The Second International Congress for Virology was held in Budapest in 1971. Meetings of the Executive Committee of the ICNV itself were held during the Congress but it was too soon after the Mexico City meeting for any new taxonomic proposals to have been generated.

A further meeting of the Executive Committee was held in London in May 1973. The Committee unanimously supported a proposal that the name of the committee should be changed to the International Committee for Taxonomy of Viruses (ICTV). The reasoning behind this decision was that the committee was inevitably concerned with wider issues than nomenclature, and that the word "taxonomy" properly described its functions. This proposal was widely supported in a postal vote of the ICNV membership. No specific taxonomic proposals were approved at this meeting.

The situation with plant viruses and plant virologists that developed following the Mexico City 1970 meeting illustrates well the kinds of difficulties that beset a democratically elected International Committee charged with the job of developing a useful virus taxonomy. Taxonomic problems involve questions of judgment and of personal opinion. Virologists are of two sorts — the majority who have no particular active interest in taxonomy but who would make use of a sensible scheme; and a minority who are interested but also have more or less strongly held views. The problem is that if a unified taxonomy for viruses is to be developed, only one view can prevail in the end for any particular issue.

Some interested plant virologists were upset about some of the decisions taken by the ICNV in Mexico City — in particular the refusal to bend rule 7 concerning new sigla, and the provisional approval by the ICNV of binomial names for type species of animal and bacterial virus genera. One of the most active plant virus taxonomists made himself no longer available to assist in the work of the ICNV. Indeed, the new President had considerable difficulty in finding a competent plant virologist to be chairman of the Plant Virus Subcommittee.

Another activity at this time illustrates the difficulty of usefully polling public opinion (i.e., virologists in general) on matters of taxonomy. The American Phytopathological Society sent out a questionnaire to members in its *Phytopathology News* (1971, 5 March, 3-5). Thirty-eight questionnaires were returned from a membership of about 200 virologists (*Phytopathology News*, 1971, 5 August, 2). There was no way of knowing how representative the views of less than one in five of the members would be. On the use now of sigla derived from vernacular names for virus groups 8 were for and 27 against. On the future use of such sigla 12 were for and 23 against. However in the period 1971 to 1975 the names became widely used and the plant virologists involved with the ICTV continued to press for the use of sigla. Subsequently, rule 7 was relaxed by the ICTV (in 1975) and all the sigla-based names originally proposed by Harrison's subcommittee were accepted. Now in 1981 these names have become generally accepted. It is not at all clear why resistance to the use of sigla persisted for so long, although there is clearly a good basis for most careful scrutiny before proposed siglas are adopted. Siglas based on English names may be strange or even ridiculous in other languages (e.g., *Picornavirus* and *Cucumovirus* in French) as was pointed out by Maurin at Budapest in 1971.

Although there were these difficulties of a ''political'' sort early in Fenner's 5-year term of office steady progress in taxonomy was made, particularly with the vertebrate viruses. This progress was greatly assisted by rapid developments in our knowledge of virus structure and replication.

As described in Section 1, all the early attempts at virus classification were doomed to failure because they relied mainly on unstable biological properties, particularly disease symptoms. A crucial factor that allowed the ICNV to make solid progress with viral taxonomy in its first 4 years was the emphasis on the properties of the virus particle as providing the most satisfactory criteria for grouping viruses.

By the early 1970s substantial information was accumulating for many viruses concerning the manner in which the genome was organized and replicated and how it functioned in virus replication — the strategy of the genome. This is of course not a single character but a multiple one. As it became available the information provided an extremely valuable new source of data for viral taxonomy.[62,63] In general most members of a family that had been formed mainly on the basis of particle properties were found to have the same genome strategy.

For example among members of the *Picornaviridae*, functional individual proteins were found to be formed by post-translational cleavage of a single polyprotein. This property gave a clear basis for excluding bacterial viruses such as MS2 from the *Picornaviridae*. As another example, viruses in the family *Retroviridae* have many structural features in common. They also have an RNA genome and a reverse transcriptase enzyme which copies the genome into DNA during replication — a feature that sets the family apart from all other known viruses.

Working by postal vote in April-May 1974 the following taxonomic proposals from the Vertebrate Virus Subcommittee and the Coordination Subcommittee were approved by the ICTV:[64] (1) a new family *Togaviridae,* including two genera — *Alphavirus* (type species sindbis virus, previously approved) and *Flavivirus* (type species, yellow fever virus); (2) a new family *Reoviridae,* including two genera — *Reovirus* (previously approved) and *Orbivirus* (type species, blue tongue virus); (3) the former genus *Poxvirus* was given family status under the name of *Poxviridae*, encompassing six genera based on host specificities — *Orthopoxvirus, Avipoxvirus, Capripoxvirus, Leporipoxvirus, Parapoxvirus, and Entomopoxvirus*.

Following 5 years of discussion by study groups, subcommittees and the Executive Committee, working by correspondence and sometimes by meetings, further substantial taxonomic developments were approved by the ICTV meeting during the Third International Congress for Virology in Madrid in 1975.[61,65] These included eight families for viruses infecting bacteria, but the suggested names were not approved.

Several changes were made to the Rules of Nomenclature. Some of these were trivial but two were important. It had become apparent over the 9 years of the ICTV's existence, that while the family and generic names for viruses were widely accepted there was currently no strong wish for a binomial system for the names of virus species. For this reason the old rule 4 ''an effort will be made towards a Latinized binomial nomenclature'' had the word ''binomial'' deleted.

The other important change concerned rule 7 and the banning of sigla. Before 1966 Animal Virologists had used sigla such as *Reovirus* (Respiratory Enteric Orphan) and these were allowed as the rule said no *new* sigla. However as outlined earlier the Plant Virus Subcom-

Table 2
MAJOR CHANGES MADE IN TAXA FOR VIRUSES INFECTING VERTEBRATES AND INVERTEBRATES IN THE PERIOD 1970 TO 1975

1970 Genus	1975 Family	Genera
Poxvirus	Poxviridae	Orthopoxvirus
		Avipoxvirus
		Capripoxvirus
		Leporipoxvirus
		Parapoxvirus
		Entomopoxvirus
Herpesvirus	Herpetoviridae	Herpesvirus:
Iridovirus	Iridoviridae	Iridovirus
Baculovirus	Baculoviridae	Baculovirus
Adenovirus	Adenoviridae	Mastadenovirus
		Aviadenovirus
Reovirus	Reoviridae	Reovirus
		Orbivirus
		(plus one unnamed genus)
Parvovirus	Parvoviridae	Parvovirus
		Densovirus
Leukovirus	Retroviridae	(subfamilies) Oncovirinae
		Spumavirinae
		Lentivirinae
		(with several un-named genera in the-Oncovirinae)
Paramyxovirus	Paramyxoviridae	Paramyxovirus
		Morbillivirus
		Pneumovirus
Orthomyxovirus	Orthomyxoviridae	Influenzavirus
Rhabdovirus	Rhabdoviridae	Vesiculovirus
		Lyssavirus
	Bunyaviridae (New family in 1975)	Bunyavirus
Coronavirus	Coronaviridae	Coronavirus
Arenavirus	Arenaviridae	Arenavirus
Alphavirus	Togaviridae	Alphavirus
		Flavivirus
		Rubivirus
		Pestivirus

mittee had submitted 12 sigla for group names in 1970 of which 9 were declined. In the event these were widely used by Plant Virologists in the period 1971 to 1975. To break the deadlock that threatened the unified international system that the ICTV was striving for, rule 7 was changed to allow sigla on the conditions that (1) they were meaningful to workers in the field, and (2) that they were recommended by International Study Groups. Following this, nine new names were approved for established plant virus groups and four new groups were approved: *Closterovirus* (beet yellows virus group); *Hordeivirus* (barley stripe mosaic virus group); *Luteovirus* (barley yellow dwarf virus group); and *Ilarvirus* (isometric labile ringspot virus group).

For viruses infecting vertebrates and invertebrates, most of the genera approved in 1971 and which had not been dealt with by postal vote in 1974 were given family status in 1975 at Madrid, each with one or more genera as in Table 2. In addition one new family group was formed. The first subfamily taxa for viruses were introduced for the *Retroviridae* (Table 2).

One decision taken at Madrid caused significant ripples, if not waves, over the next 3 years. The Executive Committee recommended to the ICTV that the family name for the former genus *Herpesvirus* should, on linguistic grounds, be *Herpetoviridae* rather than *Herpesviridae*. This name was approved without the Study Group of the Vertebrate Virus Subcommittee being given any opportunity to comment on the proposal. The Herpesvirus Study Group later voted unanimously for the retention of *Herpesviridae*, while the vertebrate Virus Subcommittee voted for *Herpesviridae* by a narrow margin. The Executive Committee recognized that it is very bad taxonomic practice to change approved names except for very sound scientific reasons. However in this instance the matter was put to the vote at the full ICTV meeting in The Hague in 1978, when the family name was changed to *Herpesviridae*.

This small piece of history, while rather trivial in itself was important in that the incident brought home clearly to the Executive Committee the need to refer back to the appropriate subcommittee any *changes* in taxonomic proposals that originate within the Executive Committee or the ICTV itself. No similar incidents have occurred since.

In retrospect perhaps the major contribution made by Fenner during his Presidency was to keep the plant virologists working within the ICTV organization. This really meant stopping the insistence of Lwoff's supporters on an hierarchical classification and Latinized binomials, and also, as noted above, deleting the rule regarding new sigla. In addition Fenner exerted pressure to ensure that following two vertebrate virologists, a plant virologist should be the next President of the ICTV.

One of my first initiatives after being elected President of the ICTV at Madrid in 1976, was to ask the chairman of the Code and Data Subcommittee to conduct a survey of the use made of the cryptogram in papers in the three leading international virological journals over the period 1971 to 1975. A total of 4,333 papers contributed by 10,611 authors contained 4,508 virus names, 174 different viruses being variously named. Plant virus names used in 1971 had accompanying cryptograms in 13% of cases and in 1975, 45%. For viruses of vertebrates the corresponding figure was 5.5% in both 1971 and 1975. The cryptogram was not used at all by bacterial virologists. These results led the Executive Committee of the ICTV to consider the future role of the cryptogram at its mid-term meeting in London in 1977. It was recognized that the cryptogram had played a valuable role in two respects, (1) introduction of the idea in 1966 at Moscow had been a material factor in preventing the premature adoption of a full-blown hierarchical classification and Latinized binomial system for viruses, (2) its use in the first report of ICNV[61] drew attention to the need for much more data on many viruses. The need was particularly marked for plant viruses.[59] However it was clear that the cryptogram had outlived its usefulness, first because of its inflexibility in the light of new and important information being reported on virus structure and replication, and second because after a trial period of 9 years it was not being used by a substantial majority of virologists.

In April 1977 the Executive Committee passed the following resolution, "That in view of the limited time and finances available the efforts of the Code and Data Subcommittee should be concentrated on the development of an effective data acquisition, storage and retrieval programme based on general virus properties, and that work on computer storage of the cryptogram should be abandoned". It was also decided to discontinue use of the cryptogram in ICTV reports.

At the meetings held in conjunction with the Fourth International Congress for Virology at The Hague in September 1978, the ICTV approved a number of taxonomic proposals. Two new bacterial virus families were approved and named: (1) *Tectiviridae* to include isometric dsDNA viruses containing lipid and two coats, (e.g., PRDI) and, (2) *Plasmaviridae* to include ds DNA viruses with pleomorphic enveloped particles with no apparent capsid (e.g., MV-L2). Names were also approved for five of the eight families that had been established in 1975. These were the *Corticoviridae, Microviridae, Inoviridae, Cystoviridae,* and *Leviviridae*. A proposal from the Vertebrate Virus Subcommittee for two new genera in the *Picornaviridae* was approved. They were *Aphthovirus* and *Cardiovirus*. Several pro posals from the Coordination Subcommittee were approved. Two subfamilies were introduced in the *Poxviridae:* the *Chordopoxvirinae* to include poxviruses of vertebrates and the *Entomopoxvirinae* to include poxviruses of insects. In addition a new genus *Suipoxvirus* was created to comprise the swinepox subgroup.

In the family *Reoviridae* the "plant reovirus group" was divided into two genera with the following approved names: -*Phytoreovirus* with wound tumor virus as the type species, and *Fijivirus* with Fiji disease virus as type species. This was an important decision because, for the first time, it brought a few plant viruses formally into the family and genus structure being developed for viruses infecting all other groups of organisms.

Another point regarding this decision is worth a mention. When a new name is proposed for some virus taxon it is quite frequently objected to on the grounds that the name is inappropriate, not relevant, or does not evoke some important property of the virus or viruses. However, as time passes a name accumulates appropriate associations and its origins may be forgotten by most people. For example, I have heard of no objections to the name *Phytoreovirus* on the grounds that a respiratory enteric orphan virus of plants is an absurd concept.

From the Plant Virus Subcommittee three new groups were approved and one of these was given an approved name. These were the Southern bean mosaic virus group, the maize chlorotic dwarf group, and the *Geminivirus* group.

The *Geminivirus* group is an interesting indicator of the general state that virus taxonomy has reached. Viruses now placed in this group cause economically important diseases and had been the subject of much study by plant pathologists since early this century. However it was not until 1974 that good evidence was provided for the existence of virus-like nucleoprotein particles consisting mainly of isometric units in pairs.[66,67]

Three years later the nucleic acid of one of these viruses was shown to be single-stranded DNA[68,69] and in 1978 the *Geminivirus* group was formally approved. This example illustrates the point that most of the families and groups of viruses approved by the ICTV are really quite distinctive, and that it is usually immediately apparent when a new family or group is discovered.

At The Hague in 1978 I was re-elected for a second 3-year term as President of the ICTV. It had become apparent during the period 1975 to 1978 that the stage was being reached for many virus families, particularly among those infecting vertebrates, where further taxonomic progress would depend on the delineation of officially approved virus species and the allocation of official international names to them. The rules of nomenclature originally approved in Moscow in 1966 were really quite inadequate as a guide to virologists wishing to delineate and suggest names for virus species. The problem was discussed at length by the Executive Committee.[71,72]

As a result, at the ICTV meeting held in Strasbourg in 1981 in conjunction with the Fifth International Congress for Virology, three Moscow rules concerning species were replaced with eight new rules. In addition guidelines for the delineation and naming of virus species were provided. The problem of virus species is discussed in more detail in the last chapter of this book.

Also at Strasbourg, the Code and Data Subcommittee took a further step in its development. Over the previous 6 years the main work of this committee under the chairmanship of Atherton had been the development of a "code for the description of virus characters" that would be applicable for all viruses, in association with computer facilities available in Brisbane, Australia. This code provides for systematic and uniform storage of data and allows for flexibility of data output. It provides inherent definition of terms to ensure standard description of virus characters.

Such a major program is not readily portable. It also requires some stability in the personnel involved. For these reasons the Executive Committee at Strasbourg resolved to replace the Code and Data Subcommittee with a "Standing Subcommittee for Virus Data". This sub-committee will have a more stable membership than was possible previously. In addition the World Data Centre for Microorganisms at Brisbane was reaffirmed as the center for storage and retrieval of data relating to all viruses.

Fifty-one new taxonomic proposals were approved by the ICTV at Strasbourg in 1981. The most important of these concerned approval of the delineation and naming of the first set of virus species. These were in the family *Adenoviridae*. They are discussed in detail in the last chapter. Other noteworthy developments were as follows:

1. The caliciviruses were excluded from the *Picornaviridae* and a new family the *Caliciviridae* was established with a single genus *Calicivirus*.
2. In the large family *Bunyaviridae* three new genera were formed in addition to the established *Bunyavirus*. These were *Nairovirus*, *Phlebovirus*, and *Uukuvirus*.
3. Two new families for viruses infecting invertebrates were formed. These were the *Nodaviridae*, a family of invertebrate small ssRNA viruses with divided genomes, and the *Nudaurelia* β virus family to include small ssRNA viruses with T = 4 symmetry.
4. In the family *Iridoviridae* four new genera were established in addition to *Iridovirus*. These were *Chloriridovirus*, *Ranavirus*, African swine fever virus genus, and lymphocystis disease virus genus.
5. The family *Parvoviridae* was given a second genus *Dependovirus* to include the helper-virus dependent parvoviruses which are known by the English vernacular name of Adeno-associated viruses.
6. For the bacterial viruses, names were approved for two further families and eight genera were established in seven families.
7. A new plant virus group, the *Dianthovirus* group was established to include viruses with bipartite ssRNA genomes in isometric particles sedimenting as a single component. After declining approval some years previously the ICTV approved the name *Sobemovirus* for the Southern bean mosaic virus group.

Thus the deliberations at Strasbourg followed those at The Hague and Madrid in providing further constructive developments in viral taxonomy, while the heated discussions and strongly worded correspondence of earlier phases had virtually disappeared.

VI. THE PRESENT STATE OF VIRAL TAXONOMY

The overall state that the taxonomy of viruses had reached following the 1981 Strasbourg meetings is summarized in Table 3 and Figures 1 to 4.

The summary of data in Table 4 shows that nearly 1400 viruses have been assigned to approved taxa, and nearly 500 more have been designated as probable or possible members of such taxa. At the present stage we have no precise estimate of the total number of distinct viruses that have been isolated, but the figure is probably of the order of a few thousand. Thus the viruses that have been allocated to taxa probably represent a significant fraction of the total that have been isolated.

Furthermore, for the viruses infecting man, domesticated animals and crop plants in particular, a high proportion of the important viruses have already received some taxonomic consideration. Thus the taxonomy is now sufficiently developed to be of substantial use to virologists, whether they be research workers, teachers, or diagnosticians.

The data summarized in Tables 3 and 4 clearly show how much more the taxonomy of viruses infecting vertebrates has developed compared with the other host groups. This is

Table 3
SUMMARY OF THE STATE OF VIRAL TAXONOMY IN 1981[a]

Particle characterization	Family or group and kinds of hosts[b]	Subfamily	Genus	Type species	Number of members	
					Definite	Probable + possible
dsDNA enveloped	Poxviridae (V,I)	Chordopoxvirinae	Orthopoxvirus	Vaccinia virus	8	3
			Parapoxvirus	Orf virus	4	0
			Avipoxvirus	Fowlpox virus	8	0
			Capripoxvirus	Sheep pox virus	3	0
			Leporipoxvirus	Myxoma virus	4	0
			Suipoxvirus	Swine pox virus	1	0
k0		Entomopoxvirinae (three probable genera)				
dsDNA enveloped	Herpesviridae (V)	Alphaherpesvirinae	Unnamed	Human (alpha) herpesvirus 1	3	0
			Unnamed	Suid (alpha) herpesvirus I (pseudorabies)	3	0
		Betaherpesvirinae	Unnamed	Human (beta) herpesvirus 5 (human cytomegalovirus)	1	0
			Unnamed	Murid (beta) herpesvirus I (mouse cytomegalovirus)	1	0
		Gammaherpesvirinae	Unnamed	Human (gamma) herpesvirus 4 (Epstein-Barr)	1	0
dsDNA enveloped	Baculoviridae (I)		Subgroup A	Nuclear polyhedrosis virus	1	0
			Subgroup B	Granulosis virus	52	0
dsDNA enveloped	Plasmaviridae (B)	—	Plasmavirus	Phage MV-L2	2	6
dsDNA nonenveloped	Iridoviridae (some members have an envelope not required for infectivity)	—	Iridovirus	Tipula iridescent virus (proposed)	19	2
			Chloriridovirus	Mosquito iridescent virus (type 3)	11	0
	(V, I)		Ranavirus	Frog virus 3	55	0
			Unnamed	African swine fever virus	1	0
			Unnamed	Lymphocystis disease virus	1	0

Table 3 (continued)
SUMMARY OF THE STATE OF VIRAL TAXONOMY IN 1981[a]

Particle characterization	Family or group and kinds of hosts[b]	Subfamily	Genus	Type species	Number of members	
					Definite	Probable + possible
dsDNA nonenveloped	Adenoviridae (V)	—	Mastadenovirus	*Mastadenovirus* h2 (human adenovirus 2) plus 55 other provisionally approved species. 34 human species are placed in five subgroups	56	0
			Aviadenovirus	*Aviadenovirus* gal 1 (fowl adenovirus 17) plus 13 other provisionally approved species	14	0
dsDNA nonenveloped	Papovaviridae (V)	—	Papillomavirus	Rabbit papilloma virus	21	0
			Polyomavirus	Mouse polyoma virus	10	0
dsDNA nonenveloped	Caulimovirus group (P)	—	—	Cauliflower mosaic virus	6	4
dsDNA nonenveloped	Tectiviridae (B)	—	Tectivirus	Phage PRD1	9	0
dsDNA nonenveloped	Corticoviridae (B)	—	Corticovirus	Phage PM2	1	2
dsDNA nonenveloped	Myoviridae (B)	—	Unnamed	Coliphage T2	56	0
dsDNA nonenveloped	Phage with long non-contractile tails (B)	—	Unnamed	Coliphage λ	4	1
dsDNA nonenveloped	Podoviridae (B)	—	—	Coliphage T7	9	0
ssDNA nonenveloped	Parvoviridae (V, I)	—	Parvovirus	Rat parvovirus (Kilham)	12	3
			Dependovirus	Adeno-associated virus type 1	7	2
			Densovirus	Densovirus of *Galleria mellonella*	2	6
ssDNA nonenveloped	Geminivirus (P)	—	—	Maize streak virus	8	6
ssDNA nonenveloped	Microviridae (B)	—	Microvirus	Phage φX 174	25	0
ssDNA nonenveloped	Inoviridae (B)	—	Inovirus	Phage fd (proposed)	1	18
			Plectrovirus	Mycoplasma virus (type 1) (proposed)	35	1
dsRNA enveloped	Cystoviridae	—	Cystovirus	Phage φ6	1	0
dsRNA nonenveloped	Reoviridae (V,I,P)	—	Reovirus	Reovirus type 1	3	0
			Orbivirus	Bluetongue virus	63	1
			Rotavirus	Human rotavirus	1	0

Genome	Family	Subfamily	Genus	Type species		
			Phytoreovirus	Wound tumor virus	1	1
			Fijivirus	Rice dwarf virus	8	0
			Unnamed	Cytoplasmic polyhedrosis virus of *Bombyx mori*	12	150
ssRNA enveloped	*Togaviridae* (V,I)	—	*Alphavirus*	Sindbis virus	25	0
			Flavivirus	Yellow fever virus	53	0
			Rubivirus	Rubella virus	1	0
			Pestivirus	Mucosal disease virus	2	0
ssRNA enveloped	*Coronaviridae* (V)	—	*Coronavirus*	Avian infectious bronchitis virus	5	7
ssRNA enveloped	*Paramyxoviridae* (V)	—	*Paramyxovirus*	Newcastle disease virus	12	0
			Morbillivirus	Measles virus	4	0
			Pneumovirus	Respiratory syncytial virus	3	0
ssRNA enveloped	*Orthomyxoviridae* (V)	—	*Influenzavirus*	Influenza virus A/WS/33/HoN1	2	0
ssRNA enveloped	*Rhabdoviridae* (V,I,P)	—	*Vesiculovirus*	Vesicular stomatitis virus	4	0
			Lyssavirus	Rabies virus	6	0
			Plant Rhabdoviruses (not allocated to genera)	—	8	67
ssRNA enveloped	*Bunyaviridae* (V, I)	—	*Bunyavirus*	Bunyamwera virus	117	28
			Phlebovirus	Sandfly fever (Sicilian) virus	31	1
			Nairovirus	Crimean-Congo haemorrhagic fever	23	4
			Uukuvirus	Uukuniemi	6	24
ssRNA enveloped	*Arenaviridae* (V)	—	*Arenavirus*	Lymphocytic choriomeningitis virus	10	3
ssRNA enveloped	Tomato spotted wilt virus group (P)	—	—	Tomato spotted wilt virus	1	0
ssRNA enveloped	*Retroviridae* (V)	*Oncovirinae*	Genus (unnamed) Type C oncovirus group			
			(subgenus) Mammalian type C oncoviruses	—	9	0
			(subgenus) Avian type C oncoviruses	—	3	0
			(subgenus) Reptilian type C oncoviruses	—	1	0

Table 3 (continued)
SUMMARY OF THE STATE OF VIRAL TAXONOMY IN 1981[a]

Particle characterization	Family or group and kinds of hosts[b]	Subfamily	Genus	Type species	Number of members	
					Definite	Probable + possible
			Genus (un-named) type B oncovirus group	—	1	0
		Spumavirinae	—	—	4	0
		Lentivirinae	—	—	1	0
ssRNA nonenveloped	Picornaviridae (V)		Enterovirus	Human poliovirus 1	102	0
ssRNA nonenveloped			Cardiovirus	Encephalomyocarditis virus	3	0
ssRNA nonenveloped			Rhinovirus	Human rhinovirus 1A	114	0
ssRNA nonenveloped			Aphthovirus	Aphthovirus 0	6	0
ssRNA nonenveloped	Caliciviridae (V)	—	Calicivirus	Vesicular exanthema of swine	18	0
ssRNA nonenveloped	Unnamed (*Nudaurelia* β virus group) (I)	—	—	*Nudaurelia* β virus	6	0
ssRNA nonenveloped	Leviviridae (B)	—	Levivirus	Phage MS2	30	1
ssRNA nonenveloped	Unnamed maize chlorotic dwarf virus group (P)	—	—	Maize chlorotic dwarf virus	1	1
ssRNA nonenveloped	Tymovirus group (P)	—	—	Turnip yellow mosaic virus	17	1
ssRNA nonenveloped	Luteovirus group (P)	—	—	Barley yellow dwarf virus	15	19
ssRNA nonenveloped	Tombusvirus group (P)	—	—	Tomato bushy stunt virus	7	2
ssRNA nonenveloped	Sobemovirus group (P)	—	—	Southern bean mosaic virus	2	4
ssRNA nonenveloped	Unnamed (tobacco necrosis virus group) (P)	—	—	Tobacco necrosis virus	1	1
ssRNA nonenveloped	Closterovirus group (P)	—	—	Beet yellow virus	11	4
ssRNA nonenveloped	Carlavirus group (P)	—	—	Carnation latent virus	23	12
ssRNA nonenveloped	Potyvirus group (P)	—	—	Potato virus Y	47	67
ssRNA nonenveloped	Potexvirus group (P)	—	—	Potato virus X	18	19
ssRNA nonenveloped	Tobamovirus group (P)	—	—	Tobacco mosaic virus	10	6
ssRNA nonenveloped	Dianthovirus group (P)	—	—	Carnation ringspot virus	2	0

ssRNA nonenveloped	*Comovirus* group	—	Cowpea mosaic virus	12	1
ssRNA nonenveloped	*Nepovirus* group (P)	—	Tobacco ringspot virus	22	5
ssRNA nonenveloped	Unnamed (pea enation mosaic virus group) (P)	—	Pea enation mosaic virus	1	0
ssRNA nonenveloped	*Tobravirus* group (P)	—	Tobacco rattle virus	2	1
ssRNA nonenveloped	*Nodaviridae* (I)	—	Nodamura virus	5	0
ssRNA nonenveloped	*Cucumovirus* group (P)	—	Cucumber mosaic virus	3	1
ssRNA nonenveloped	*Bromovirus* group (P)	—	Brome mosaic virus	3	1
ssRNA nonenveloped	*Ilarvirus* group (P)	—	Tobacco streak virus	11	0
ssRNA nonenveloped	Unnamed (alfalfa mosaic virus group) (P)	—	Alfalfa mosaic virus	1	0
ssRNA nonenveloped	*Hordeivirus* group (P)	—	Barley stripe mosaic virus	3	0

[a] Data taken from Matthews.[72] Only approved taxa have been included. For families containing subfamilies or genera only members definitely allocated to such subfamilies or genera have been included. This means that for a few genera the numbers in the table substantially underestimate the number of viruses already known but not allocated (e.g., the *Rotavirus* genus and the *Baculoviridae*). As a matter of convenience the plant virus groups are included in the same column as virus families. The ICTV has not yet made any decisions as to the family or genus status of any of the plant virus groups.

[b] V = Vertebrates; I = Invertebrates; B = Bacteria; P = Plants.

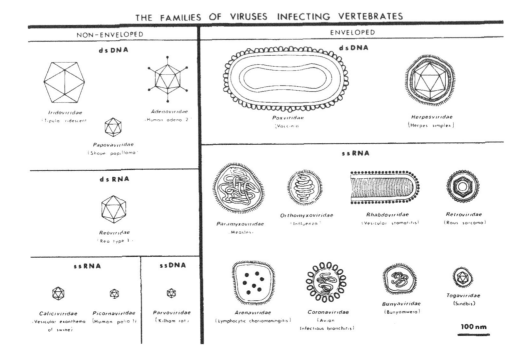

FIGURE 1.   The families of viruses infecting vertebrates. (From Matthews, R. E. F., *Intervirology*, 17(1-3), Karger, Basel, 1982. With permission.)

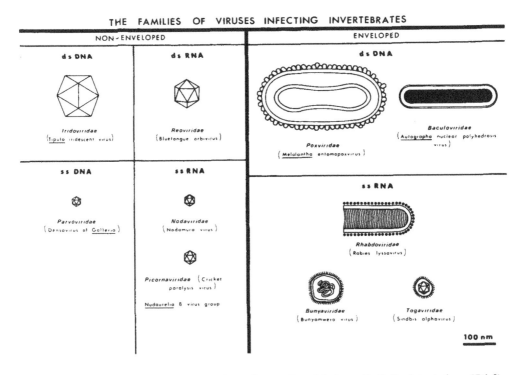

FIGURE 2.   The families of viruses infecting invertebrates. (From Matthews, R. E. F., *Intervirology*, 17(1-3), Karger, Basel, 1982. With permission.)

THE FAMILIES OF VIRUSES INFECTING BACTERIA

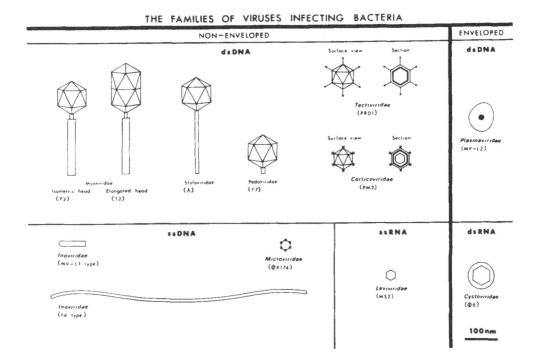

FIGURE 3. The families of viruses infecting bacteria. (From Matthews, R. E. F., *Intervirology,* 17(1-3), Karger, Basel, 1982. With permission.)

THE FAMILIES AND GROUPS OF VIRUSES INFECTING PLANTS

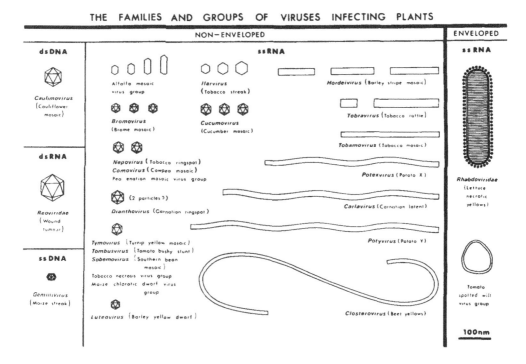

FIGURE 4. The families and groups of viruses infecting plants. (From Matthews, R. E. F., *Intervirology,* 17(1-3), Karger, Basel, 1982. With permission.)

Table 4
NUMBERS OF VIRUSES INCLUDED IN
INTERNATIONALLY APPROVED TAXA

Host group	A Definite members	B Probable and possible members	C Total	A as a % of C
Vertebrate	855	76	931	92%
Invertebrate	108	158	266	41%
Bacterial	173	29	202	86%
Plant	236	223	459	51%
Fungal	0	0	0	—
Total	1372	486	1858	74%

Note: Data for this table were taken from Table 3.

further emphasized in the data of Figure 5 which shows the growth in numbers of families and genera for the various host groups of viruses following the four plenary sessions of the ICTV that have been held to date. The Fungal Virus Subcommittee has been in existence since 1975, but as yet no taxonomic proposals from this subcommittee have been approved.

There have been three main reasons why the taxonomy of viruses infecting vertebrates has become the most developed: (1) the importance of these viruses for human and veterinary medicine, and their consequent economic importance means that vastly more money has been invested in vertebrate virology over many decades and in many countries. Thus there are many more working vertebrate virologists to deal with a not too dissimilar number of viruses; (2) Enders et al.[73] first reported that poliovirus could be grown in cultured non-neural cells. Since then the development of cell culture techniques for the study of the molecular biology and biochemistry of vertebrate virus replication has increased our knowledge about previously known viruses and led to the discovery of many more. Invertebrate cells are much more difficult to culture in quantity. The protoplast systems used over the last 12 years to study plant virus replication, while making a useful contribution have significant limitations compared to the established cell lines of the animal virologist, and (3) plant virologists generally have taken a rather cautious approach to taxonomic development. Some individuals may have been rather too resistant to any change. As noted earlier most of the premature and ineffective attempts at a comprehensive viral taxonomy were put forward by plant virologists. This led to a "once bitten, twice shy" attitude among plant virologists. However, there has also been good scientific reason to proceed slowly. The virions of the families of vertebrate viruses have diverse and distinctive morphology (Figure 1) whereas the plant viruses display much less morphological diversity (Figure 4). Most of the viruses infecting plants are small, nonenveloped with ssRNA genomes. Furthermore 14 of the 26 approved groups have small icosahedral particles. For some of these it is still not apparent whether they equate best with a vertebrate virus genus or family.

The difficulties that have faced and will continue to face virologists working with bacterial, insect, and fungal viruses are discussed in later chapters. However, as with the vertebrate and plant viruses, the present state of their taxonomy basically reflects the financial input, worldwide over recent decades.

In conclusion, if we look back over the historical development of viral taxonomy we can see that the major difficulties stemmed from the fact that the need for a useful taxonomy arose about the 1920s but the information required to construct a meaningful and stable taxonomy began to accumulate only some 30 years later. Even then, development was very uneven between host groups, a significant cause of the tensions that existed in the early

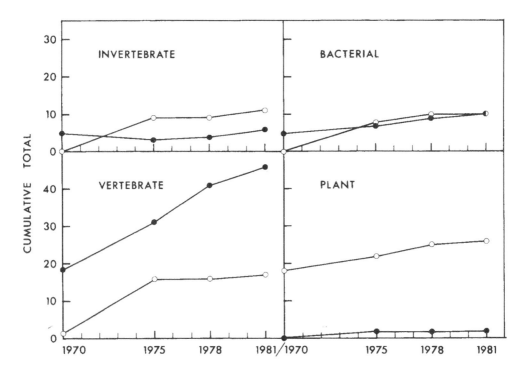

FIGURE 5. Cumulative totals for the numbers of families (0-0) and genera (●-●) for viruses infecting four host-groups approved by the ICTV at its meetings in 1970, 1975, 1978, and 1981. For the purposes of this figure the 24 plant groups have been considered as families. In 1975 many taxa that had been given generic status in 1970 were elevated to family status.

years of international collaboration. The ICTV has now achieved a substantial and essentially stable framework for the taxonomy of viruses. A major feature of the taxonomy developed so far is that it brings out the essential unity of virology as a field of study. This is particularly evidenced by those virus families which have members that infect more than one kind of host.

Problems and prospects for the future are discussed in the last chapter.

VII. THE RULES OF VIRAL NOMENCLATURE, 1981

Rule 1	The code of bacterial nomenclature shall not be applied to viruses.
Rule 2	Nomenclature shall be international.
Rule 3	Nomenclature shall be universally applied to all viruses.
Rule 4	An effort will be made towards a Latinized nomenclature.
Rule 5	Existing latinized names shall be retained whenever feasible.
Rule 6	The law of priority shall not be observed.
Rule 7	Sigla may be accepted as names of viruses or virus groups, provided that they are meaningful to workers in the field and are recommended by international virus study groups.
Rule 8	No person's name shall be used.
Rule 9	Names should have international meaning.
Rule 10	The rules of orthography of names and epithets are listed in Chapter 3, Section 6 of the proposed international code of nomenclature of viruses [Appendix D; Minutes of 1966 Moscow meeting].

New Rule 11   A virus species is a concept that will normally be represented by a cluster of strains from a variety of sources, or a population of strains from a particular source, which have in common a set or pattern of correlating stable properties that separate the cluster from other clusters of strains.

New Rule 12   The genus name and species epithet together with the strain designation must give an unambiguous identification of the virus.

New Rule 13   The species epithet must follow the genus name and be placed before the designation of strain, variant, or serotype.

New Rule 14   A species epithet should consist of a single word, or if essential a hyphenated word. The word may be followed by numbers or letters.

New Rule 15   Numbers, letters, or combinations thereof may be used as an official species epithet where such numbers or letters already have wide usage for a particular virus.

New Rule 16   Newly designated serial numbers, letters, or combinations thereof are not acceptable alone as species epithets.

New Rule 17   Artificially created laboratory hybrids between different viruses will not be given taxonomic consideration.

New Rule 18   The approval by ICTV of newly proposed species, species names and type species will proceed in two stages. In the first stage provisional approval may be given. Provisionally approved proposals will be published in an ICTV report. In the second stage, after a 3-year waiting period, the proposals may receive the definitive approval of ICTV.

Rule 19   The genus is a group of species sharing certain common characters.

Rule 20   The ending of the name of a viral genus is " . . . virus".

Rule 21   A family is a group of genera with common characters, and the ending of the name of a viral family is " . . . viridae".

New Rule 22   Approval of a new family must be linked to approval of a type genus; approval of a new genus must be linked to approval of a type species.

Guidelines for the delineation and naming of species:

1.   Criteria for delineating species may vary in different families of viruses.

2.   Wherever possible duplication of an already approved virus species name should be avoided.

3.   When a change in the type species is desirable this should be put forward to ICTV in the standard format for a taxonomic proposal.

4.   Subscripts, superscripts, dashes, underlining, italics, oblique bars, or Greek letters should be avoided in future virus nomenclature.

5.   When designating new virus names, study groups should recognize national sensitivities with regard to language. If a name is universally used by virologists (those who publish in scientific journals) that name or a derivative of it should be used regardless of national origin. If different names are used by virologists of different national origin, the study group should evaluate relative international usage and recommend the name that will be acceptable to the majority and which will not be offensive in any language.

6.   ICTV is not concerned with the classification and naming of strains, variants or serotypes. This is the responsibility of specialist groups.

7.   Viral taxonomy at its present stage has no evolutionary or phylogenetic implications.

REFERENCES

1. **d'Herelle, F.,** Technique de la recherche du microbe filtrant bactériophage *(Bacteriophagum intestinale),* *C. R. Soc. Biol. Paris,* 81, 1160, 1918.
2. **Johnson, J.,** The classification of plant viruses, *Wisconsin Univ. Agr. Exptl. Sta. Res. Bull.,* 76, 1, 1927.
3. **Johnson, J. and Hoggan, I. A.,** A descriptive key for plant viruses, *Phytopathology,* 25, 328, 1935.
4. **Smith, K. M.,** *A Textbook of Plant Virus Diseases,* Churchill, London, 1937, 615.
5. **Holmes, F. O.,** *Handbook of Phytopathogenic Viruses,* Burgess, Minneapolis, 1939, 221.
6. **Valleau, W. D.,** Classification and nomenclature of tobacco viruses, *Phytopathology,* 30, 820, 1940.
7. **Thornberry, H. H.,** A proposed system of virus nomenclature and classification, *Phytopathology,* 31, 23, 1941.
8. **Fawcett, H. S.,** Virus nomenclature, *Chronica Botanica,* 7, 7, 1942.
9. **McKinney, H. H.,** Genera of the plant viruses, *J. Wash. Acad. Sci.,* 34, 139, 1944.
10. **McKinney, H. H.,** Descriptions and revisions of several species of viruses in the genera Marmor, Fractilinea and Galla, *J. Wash. Acad. Sci.,* 34, 322, 1944.
11. **Limasset, P.,** Nomenclature des virus phytopathogènes, *Ann. Epiphyties,* 12, 317, 1946.
12. **Carsner, E., Holmes, F. O., Johnson, J., McKinney, H. H., Thornberry, H. H., Weiss, F., and Bennett, C. W.,** Report of the committee on nomenclature and classification of plant viruses, *Phytopathology,* 33, 424, 1943.
13. **Holmes, F. O.,** Order Virales. The filterable viruses, in *Bergey's Manual of Determinative Bacteriology,* 6th ed., Bailli⊰ere Tindall and Cox, London, 1948.
14. **Hansen, H. P.,** Correlations and interrelationships in viruses and in organisms. I. Classification and nomenclature of plant viruses, *K. Vet. Landbohöjsk. Arsskr.,* 108, 1956.
15. **Hansen, H. P.,** On the general nomenclature of viruses, *K. Vet. Landbohöjsk. Arsskr.,* 191, 1966.
16. **Hansen, H. P.,** Contribution to the systematic plant virology, *K. Vet. Landbohöjsk. Arsskr.,* 110, 1970.
17. **Bawden, F. C.,** *Plant Viruses and Virus Diseases,* Chronica Botanica, Leiden, Holland, 1939, 272.
18. **Bawden, F. C.,** Nomina ad infinitum, *Chronica Botanica,* 6, 385, 1941.
19. **Bawden, F. C.,** *Plant Viruses and Virus Diseases,* 2nd ed., Chronica Botanica, Waltham, Mass, 1943, 294.
20. **Bawden, F. C.,** *Plant Viruses and Virus Diseases,* 3rd ed., Chronica Botanica, Waltham, Mass, 1950, 335.
21. **Valleau, W. D.,** The binomial system of nomenclature for plant viruses, *Chronica Botanica,* 6, 223, 1941.
22. **Brandes, J. and Wetter, C.,** Classification of elongated plant viruses on the basis of particle morphology, *Virology,* 8, 99, 1959.
23. **Andrewes, F. W.,** *A System of Bacteriology,* Vol. 1, H.M.S.O., London, 1930, 292.
24. **Levaditi, C. and Lepine, P.,** *Les ultravirus des maladies humaines,* Narbet Maloine, Paris, France, 1937, 1182.
25. **Goodpasture, E. W.,** Borreliotoses: fowl-pox, molluscum contagiosum, variola-vaccinia, *Science,* 77, 119, 1933.
26. **Andrewes, C. H.,** The classification of viruses, *J. Gen. Microbiol.,* 12, 358, 1955.
27. **Zhdanov, V. M.,** *Guide to Viruses Human and Animal,* Academy of Medicine and Science, USSR, Izdatelstro, Moscow, 1953, 1.
28. **Breed, R. S. and Petraitis, A.,** Some Russian contributions to taxonomy and nomenclature of the viruses. A review, *Int. Bull. Bacteriol. Nomenclature and Taxonomy,* 4, 189, 1954.
29. **van Rooyen, C. E.,** A revision of Holmes's classification of animal viruses, suborder III (Zoophagineae), *Canad. J. Microbiol.,* 1, 227, 1954.
30. **Cooper, P. D.,** A chemical basis for the classification of animal viruses, *Nature (London),* 190, 302, 1961.
31. **Lwoff, A., Horne, R., and Tournier, P.,** A system of viruses, *Cold Spring Harbor Symp. Quant. Biol.,* 27, 51, 1962.
32. **Johnson, J. and Hoggan, I. A.,** The challenge of plant virus differentiation and classification, *Proc. 5th Intern. Botan. Congr.,* Cambridge, England, 225 (abstr.), 1930.
33. **Johnson, J.,** Virus nomenclature and committees, *Chronica Botanica,* 7, 65, 1942.
34. **Andrewes, C. H.,** Classification and nomenclature of viruses, *Ann. Rev. Microbiol.,* 6, 119, 1952.
35. **Buchanan, R. E.,** Conference on virus and rickettsial classification and nomenclature. A resumé, *Int. Bull. Bacteriol. Nomenclature and Taxonomy,* 2, 71, 1952.
36. **Andrewes, C. H.,** Report of the Subcommittee on Viruses (1953), *Int. Bull. Bacteriol. Nomenclature and Taxonomy,* 4, 109, 1954.
37. **Andrewes, C. H.,** Nomenclature of viruses, *Nature (London),* 173, 620, 1954.
38. **Andrewes, C. H. and Sneath, P. H. A.,** The species concept among viruses, *Nature (London),* 182, 12, 1958.
39. **Lwoff, A.,** The concept of virus, *J. Gen. Microbiol.,* 17, 239, 1957.

40. **Andrewes, C. H., Burnet, F. M., Enders, J. F., Gard, S. Hirst, G. K., Kaplan, M. M., and Zhdanov, V. M.,** Taxonomy of viruses infecting vertebrates: present knowledge and ignorance, *Virology,* 15, 52, 1961.

41. **Andrewes, C. H.,** Classification of viruses of vertebrates, *Adv. Virus Res.,* 9, 271, 1962.

42. **Andrewes, C. H.,** *Viruses of vertebrates,* Baillière, Tindall and Cox, London, 1964, 401.

43. **Brenner, S. and Horne, R. W.,** A negative staining method for high resolution electron microscopy of viruses, *Biochim. Biophys. Acta.,* 34, 103, 1959.

44. **Horne, R. W. and Wildy, P.,** Symmetry in virus architecture, *Virology,* 15, 348, 1961.

45. **Wildy, P. and Watson, D. H.,** Electron microscopic studies on the architecture of animal viruses, *Cold Spring Harbor Symp. Quant. Biol.,* 27, 25, 1962.

46. **Horne, R. W. and Wildy, P.,** An historical account of the development and applications of the negative staining technique to the electron microscopy of viruses, *J. Microscopy,* 117, 103, 1979.

47. **Andrewes, C. H.,** Minutes of meeting of the Subcommittee on Taxonomy of the Viruses, Montreal, 1962, *Int. Bull. Bacteriol. Nomenclature and Taxonomy,* 13, 217, 1963.

48. **Lwoff, A.,** The new provisional committee on nomenclature of viruses, *Int. Bull. Bacteriol. Nomenclature and Taxonomy,* 14, 53, 1964.

49. **P.C.N.V.,** Proposals and recommendations of the Provisional Committee for Nomenclature of Viruses (P.C.N.V.), *Ann. Inst. Pasteur,* 109, 625, 1965.

50. **Lwoff, A. and Tournier, P.,** The classification of viruses, *Ann. Rev. Microbiol.,* 20, 45, 1966.

51. **Gibbs, A. J., Harrison, B. D., Watson, D. H., and Wildy, P.,** What's in a virus name? *Nature (London),* 209, 450, 1966.

52. **Adanson, M.,** *Familles des Plantes,* Vol. 1, Vincent, Paris, 1763.

53. **Gibbs, A. J. and Harrison, B. D.,** Realistic approach to virus classification and nomenclature, *Nature (London),* 218, 927, 1968.

54. **Gibbs, A.,** Plant virus classification, *Adv. Virus Res.,* 14, 263, 1969.

55. **Lwoff, A.,** Principles of classification and nomenclature of viruses, *Nature (London),* 215, 13, 1967.

56. **Harrison, B. D.,** Comments on the proposals and recommendations of the provisional committee for nomenclature of viruses, in *Proceedings of IX International Congress of Microbiology,* Moscow, 1966, 453.

57. **Wildy, P., Ginsberg, H. S., Brandes, J., and Maurin, J.,** Virus classification nomenclature and the International Committee on Nomenclature of viruses, *Prog. Med. Virol.,* 9, 476, 1967.

58. **Harrison, B. D., Finch, J. T., Gibbs, A. J., Hollings, M., Shepherd, R. J., Valenta, V., and Wetter, C.,** Sixteen groups of plant viruses, *Virology,* 45, 356, 1971.

59. **Martyn, E. B.,** *Plant Virus Names,* Commonwealth Mycological Institute, Phytopathological paper number 9, 1968, 204.

60. **Fenner, F.,** *Classification and Nomenclature of Viruses,* Second report of the International Committee on Taxonomy of Viruses, S. Karger, Basel, 1976, 115.

61. **Wildy, P.,** *Classification and Nomenclature of Viruses. First Report of the International Committee on Nomenclature of viruses,* Monographs in Virology, Vol. 5, S. Karger, Basel, 1971, 81.

62. **Baltimore, D.,** Expression of animal virus genomes, *Bact. Rev.,* 35, 235, 1971.

63. **Cooper, P. D.,** Towards a more profound basis for the classification of viruses, *Intervirology,* 4, 317, 1974.

64. **Fenner, F., Pereira, H. G., Porterfield, J. S., Joklik, W. K., and Downie, A. W.,** Family and generic names for viruses approved by the International Committee on Taxonomy of Viruses, June 1974, *Intervirology,* 3, 193, 1974.

65. **Fenner, F.,** The classification and nomenclature of viruses. Summary of results of meetings of the International Committee on Taxonomy of Viruses in Madrid, September 1975, *Virology,* 71, 371, 1976.

66. **Mumford, D. L.,** Purification of curly top virus, *Phytopathology,* 64, 136, 1974.

67. **Bock, K. R., Guthrie, E. J., and Woods, R. D.,** Purification of maize streak virus and its relationship to viruses associated with streak diseases of sugar cane and *Panicum maximum, Ann. Appl. Biol.,* 77, 289, 1974.

68. **Goodman, R. M.,** Infectious DNA from a white fly-transmitted virus of *Phaseolus vulgaris, Nature (London),* 266, 54, 1977.

69. **Goodman, R. M.,** Single-stranded DNA genome in a whitefly-transmitted plant virus, *Virology,* 83, 171, 1977.

70. **Matthews, R. E. F.** *Classification and Nomenclature of Viruses. Third Report of the International Committee on Taxonomy of Viruses,* S. Karger, Basel, 1979, 160.

71. **Matthews, R. E. F.,** The classification and nomenclature of viruses. Summary of results of meetings of the International Committee on Taxonomy of Viruses in Strasbourg, August, 1981, *Intervirology,* 16, 53, 1981.

72. **Matthews, R. E. F.**, *Classification and Nomenclature of Viruses. Fourth Report of the International Committee on Taxonomy of Viruses*, S. Karger, Basel, 1982, 199.
73. **Enders, J. F., Weller, T. H., and Robbins, F. C.**, Cultivation of the Lansing strain of poliomyelitis virus in cultures of various human embryonic tissues, *Science*, 109, 85, 1949.

Chapter 2

CURRENT PROBLEMS IN VERTEBRATE VIRUS TAXONOMY

Frederick A. Murphy

TABLE OF CONTENTS

''The world, said Paul Valery, is equally threatened with two catastrophies: order and disorder. So is virology.''

I. PERSPECTIVE

When, in 1966, the International Committee on Nomenclature of Viruses (ICNV) was established at the International Congress for Microbiology in Moscow, there was already a sense among vertebrate virologists that a universal taxonomic scheme was needed.[2] In fact, several individuals and groups had already been prompted by a rapidly growing mass of physicochemical data on many human and animal viruses to advance their own classification schemes. This overall history is detailed by Matthews in Chapter 1. The coming of age of vertebrate virology occurred logarithmically through the preceding 30 years starting with the first substantial studies of the nature of viruses in the 1930s. If one were to plot in some way efforts made to cluster viruses on the basis of similar pathogenic properties (e.g., ''hepatitis viruses'' to include hepatitis A virus, hepatitis B virus, yellow fever virus), organ tropisms (e.g., ''respiratory viruses'' to include influenza viruses, rhinoviruses, adenoviruses), or ecologic characteristics (e.g., ''arboviruses'' to include togaviruses, bunyaviruses, rhabdoviruses) against the date, the result would be a descending curve from a peak in the 1930s to a null point today. If, on the other hand, one were to plot efforts made to cluster viruses on the basis of similar virion physicochemical and structural properties, the result would be an ascending curve from a null point in the 1930s to a point of unanimity today. If one were to overlap these two hypothetical plots it would be clear that it was not until about 1950 that the two curves cross and the influence of biochemical and morphologic data could be seen as the wave of the future. From about this same time, that is 1950, there was also an explosion in the discovery of new viruses of man, and domestic and wild animals.

So, in 1966, when the ICNV was established, there was less controversy about whether the hundreds of viruses isolated from vertebrates should be classified than on the issue of just how this should be done. The central argument concerned the question of hierarchial relationships among primary groupings of viruses. Lwoff, Horne, and Tournier[1] stated that ''a system of viruses should, as any other, imply a hierarchy of characters and a hierarchy implies a choice'' (choice among criteria and weighting of criteria). Even with the disclaimer that no phylogenetic relationships were implied, this scheme would have placed all the viruses in one ''kingdom'' and would have based descending hierarchial divisions upon nucleic acid type, strategy of replication, capsid symmetry, presence or absence of envelope, and further capsid structural details. The hierarchial issue was defeated by the ICNV on the basis that it was presumptous and implied phylogenetic relationships, irrespective of disclaimers to the contrary (see Chapter 1). This early decision set into place within the ICNV the overall scheme which has been built upon ever since. In this scheme, as many virion characteristics as possible are considered and weighted as criteria for making divisions. The relative weight of each characteristic is in fact set arbitrarily, and influenced by prejudged relationships which ''we would like to believe but are unable to prove.''[3] The scheme does not involve any hierarchial levels beyond families, and the scheme does not imply any phylogenetic relationships beyond those proven experimentally (e.g., by nucleic acid hybridization, nucleotide sequencing, and gene reassortment experiments).

Using this scheme, the First Report of the ICNV,[4] covering the term 1966 to 1970 under the Presidency of Wildy, included the following ''genera'' of vertebrate viruses: *Poxvirus, Iridovirus, Herpesvirus, Adenovirus, Papillomavirus, Polyomavirus, Parvovirus, Reovirus, ''Leukovirus'', Paramyxovirus, Orthomyxovirus, Rhabdovirus, Alphavirus, ''Flavovirus'', Calicivirus, Rhinovirus, Enterovirus, Coronavirus, Arenavirus*. Already at the time of this First Report there was an intention to elevate all major groupings to the level of families; two had already been approved. *Papovaviridae* and *Picornaviridae*. These constructions came from the First Vertebrate Virus Subcommittee of ICNV under the chairmanship of Sir Christopher Andrewes (members: H. G. Pereira, J. Maurin, D. Blaskovic, J. B. Brooksby, J. Casals, H. S. Ginsberg, M. M. Kaplan, J. L. Melnick, R. Rott, P. Tournier, C. J. York, V. Zhdanov). Some of the constructions were based upon long accepted groupings, others were completed very shortly after discovery of the unique nature of member viruses (e.g., *Coronavirus, Arenavirus*).

By the time of the Second Report of the International Committee on Taxonomy of Viruses (ICTV)*,[5] covering the term 1971 to 1975 under the Presidency of Fenner, the Vertebrate Virus Subcommittee under the chairmanship of Pereira had assimilated the vast amounts of virus characterization data which were then coming forth on many viruses of man, domestic animals, and other vertebrates. Fewer and fewer unclassifiable viruses remained, while hundreds of viruses were assigned to one or another of 16 families. It was already being said that "it seems unlikely that many new families of viruses of vertebrates remain to be discovered, although there are probably very large numbers of as yet undiscovered species of viruses among vertebrate species other than man and domesticated animals".[5] During the 1971 to 1975 term of the Vertebrate Virus Subcommittee, 10 Study Groups were set up, leading to the approval of several new virus family constructions by the end of the term. The families containing vertebrate viruses as of 1975 were *Poxviridae, Iridoviridae, Herpesviridae, Adenoviridae, Papovaviridae, Parvoviridae, Reoviridae, Retroviridae, Orthomyxoviridae, Paramyxoviridae, Rhabdoviridae, Togaviridae, Picornaviridae, Coronaviridae, Bunyaviridae, Arenaviridae*.[6] At this time five of the families contained one genus, and the rest more than one, but there was expectation that many more genera would soon be created. At the same time it was already clear which families represented difficult taxonomic problems; the *Herpesvirus* and *Retrovirus* Study Groups were seen to have particularly difficult problems, involving the need for complex substructuring within these families.

The Third Report of the ICTV, covering the period 1976 to 1978, under the Presidency of Matthews, brought the number of approved families and "groups"** to 50 (with three additional families proposed), but the number of families containing viruses of vertebrates remained at 16, the same as in 1975.[7] The Fourth Report of the ICTV, covering 1979 to 1981 and continuing under the Presidency of Matthews, added a few new families, but only one, the family *Caliciviridae*, contains vertebrate viruses.[8] Many of the issues facing the Vertebrate Virus Subcommittee during these two most recent terms, covering the period 1975 to 1981, are still unresolved. These will be considered in some detail. During this period the Subcommittee, under the chairmanship of Murphy (members: F. Brown, C. Andrewes, P. Bres, J. Casals, J. H. Gillespie, I. H. Holmes, M. C. Horzinek, W. K. Joklik, R. Kono, J. Maurin, J. L. Melnick, P. K. Vogt, V. M. Zhdanov), switched most primary taxonomic responsibility to 18 Study Groups (12 Study Groups representing virus families containing vertebrate viruses, plus six Study Groups representing virus families containing viruses of vertebrates, plants and invertebrates). The Study Groups and their chairmen were

Study group	Chairman
1. *Poxviridae****	K. Dumbell
2. *Iridoviridae****	A. Granoff/D. Willis
3. *Herpesviridae*	B. Roizman

---

* The International Committee on Nomenclature of Viruses (ICNV) became the International Committee on Taxonomy of Viruses (ICTV) in 1973.

** The term "group" rather than family is used by plant virologists; see Chapter 3.

***Families containing viruses of vertebrates, invertebrates and/or plants; therefore managed by the Coordination Subcommittee of the ICTV (see Chapters 1 and 3).

4. *Adenoviridae*	R. Wigand
5. *Papovaviridae*	J. Melnick
6. *Parvoviridae**	P. Bachmann/G. Siegl
7. *Reoviridae**	W. Joklik
8. *Retroviridae*	P. Vogt
9. *Paramyxoviridae*	D. Kingsbury
10. *Orthomyxoviridae*	W. Dowdle
11. *Rhabdoviridae**	F. Brown
12. *Togaviridae*	J. Porterfield/M. Horzinek
13. *Picornaviridae*	P. Cooper/R. Rueckert
14. *Caliciviridae*	F. Schaffer
15. *Coronaviridae*	D. Tyrrell
16. *Bunyaviridae*	D. Bishop
17. *Arenaviridae*	W. Rowe/C. Pfau
18. Unnamed family comprising the viruses with 2 segments of DS RNA*	P. Dobos

More than 215 virologists served as members of these Study Groups during this period; this broad democratic representation has tied taxonomic decisions to the leading edge of virological research and has assured that the universal scheme continues to be useful and used.

II. CURRENT ISSUES IN VERTEBRATE VIRUS TAXONOMY

A. Interrelationships of the ICTV with Other Organizations and Specialty Groups

Throughout its early years, the ICTV competed with other organizations seeking to develop classification and nomenclature systems; usually these organizations restricted their activities to viruses of special interest. For example, the Western Hemisphere Animal Virus Characterization Committee (organized under the United States Animal Health Association) continually updated a system for classifying veterinary pathogens.[9] One of the objectives of the WHO/FAO Comparative Virology Programme was to do the same.[10] Some of these organizations dissolved spontaneously and others shifted their objectives as the ICTV exerted its dominance over all virus taxonomy. For example, in 1976, the Working Teams of the WHO/FAO Comparative Virology Programme arranged overlapping membership with parallel ICTV Study Groups and then dropped their separate taxonomic activities. This kind of complete deferral to the ICTV has left some important voids.

There has been a decline in specialty group activities concerned with relationships among virus species, subspecies, and strains. Specialty groups had been led to believe that the ICTV would deal with such matters, but because of its broadness, ICTV and even its Study Groups lack necessary expertise. A formal statement was made in 1980 on behalf of the ICTV by its President, Matthews, clarifying to some extent the line of responsibility between the ICTV specialty groups.[8] This line was set between the species and subspecies levels for the time being — that is, the ICTV retains primary responsibility for placement of viruses at the species level and defers to specialty groups at the subspecies and strain level. This decision pertains to several matters of practical importance, such as the means of deciding whether an isolate represents a new virus species (or a new strain), and the means of choosing type species (and type strains).

---

* Families containing viruses of vertebrates, invertebrates, and/or plants; therefore managed by the Coordination Subcommittee of the ICTV (see Chapters 1 and 3).

The matter of deciding whether a particular isolate represents a new species or a new strain remains unresolved. Surely the ICTV must make this decision, but evidence and opinion must derive from specialty group proceedings. The virologist who makes a candidate isolation is biased toward wanting it to stand as a new species; checks and balances must counter this influence. (See also Chapter 9 for a discussion of this problem).

The matter of deciding how type species (and type strains) are chosen also remains unresolved. In some cases the personal interests of individuals conflict with the scientific bases for choosing prototypes. This is especially the case where prototypes become valuable as substrates for vaccines, diagnostic reagents, etc. In this regard, some confusion may have been caused by the ICTV's identification of type viruses as part of the definition of genera. However, the ICTV has never identified type viruses with the kind of precision that is used by specialty groups or culture collections (isolate identification, passage history, dates, etc.). The best model for the kind of precise description necessary to avoid all unambiguity is that of the American Type Culture Collection in its constantly updated Catalogue of Strains of Viruses. For example, St. Louis encephalitis virus is listed as:

ATCC VR-80 St. Louis encephalitis virus
Strain: Hubbard. Original source: Brain of patient, Missouri, 1937. Reference: McCordock, H. A., et. al., *Proc. Soc. Exp. Biol. Med.*, 37:288,1937. Passage History: M/103 (Lot #3). Preparation: 20% SMB in 50% NIRS infusion broth; supernatant of low speed centrigugation. (additional information concerns growth parameters)

Clearly, this kind of description goes beyond the ICTV's mandate, into the area of responsibility of specialty groups. There is need for the ICTV to take new initiatives to coordinate these and other matters which affect taxonomy at the species level. These initiatives must be organized with close communication among internationally recognized specialty groups. In some cases the specialty groups must first be identified and encouraged to undertake necessary activities. Specialty groups must represent working virologists in the field and must coordinate their activities through the ICTV, its Subcommittees and its Study Groups. Specialty groups should consider publishing their proceedings and recommendations in *Intervirology*, the official journal of the Virology Division, International Union of Micro biological Societies. Specialty groups should also develop a working relationship with the WHO Center for the Collection and Evaluation of Data on Comparative Virology (Munich). This Center, which is tied to the ICTV, to the WHO/FAO Comparative Virology Programme, and to the World Data Center for Microorganisms (Brisbane), provides wide access and distribution of the information stemming from ICTV and specialty group proceedings.

There are a few specialty groups operating now which may serve as examples of ways that various virologic constituencies may be served. One example is WHO's Committee on Influenza. This specialty group has become internationally recognized as the controlling influence upon orthomyxovirus activities world-wide. The group has ties to WHO International Reference Center laboratories, Regional Reference Center laboratories, and to the WHO Priority Programme on Acute Respiratory Infections (ARI). Through these interrelationships, the Committee has access to information from most countries of the world pertaining to new strain emergence, new research findings, and other matters affecting public health. There is a close working relationship between this specialty group and the ICTV *Orthomyxoviridae* Study Group. This relationship is achieved primarily by overlapping memberships. The primary need of the specialty group is to identify influenza virus isolates with extreme precision. The cryptic strain nomenclature system developed over many years by the group allows unambiguous communication regarding the emergence and movement of particular strains.[12] The practical needs of influenza epidemiologists have been melded into the ICTV's universal taxonomy scheme quite satisfactorily: The ICTV *Orthomyxoviridae* Study Group has divided the family into a genus *Influenzavirus* (comprising influenza viruses

A and B) and a probable genus, unnamed, comprising influenza virus C. The ICTV Study Group has the species issue under review, and the specialty group, under WHO auspisces, has modified the strain nomenclature scheme several times in the past 10 years. This is an excellent example of specialty group — ICTV Study Group interrelationship with benefit for all.

Another example of specialty group involvement in taxonomy concerns the WHO Expert Committee on Viral Hepatitis and its assumed leadership role in developing an unambiguous nomenclature for hepatitis B virus (HBV). This committee's terminology scheme has replaced all conflicting proposals from individual virologists.[13] This had been done without interaction with the ICTV, but now, this shortcoming is being resolved. An interim ICTV Hepatitis Study Group, chaired by Gust, was established in 1979, and a formal HBV Study Group is being established in the present term. There is now overwhelming physicochemical evidence that HBV and related viruses deserve placement in a new family. The charge to the new Study Group is to develop a formal basis for this construction. For these viruses as with some others, ICTV action may seem too slow. However, the debate and experimentation over terminology which was fought out in the specialty group's proceedings has allowed the ICTV to continue to maintain a valuable principle — that is, taxonomic constructions and terms, once set in place by the ICTV, shall not be changed except for the most compelling reasons.

In the case of the arboviruses, the WHO defers to the American Committee on Arthropod-borne Viruses and its Subcommittee on Interrelationships Among Catalogued Arboviruses (SIRACA) in matters pertaining to serologic relationships and nomenclature. Despite its name, this specialty group has international representation, and unlike many specialty groups, this one actually supports and coordinates research projects aimed at establishing quantitative and hierarchial relationships among the large numbers of related arboviruses. This work is financially supported by national agencies because it is a key to epidemiologic investigation of disease trends; nevertheless, it also extends significantly the base for formal taxonomic decisions. This group has published in *Intervirology* multitiered substructures for the genus *Alphavirus* and the genus *Bunyavirus,* and is working on others.[14,15] These substructures, falling at subgenus, species, subspecies, strain, variant, etc., levels, are shared with ICTV *Togaviridae* and *Bunyaviridae* Study Groups, so that the results fit the universal taxonomy scheme while at the same time providing value to arbovirus epidemiologists. This is a good model for other specialty groups working independently of the ICTV.

One specialty group, the Simian Virus Working Team of the WHO/FAO Comparative Virology Programme (S. S. Kalter, chairman) has undertaken activities which involve many virus families and, therefore, relate to many ICTV Study Groups. This specialty group controls reference identification of viruses isolated from primates, and controls reference diagnostic reagents and a master collection of virus stocks. Such activities have involved the specialty group in the naming of viruses and in the publication of a comprehensive scheme for simian virus taxonomy.[11] In its efforts to serve its constituency, the specialty group has moved faster than several of the ICTV Study Groups concerned with virus families containing simian viruses. Discrepancies have involved herpesviruses, adenoviruses, and some others. Certainly, such discrepancies are resolvable under guidelines which should apply to all specialty groups: (1) the ICTV, its officers and its Study Group chairmen, and specialty group chairmen must maintain necessary communication no matter how bothersome this may be; (2) ICTV deliberateness (usually stemming from the difficulty of taxonomic decisions) must not be bypassed in the specialty group's sense of urgency; and (3) ICTV must be accepted as having final authority for all virus taxonomy — specialty groups must in the end bend to this position of seniority.

B. Parallel Taxonomic Structure Among the Families Containing Vertebrate Viruses

Because the taxonomic structures used to organize relationships between member viruses within vertebrate virus families often predate the ICTV, and because the Study Groups now concerned with these families work rather independently, there has been a drift toward increasing differences in structural detail between families. This drift has accelerated recently and it seems likely that it will continue, since it suits the research virologist working in a narrow subject area (usually in an area much narrower than a whole virus family). That is, accomodation of the taxonomy of a family, or even a genus, to the data base of the day, suits the research worker who otherwise would see taxonomic structural change as an unnecessary bureaucratic exercise. The data base and emphases used in making physico-chemical comparisons vary greatly between virus families. On the other hand, the other users of the universal taxonomic scheme, that is teachers, students, diagnosticians, and specialists in related fields such as infectious diseases, entomology, ecology, etc., have an increasing burden in trying to keep up with the growth in the size and diversity of virus taxa, and this is not helped by the use of different structural details within each family. The formal "Rules of Nomenclature of Viruses" published in each ICTV Report provide the overall influence upon common structuring within all families, but these rules are necessarily very broad in order to be equally valid in plant, invertebrate, and bacterial virology. These rules alone will not be too effective in supporting the solution of detailed problems within vertebrate virology.

Some examples may help make the point that new ICTV initiatives toward achieving structural parallelism between virus families are needed. Such examples also serve as a reminder of the great differences in the relationships of member viruses within the various families. Consider the basis for subdividing the family *Poxviridae*, wherein "host specific-ity" is used to subdivide the family into two subfamilies, the *Chordopoxvirinae* and the *Entomopoxvirinae* and, is also used for making subdivisions into genera such as *Avipoxvirus, Capripoxvirus*, etc. In this scheme the major physicochemical differences between the mem-ber viruses of the genus *Parapoxvirus* and all other poxviruses is given too little weight, and other physicochemical data have not been used comprehensively. The result is that *Poxviridae* taxonomy seems out of date.

Consider the bases for subfamily structuring in the families *Herpesviridae* and *Retrovir-idae*. Both are extremely complex and require further explanation to the user of virus taxonomy. Perhaps the bases for substructuring in these two families represent another wave of the future, but further efforts must be made to complete the genus (and species) con-structions in these families so that the issue of parallelism with other families may be addressed.

Consider the early justification of placing papillomaviruses and polyomaviruses together in the family *Papovaviridae* despite the great difference in the size of the particles and genome of member viruses. Consider the early justification of placing the alphaviruses and the flaviruses (and the rubiviruses and pestiviruses) together in the family *Togaviridae* despite major differences in protein composition and strategy of replication of member viruses. More parallelism is needed. At a lower level in the taxonomic hierarchy consider the present wish to delay placement of hepatitis A virus in the genus *Enterovirus* pending demonstration of post-translational cleavage and other evidence of replication strategy. Parallelism must relate to new as well as old taxonomic constructions, and must relate to all hierarchial levels.

Consider the many incongruities in the definition of the term "genus" and consider the ease in defining new genera (and species) in some families compared with difficulty in others. New species of adenoviruses and enteroviruses emerge from reference center labo-ratories by established criteria. New species of alphaviruses, flaviviruses, and bunyaviruses emerge similarly, but with criticism from other virologists. Meanwhile, there are many problems in deciding the taxonomic status of most rhabdoviruses, orbiviruses, rotaviruses, and coronaviruses. Some effort toward considering parallel constructions might help solve these problems.

Clearly the issue of parallelism in taxonomic constructions centers on the subfamily and genus levels, but the issue also influences considerations regarding the definition of species (see Section III and Chapter 9). The achievement of more parallelism in vertebrate virus taxonomy requires, first, a commitment by ICTV leadership, and then a plan for action. The circumstances for addressing the issue are not present in the regular meetings of the ICTV, its Subcommittees or its Study Groups. Such meetings are consumed with handling specific taxonomic proposals, many of which further the drift away from parallelism. Perhaps progress would best be made at a special workshop attended by members of the Vertebrate Virus Subcommittee, Study Group chairmen and ICTV Executive Council members.

C. The Adequate Description of "New" Viruses

Over the years, as thousands of viruses have been isolated from vertebrate hosts, there have been many errors of duplication — that is, viruses isolated in different laboratories have been given different names and thereby the chance for coexistence in various listings. There have also been placements of viruses in wrong listings — the most notable instance involving the emergence of the family *Reoviridae* from the initial hiding place of its prototype member in the long listing of the enteroviruses. The basis for these and similar problems has been inadequate characterization and inadequate review and oversight by appropriate bodies. The number of such problems is declining in vertebrate virology, but because taxonomic matters are affected by errors in identification or placement, continuing attention is warranted.

When an "unknown" is first worked on in a laboratory, its initial characterization may be considered as just an extension of diagnostic protocols. That is, very few characteristics are determined as leads for specific (usually serologic) identification procedures. Only when an "unknown" fails to yield to the diagnostic system is there call for comprehensive study of many physicochemical properties. One key to simplifying such study is to set useful techniques in proper sequence. This is easier said than done; many laboratories handling "unknowns" do not use techniques which represent "short-cuts", but instead continue to run all standard tests so as to "fill in all the boxes". In fact, the only situation where there should be no "short-cuts" is in the characterization of a virus which is to stand as the type virus of a new family. In all other instances techniques and their sequence of use should be reviewed and streamlined. For example, electron microscopy is a great "short-cut" for placing unknowns which emerge from diagnostic protocols. If an unknown is shown to be a rhabdovirus by electron microscopy what is the value of routinely checking whether its genome is DNA or RNA, or whether it is sensitive to low pH or solvents or particular enzymes? What is the value of serologic testing of the "unknown" against viruses other than the known rhabdoviruses? Perhaps comprehensive testing of virus characteristics is a key to the discovery of novel viruses (such as a hypothetical bullet-shaped virus with a double-stranded DNA genome), but the system is not productive and is so demanding of resources that it detracts from needed, focused study of viruses exhibiting some characteristic indicative of an unusual nature. Adequate description of *new* viruses deserves more of the reference laboratory's resources.

It is impossible to generalize on approaches which would best lead to adequate description of the first member of a new species, genus, or family. The differences among the known viruses of vertebrates are so great that such an attempt to describe techniques and the sequence of techniques would be naive. The key to developing an adequate description of a new virus lies in *expert* use of discriminating techniques. Clearly, this is the role of the reference laboratory or the international reference center for the virus family in question. The procedures always require accomplished expertise in working with the particular virus type. There is always need for reference materials, that is a prototype virus collection and a serologic reagent collection (antigens, antisera). There is always need for broad communication between laboratories making preliminary identifications, and there is need for influencing quality control. For all those virus families where international reference centers exist, problems of duplication (one virus, two names) or other failings due to inadequate characterization have been reduced to a minimum. Taxonomic problems have likewise been reduced to a minimum. For those virus families where independent reference centers do not exist, there needs to be extra review conducted by the appropriate ICTV Study Group. Where a ''new'' virus does not fit into an established family, additional review must be conducted by the ICTV Executive Committee, itself.

D. Issues Pertaining to Individual Virus Families

For many reasons, different issues are of paramount concern in each of the 18 ICTV Study Groups dealing with vertebrate viruses. The following is an attempt to review some of these issues:

Poxviridae — Two subfamilies, the *Chordopoxvirinae* and the *Entomopoxvirinae* were proposed by the ICTV Study Group under the chairmanship of Dumbell; these were approved by the ICTV in 1978 in recognition of major differences in the poxviruses of vertebrates and insects. However, this division certainly does not mark physicochemical differences as great as those between orthopoxviruses and parapoxviruses; a complete review of the substructure within this family is needed as a basis for reemphasizing virion physicochemical characteristics, rather than host species or serological relationships, as critiera for the delineation of subfamilies and genera.

Iridoviridae — African swine fever virus is the only iridovirus which infects vertebrates, so it is appropriate that most taxonomic matters concerning this family remain in the hands of invertebrate virologists. In 1981, genera were constructed by the ICTV Study Group under the chairmanship of Willis to accommodate all the well known iridoviruses (see Chapter 5); a separate genus was reserved for African swine fever virus. This genus has not been named. It is important that virologists working on African swine fever be invited to become involved in the naming of this genus. Ongoing physicochemical studies on this virus will go far to formalize the definition of the genus.

Herpesviridae — Three subfamilies were approved in 1978, the *Alphaherpesvirinae* comprising viruses similar to herpes simplex virus of man, the *Betaherpesvirinae* comprising viruses like the cytomegalovirus of man, and the *Gammaherpesvirinae* comprising viruses like EB virus of man. This construction was the tip of the iceberg with respect to the activity of the ICTV Study Group under the chairmanship of Roizman. It provides a framework for the creation of genera in the near future — a matter which had been impossible previously because of differences in opinion as to bases for genus definition. The Study Group has recently published a report in *Intervirology*[16]; this report sets the stage for species definition as well as genus construction.

Adenoviridae — Construction of two genera, *Mastadenovirus* and *Aviadenovirus,* was complete by the time of the First Report of the ICTV Study Group published in *Intervirology* in 1976.[17] Recently, proposals for species nomenclature were developed by the Study Group under the chairmanship of Wigand and were provisionally approved by the ICTV (Table 1): these are the first species terms ever given such formal approval.[18] Species have been defined precisely by serologic and physicochemical criteria and have been given names consisting of a three-letter code for the host genus followed by a sequential arabic number.[8,17] In addition, a scheme for creating five subgenera of the human adenoviruses has been approved. Bases for these divisions are primarily physicochemical. Thus, this family becomes one model for naming species. It will serve as the model for those families where individual viruses have been indentified in the past by serial numbers or number-letter terms.

Papovaviridae — Constructions within this family were formalized in 1974 in a published report of the ICTV Study Group under the chairmanship of Melnick.[19] One recent issue has

Table 1
GENERA AND SPECIES IN THE *ADENOVIRIDAE*

| Host | | Virus | |
English name	Zoological name	Genus	Species
Man	*Homo sapiens*		h1-h37
Cattle	*Bos taurus*		bos1-bos9
Pig	*Sus domesticus*	*Mastadeno-*	sus1-sus4
Sheep	*Ovis aries*	*virus*	ovi1-ovi5
Horse	*Equus caballus*		equ1
Dog	*Canis familiaris*		can1
Goat	*Capra hircus*		cap1
Mouse	*Mus musculus*		mus1
Fowl	*Gallus domesticus*		gal1-gal9
Turkey	*Neleagris gallopavo*		mel1-mel2
Goose	*Anser domesticus*	*Aviadeno-*	ans1
Pheasant	*Phasianus colchicus*	*virus*	pha1
Duck	*Anas domestica*		ana1

concerned the propriety of having the genus containing the human viruses BK and related isolates named *Polyomavirus*. The inference that these human isolates cause multiple tumors in man prompted the Study Group to recommend changing the genus name. After repeated discussions by the ICTV, this proposal for change was rejected on the grounds that anthropocentric considerations should not be a major influence in taxonomic decisions. This rejection was also influenced by the principle that terms which have been in use for some time must not be changed except for the most compelling reasons.

Parvoviridae — For many years the construction of two genera, the *Parvovirus* and *Densovirus* genera, had been considered complete, but no agreement could be reached for naming the genus comprising the adeno-associated viruses.[20] This impasse was resolved in 1981 with the approval by the ICTV of a recommendation of the Study Group, under the chairmanship of Siegl, for the name *Dependovirus*. An ongoing problem facing the Study Group concerns biological differences of great practical importance between parvoviruses which are virtually identical physicochemically. In particular, the etiologic agents of feline panleukopenia and canine parvovirus infection exhibit 90% DNA homology and similar near identity in other assays, but cause diseases in different host species in nature. This matter was submitted to the ICTV in 1981 by the Study Group in the form of a motion to divide one species (presently termed feline panleukopenia virus) into three subspecies (feline panleukopenia virus, canine parvovirus, mink enteritis virus). This proposal which did not include official names was deferred pending further species definition in this family, and pending the recommendation of an international specialty group.

Reoviridae — This family contains six genera, five with officially approved names and one, comprising the cytoplasmic polyhedrosis viruses of insects, with the proposed name, name, *Cypovirus*. This name will be submitted to the ICTV for final approval in the present term. Of these six genera, two *(Phytoreovirus and Fijivirus)* represent the first approved genera for any plant viruses; as such they set a precedent and become a model for cooperation in taxonomic decisions by plant, invertebrate and vertebrate virologists. This Study Group, under the chairmanship of Joklik, also served as a model for the democratic process in taxonomic decision-making by its handling of the creation of the genus *Rotavirus*. At present the same approach is being used by the Study Group to manage detailed substructuring in the genus *Orbivirus* so as to maintain the definition of the genus and at the same time settle problems stemming from the economic impact of newly identified bluetongue virus types.*

Retroviridae — Many compromises were made in 1975 in order to establish a general structure for this family; these compromises have been rather well accepted and the subfamily structure is being used (three subfamilies, *Oncovirinae, Spumavirinae, Lentivirinae*). However, progress has been slow on subdivision at the genus and species level. The ICTV Study Group chairman, Vogt, published a paper on definitions and characteristics of the subfamily *Oncovirinae* in 1976,[21] but review of the whole family substructure with the aim of defining and naming genera is now needed.

Paramyxoviridae — The three genera comprising this family were approved in 1975 *(Paramyxovirus, Morbillivirus, Pneumovirus).*[22] In the absence of pressing problems, the ICTV Study Group under the chairmanship of Kingsbury, starting in 1979, undertook to propose species nomenclature as a model for those families where individual viruses are named rather than numbered. The Study Group's choice of terms included *Paramyxovirus newcastle, Paramyxovirus mumps, Paramyxovirus parainfluenza hl,* (. . . h2, h3, h4), *Paramyxovirus yucaipa, Morbillivirus measles, Morbillivirus distemper, Morbillivirus rinderpest, Pneumovirus syncytium h1, Pneumovirus bov1, Pneumovirus mus1.* Terms for other viruses, such as peste-des-petits-ruminants virus, were debated but in the absence of further guidelines were not decided. When the members of this Study Group reviewed the letter/ number species scheme developed by the Adenoviridae Study Group (see Section II.D.) in 1980, they developed a parallel scheme as a working exercise. In this scheme, Newcastle disease virus became *Paramyxovirus gall*, mumps virus became *Paramyxovirus h5* (following *Paramyxovirus h1 to h4* for the human parainfluenza viruses), measles virus became *Morbillivirus h1*, canine distemper virus became *Morbillivirus can 2*, rinderpest virus became *Morbillivirus bos1*, respiratory syncytial viruses became *Pneumovirus h1, Pneumovirus bos1* and *Pneumovirus mus1.* This system allowed the Study Group to solve problem of species terms, e.g., peste-des-petits-ruminants virus as *Morbillivirus cap1.* The two models, the species name model and the species letter/number code model, are valuable examples — they highlight the issue of whether virologists want to keep well-known vernacular (mostly English) terms, or want to develop a uniform coding system for the formal species epithet in the universal taxonomy scheme. At this point this, and other, Study Groups await ICTV action.

Orthomyxoviridae — As stated above (Section II.A), the taxomonic structure of this family is consistent with the nomenclature code developed by WHO. The ICTV Study Group, under the chairmanship of Dowdle, has published a formal report,[23] and keeps close ties with WHO.

Rhabdoviridae — Taxonomic structure within this family was considered in a formal report of the ICTV Study Group, under the chairmanship of Brown, published in 1979.[24] Two approved genera *(Vesiculovirus* and *Lyssavirus)* have been recognized by ICTV, covering about 14 viruses, but all of the other rhabdoviruses are listed without regard to further subdivision. The question had been asked whether there was any value in dividing the more than 80 other rhabdoviruses according to host— plant, invertebrate, and vertebrate rhabdoviruses, but this was dismissed in the hope that better criteria would emerge. This matter should be addressed in the present term of the ICTV, if for no other reason than to help in teaching the virology of this large and important family.

Togaviridae — This was one of the first families created by the ICTV, but its status is far from settled. The ICTV Study Group, under the co-chairmanship of Porterfield and Horzinek, published a formal report in 1978[25] which made it clear that virion composition, construction and genome strategy differ substantially between the member viruses of the

---

* *Reoviridae* Study Group subcommittee established in 1981, B. M. Gorman, chairman.

genera *Alphavirus* and *Flavivirus* (and are also likely between the genera *Rubivirus* and *Pestivirus*). This matter must be readdressed and a consensus reached as to whether the family should remain intact or be divided. That is, should the similarities of member viruses of the four genera, which justified the formation of the family in the first place, be weighted more than the more recently discovered differences, or not? This matter is an example of the arbitrary nature of many of the key decisions in virus taxonomy, especially those made early and on rather incomplete evidence. Clearly, it will be necessary throughout virus taxonomy to reconsider older structures from time to time in the light of continuing study and to restructure where necessary and appropriate.

Picornaviridae — The structure of this family was refined by division into four genera in 1978 *(Enterovirus, Cardiovirus, Rhinovirus, Aphthovirus)*. This was justified in the formal report of the ICTV Study Group, under the chairmanship of Cooper (succeeded in 1980 by Rueckert).[26] This resulted in a precise definition of criteria for membership in the family, and hastened the removal of the caliciviruses and the many small RNA viruses of bacteria, plants and insects whose virion structure and/or genomes strategy were known or thought to be different from the picornaviruses. A task for the Study Group in the future will be to consider the placement back in the family of those few excluded small RNA viruses which are now proven to match the vertebrate virus members of the family in relevant taxonomic characteristics. Another task concerns the placement of hepatitis A virus (HAV) in the genus *Enterovirus* now that data are accumulating to indicate that such placement is appropriate.

Caliciviridae — Final approval for this family was given by the ICTV in 1981, based upon the published report of the ICTV Study Group, under the chairmanship of Schaffer.[27] At this time, there is no real basis for subdividing this monogeneric family.

Coronaviridae — There has been much recent progress toward understanding the composition and genome strategy of the coronaviruses; these new data reinforce the decision to maintain the monogeneric structure of the family which was made in the 1978 report of the ICTV Study Group, under the chairmanship of Tyrrell.[28] In addition, this Study Group has found it necessary to make it known that comprehensive data must be presented to advance the candidacy of isolates as "new" coronaviruses (see II.C). This was prompted by claims in the literature for discovery of new coronaviruses based solely upon negative contrast electron microscopic findings of "coronovirus-like particles" in stool specimens from various animal species. The problem of adequately based descriptions for "new" viruses is a general one, but is particularly acute for Study Groups dealing with uncultivatable viruses and "virus particles" found in feces.

Bunyaviridae — In 1981, the structure of this family was filled out by ICTV approval of three new genera *(Phlebovirus, Nairovirus, Uukuvirus)* complementing the original genus, *Bunyavirus*. This construction, based upon the 1981 report of the ICTV Study Group, under the chairmanship of Bishop,[29] has reduced to very few viruses the former list of more than 150 "other possible bunyaviruses". The construction has also resulted in placement of important pathogens of man and domestic animals in taxa together with viruses with well-known arthropod vector cycles. This has led to several important ecologic extrapolations concerning viruses such as Rift Valley fever virus, Nairobi sheep disease virus, and Crimean hemorrhagic fever virus.[30] For example, when it was shown by Shope that Rift Valley fever virus was serologically related to members of the *Phlebovirus* genus, new questions regarding the unknown interepidemic cycle of this important pathogen were asked. Rift Valley fever virus is spread in its epidemic cycle by mosquitoes, but its serologic relationships beg the question of whether phlebotomus flies play a role in its overall natural history. Studies to answer this question have been proposed. Similarly, when a virus from India called Ganjam virus was shown to be serologically indistinquishable from Nairobi sheep disease virus of East Africa, epidemiologic and natural history questions were framed by knowledge of other members of the genus *Nairovirus*. Similar epidemiological questions have been framed in regard to the distiribution of another member of the genus *Nairovirus,* the human pathogen Crimean hemorrhagic fever virus, as it emerged from diverse sources in the Soviet Union, Pakistan, the Persian Gulf, and South Africa. These examples of discoveries of unexpected taxonomic relationships between some of the many viruses in the family *Bunyaviridae* confirm the practical value of the universal virus classification scheme. These discoveries have allowed extrapolations and predictions of epidemiologic and ecologic characteristics in human and animal diseases. As the taxonomy scheme is used more, we can expect more of these kind of discoveries in this family and in other families as well.

Arenaviridae — This family's original construction, based upon a formal ICTV Study Group report completed in 1974 under the chairmanship of Rowe (later succeeded by Pfau) has not changed.[31] Its monogeneric structure still seems appropriate, although it has been suggested by some investigators that the family should be divided according to the western hemisphere and eastern hemisphere origin of member viruses.

Proposed unnamed family comprising the viruses with 2-segments of DS RNA — That the viruses like infectious pancreatic necrosis virus of fish and bursal disease virus of fowl deserve separate, new family status has been substantiated by numerous studies of virion physicochemical properties. A Study Group, under the chairmanship of Dobos, was established by the ICTV in 1980 to develop the formal basis for such action. Initial proposals were considered by the ICTV at its meetings in Strasbourg in 1981, but formalization of the family was deferred until 1984. The question of an appropriate name for the family has not yet been resolved; the term Birnaviridae has been advanced and is being used vernacularly by some virologists in the field.

Unnamed and as yet only informally proposed family comprising hepatitis B virus (HBV) and related viruses from woodchucks and other mammals — As stated above, there is overwhelming physicochemical evidence that these viruses warrant separate, new family status. This matter is being addressed by the ICTV in its present term.

Unnamed and as yet only informally proposed family comprising Marburg and Ebola viruses — The unusual nature of these viruses, which are the most lethal human pathogens known other than rabies virus, was appreciated from their first isolation and morphologic characterization. In anticipation of the establishment of a new family, the *Rhabdoviridae* Study Group of the ICTV, under the chairmanship of Brown, was given "watching brief" responsibility in 1977, while RNA and protein characterization studies were in progress. Recently, an informal group of virologists, under the chairmanship of Kiley, has gathered data emerging from these studies and has proposed the immediate construction of a new family.[32] This will be considered by the ICTV in the present term.

Unclassified viruses — There are viruses which remain unclassified or unclassifiable — the leftovers of the universal taxonomy scheme. Some remain in this category for lack of comprehensive study (usually because they have not been grown in the laboratory, e.g., Borna disease virus the agent of roseola infantum [also called exanthum subitum] and the agent of cat-scratch disease). Others remain unclassified because their unusual or mysterious nature requires new decisions by the ICTV (e.g., the agents of the spongiform encephalo-pathies [Kuru, Creutzfeldt-Jakob disease, scrapie, mink encephalopathy, and wasting disease of deer and elk]). There are others, and more are certain to be isolated in the future.

III. THE SPECIES ISSUE AS IT PERTAINS TO VERTEBRATE VIRUSES

From the first meeting of the ICNV in 1966, there was an intention that the universal taxonomic scheme would, one day, be completed to the species level (species being defined as synonomous with "virus"). In the first years of the ICNV, the species issue was properly deferred — it was a matter of "first things first" and family and genus structuring appeared to many vertebrate virologists to be of primary importance. As an interim solution to the need to identify and name species, ICTV reports listed viruses under a heading "English Vernacular Name" (usually without mention of vernacular synonyms) and left blank blank the heading "International Name". Starting in 1977, correspondence was exchanged on the species issue among individuals associated with the development of vertebrate virus taxonomy; there was agreement that the time was ripe for reconsideration of the issue, but there was no agreement as to the best scheme for naming species. A consensus wish was expressed for developing one ultimate scheme — a "final solution," which would be widely used. However, ultimate schemes were envisioned by some as Latin binomials, and by others as various adaptation of vernacular usages. This topic was addressed by the ICTV in 1978, and was brought into sharper focus by Matthews, the ICTV's President, in the Third Report of the Committee.[7] The topic has been discussed at every ICTV meeting since then, and the overall present situation is reviewed by Matthews in Chapter 9 of this volume.

One purpose of a complete, official species nomenclature for the vertebrate viruses is to assure unambiguous identification in literature citations and in data retrieval systems. This would require use somewhere in every paper of official family, genus and species terms (sometimes subfamily and subgenus terms also) and in most cases strain or variant description (e.g., isolate number or term from specialty group or culture collection). As an indication of the interest in this kind of precision, Matthews found that most of the editors of major virological journals questioned would enforce such usage, once it was set in place by the ICTV. This, then, is one reason why the listings in ICTV reports of member viruses of each family should be up to date and complete; likewise it is a reason why species nomenclature should be officially established as soon as possible for *all* viruses — even those which represent particular nomenclature problems.

With this motivation, Matthews asked the Vertebrate Virus Subcommittee and its Study Groups in 1978 to begin to consider the issue and put forward definite proposals for species nomenclature. At a meeting of the Subcommittee and its Study Group chairmen, held at the Fourth International Congress of Virology in The Hague the same year, discussion of this issue led to a "straw vote" on the alternative bases for developing a nomenclature. The result was a unanimous preference for starting from "English Vernacular Name" terms, whether these be disease or place names, or numbers, or combinations of these. There was no support for any Latinized or other system (20 people voting). Further influence toward consistency in the development of proposals came in the form of new rules and "guidelines" from the ICTV[18] (see also Chapter 1): most important were *rules* that "a species epithet should consist of a single word or if essential, a hyphenated word", . . . "the species word may be followed by numbers or letters," . . . "numbers and letters or combinations thereof may be used as the species epithet where they already have wide usage," . . . "newly designated serial numbers, letters, or combinations of these are not acceptable as species epithets," and *guidelines* that " . . . if a name is universally used by virologists (those who publish in scientific journals) that name or a derivative of it should be used, regardless of national origin." The ICTV made it clear that there would be no transliteration of formal terminology into various languages.

Most Study Groups dealing with vertebrate viruses addressed the issue with energy and care; their results were presented informally to the ICTV in 1981 at the Fifth International Congress of Virology in Strasbourg. The scheme of the *Adenoviridae* Study Group was formally approved (see Section II.D.) but there was difficulty in achieving any sense of uniformity among the families where member viruses are named rather than numbered. Clearly, by 1981 the Study Groups had reached a point where further instruction from the ICTV and further overall influence upon uniformity was necessary. They also had reached a point where much more communication with virologists was necessary; most virologists who are close to the ICTV and understand the historic basis for family and genus construction seem willing to extrapolate to the species level, but others become concerned that at the species level arbitrary, nonevolutionary criteria for division are mixed with real phylogenetic relationships. These conflicting concerns influence opinion on the development of species nomenclature. In the 1981 to 1984 ICTV term, under the Presidency of Brown, the Vertebrate Virus Subcommittee, under the chairmanship of Kingsbury, is reconsidering the whole species issue.

The following exercise (Table 2) is not an attempt to prejudge an issue which can only be resolved internationally and by the democratically constituted ICTV, but rather it is a personal attempt to show what a species nomenclature scheme would look like if it were bent to the following guidelines: (1) English vernacular virus names used as they are in the virology literature (changed less than would be required by ICTV rules and guidelines), (2) extrapolations made from some commonly agreed upon examples of "good species terms", and (3) only minimal attempts to reduce awkward multiple word names which are entrenched in vernacular usage. In any exercise like this it is necessary to address difficult nomenclature problems, such as multiple word terms. Perhaps the greatest difficulty seen in this exercise concerns the inclusion of host species terms in virus species epithets. Attempts to maintain entrenched vernacular virus names which include host species lead to quite unwieldy names often with terms duplicated in the genus and species terms (e.g., *Paramyxoviridae, Paramyxovirus parrot paramyxovirus*, or *Paramyxoviridae, Pneumovirus bovine respiratory syncytial virus*). The coded host species epithets, as used in the adenovirus nomenclature scheme, avoid this problem, but their extension to taxa where viruses have been given uncoded names is difficult.

IV. CONCLUSIONS

Within vertebrate virus taxonomy, it would appear that the most difficult issues have been saved for last. Many families have been established and family names are being used. For many families, the construction of genera is well established and, again, generic names are being used. The patterns which have evolved for establishing and justifying these constructions provide a good framework for establishing further new families and genera as warranted by continuing physicochemical and molecular biological studies. The most difficult, unresolved issues which remain are the same as they were several years ago — structuring and nomenclature within the families *Herpesviridae* and *Retroviridae* (and to a lesser extent in the families *Poxviridae, Rhabdoviridae* and *Togaviridae*), and formal species definition and nomenclature for all vertebrate viruses.

With respect to the herpesviruses, great progress has been made by the ICTV *Herpesviridae* Study Group under the chairmanship of Roizman, and there is confidence that a system of genera and species definitions and terms will be completed soon. The delay has been more attributable to the complexity and size of this virus family, than to original disagreement as to bases for divisions. Even so, the complex intertwining of different bases for defining subfamilies, genera, and species will make it difficult to extend the scheme to include all herpesviruses, including those which have not been studied extensively. Unlike the simpler schemes used with other families which require minimal data to place viruses within suitable taxa, the herpesvirus scheme is very demanding of data. It remains to be seen how the scheme will work in this regard.

The progress which has been made in the past several years in retrovirus taxonomic constructions is due to the efforts of Vogt, chairman of the ICTV *Retroviridae* Study Group from 1975 to 1981. Some lack of concern for taxonomy and disagreement with proposals when put forth seem to have characterized many retrovirologists, making efforts to complete the family structure most difficult. Superimposed upon these human and organizational problems has been the extreme complexity of relationships among the retroviruses. As with the herpesviruses, this complexity is evidenced by the need to introduce additional hierarchial

Table 2

VERTEBRATE VIRUS SPECIES NOMENCLATURE — A PERSONAL EXERCISE IN FORMALIZING SPECIES NAMES

Family[a]	Genus	Species[b] (examples-not complete listings)	Comments[c]
Poxviridae[d]			
	Orthopoxvirus	Vaccinia	
		Variola	
		Ectromelia	
		Cowpox	
		Monkeypox	
		Rabbitpox	
	Parapoxvirus	Orf	
		Bovine papular stomatitis }	Awkward multiple word names; entrenched usage; what are alternatives!
		Milker's nodule	
	Avipoxvirus	Fowlpox	
		Canarypox	
		Pigeonpox	
	Capripoxvirus	Goatpox	
		Sheeppox	
	Leporipoxvirus	Myxoma	
		Hare fibroma	
		Rabbit fibroma	
		Squirrel fibroma	
Iridoviridae	[Unnamed genus]	African swine fever	Awkward multiple word name; precisely defined; entrenched usage; what is alternative?
Herpesviridae (Subfamily: *Alphaherpesvirinae*) [*Herpesvirus*][e]		Human herpesvirus 1	(HSV 1)[f]
		Human herpesvirus 2	(HSV 2)
		Bovine herpesvirus 2	(Mammillitis)

	Species	
[Unnamed genus]	Suid herpesvirus 1	(Pseudorabies)
	Equid herpesvirus 1	(Equine abortion)
(Subfamily: *Betaherpesvirinae*)	Human herpesvirus 3	(Varicella-zoster)
[Dermatovirus]		
[Cytomegalovirus]	Human herpesvirus 5	(Human cytomegalovirus)
(Subfamily: *Gammaherpesvirinae*)	Murid herpesvirus 1	(Murine cytomegalovirus)
[Unnamed genus]		
[Lymphovirus]	Human herpesvirus 4	(EB virus)
	Cercopithecine herpesvirus 12	(Baboon herpesvirus)
[Folliculovirus]	Gallid herpesvirus 1	(Marek's disease)
	Meleagrid herpesvirus 1	(Turkey herpesvirus)
[Rhadinovirus]	Ateline herpesvirus 1	(Herpesvirus atales)
	Cebine herpesvirus 1	(Herpesvirus Saimiri)
Adenoviridae		
See Section II.D. for approved species terms for 2 genera		No designation of species for adenoviruses of nonhuman primates has been approved
Papovaviridae		
Papillomavirus	syl 1 (Shope)	*Adenovirus* species nomenclature system tried for this genus
	h 1	
	h 2 etc	
	bos 1	
	can 1	
Polyomavirus	polyoma	Traditional vernacular terms tried for this genus
	SV 40	
	BK	
	JC	
	K	
	SA 12	
Parvoviridae		
Parvovirus	rat 1 (Killam)	*Adenovirus* species nomenclature system tried for some members of this genus, not for members with letter epithets
	bos 1	
	fel 1	
	ans 1	
	Lu III	
	MVM	
	RT	
	TVX	

Table 2 (continued)
VERTEBRATE VIRUS SPECIES NOMENCLATURE — A PERSONAL EXERCISE IN FORMALIZING SPECIES NAMES

Family[a]	Genus	Species[b] (examples-not complete listings)	Comments[c]
	Dependovirus	[H 1]	Unaccepted term; implies human source because there is confusion with the adenovirus species nomenclature system (AAV 1) (AAV 2) etc
		1	
		2 etc	
Reoviridae	*Reovirus*	1	
		2 etc	
	Orbivirus[g]	Bluetongue 1	Many species have place names many species identified by isolate number
		Bluetongue 2, etc.	
		Eubenangee	Awkward multiple word name; precisely defined
		BeAr 35646	
		Colorado tick fever[h]	
		Epizootic hemorrhagic disease of deer	
		African horsesickness 1	Awkward multiple word names; what are alternatives?
		African horsesickness 2, etc.	
		Equine encephalosis	
	Rotavirus	h 1	
		h 2, etc.	
		bos 1	
		mus 1 (EDIM)	
		cav 1	
		[SA 11]	

Retroviridae
(Subfamily: *Oncovirinae*)
["Type C oncovirus"]

(Subgenus: "mammalian type C oncovirus")
  Murine sarcoma
  Murine leukemia
  Feline sarcoma
  Feline leukemia
    Subspecies and laboratory strains identified

(Subgenus: "avian type C oncovirus")
  Fowl sarcoma and leukosis
  Avian reticuloendotheliosis
    Subspecies and strains identified

(Subgenus: "reptilian type C oncovirus")
Viper type C virus

["Type B oncovirus"]
["Type D oncovirus"]
  Murine mammary tumor virus
    Awkward multiple word term

(Subfamily: *Spumavirinae*)
  [Spumavirus]
  Bovine syncytium
  Feline syncytium
  Simian foamy 1
  Simian foamy 2, etc.

(Subfamily: *Lentivirinae*)
  [Lentivirus]
  Visna

Paramyxoviridae[j]

Paramyxovirus
  Newcastle
  Mumps
  Parainfluenza mus 1 (Sendai)
  Parainfluenza h 2
  Parainfluenza can 2 (SV5)
  Parainfluenza h 3
  Parainfluenza bos 3
  Yucaipa
  Nariva
    Combination of traditional vernacular terms and adenovirus species nomenclature terms tried for these viruses.

  Parrot paramyxovirus
  Finch paramyxovirus
    Awkward usage whenever host term is included without coding; perhaps adenovirus species nomenclature system appropriate here

Morbillivirus
  Measles
  Canine distemper
  Rinderpest
  Pest-des-petits-ruminants
    Awkward multiple word name

Table 2 (continued)

VERTEBRATE VIRUS SPECIES NOMENCLATURE — A PERSONAL EXERCISE IN FORMALIZING SPECIES NAMES

Family[a]	Genus	Species[b] (examples-not complete listings)	Comments[c]
	Pneumovirus	Respiratory syncytium (human) Bovine respiratory syncytium Pneumonia of mice	Awkward usage whenever host term is included without coding; entrenched names — what are alternatives?
Orthomyxoviridae	*Influenzavirus*	A B	Whole WHO strain nomenclature system would be used in most instances
	[Unnamed, proposed] genus	[Influenza virus C]	
Rhabdoviridae	*Vesiculovirus*	Vesicular stomatitis Indiana Vesicular stomatitis New Jersey Piry	
	Lyssavirus	Rabies Mokola Logos bat Duvenhage Flanders	
	[No genera established]	Bovine ephemeral fever Infectious hematopoietic necrosis Red disease of pike Spring viremia of carp Rhabdovirus of eels Rhabdovirus of blue crab	Many species have place names Awkward multiple word names; awkward usage whenever host term is included without coding
Togaviridae	*Alphavirus*[k]	Sindbis Semliki forest Ross River Eastern equine encephalomyelitis Venezuelan equine encephalomyelitis	Many species have place names Awkward multiple word names; entrenched usage; what are alternatives?

Genus	Species	Notes
Flavivirus	Yellow fever Dengue 1 Dengue 2, etc. Edgehill Japanese encephalitis St. Louis encephalitis Murray Valley encephalitis Israel turkey encephalitis	Many species have place names; but entrenched usage and not as awkward as multiple word term including host name
Rubivirus	Rubella	
Pestivirus	Muscosal disease Hog cholera	
Picornaviridae		
Enterovirus	Polio 1 Polio 2, etc. Coxsackie A1 Coxsackie B2, etc. Echo 1 Echo 2, etc. 68 69, etc. bos 1 bos 2, etc. sus 1 sus 2, etc. mus 1 Hepatitis A	
Cardiovirus	Encephalomyocarditis Mengo	*Adenovirus* species nomenclature system tried for some members of genus, not for members with entrenched epithets
Rhinovirus	h 1A h 2, etc. bos 1 bos 2, etc.	Traditional terms used as species epithets for this genus *Adenovirus* species nomenclature system used for this genus
Aphthovirus	0 A C SAT 1 SAT 2, etc.	Traditional terms used as species epithets for this genus

Table 2 (continued)
VERTEBRATE VIRUS SPECIES NOMENCLATURE — A PERSONAL EXERCISE IN
FORMALIZING SPECIES NAMES

Family[a]	Genus	Species[b] (examples-not complete listings)	Comments[c]
Caliciviridae	*Calicivirus*	Vesicular exanthema A Vesicular exanthema B, etc. fel 1 fel 2, etc. San Miguel sea lion 1 San Miguel sea lion 2	
Coronaviridae	*Coronavirus*	Infectious bronchitis h 1 h 2, etc. bos 1	*Adenovirus* species nomenclature system tried for unnamed viruses—but designation which serial numbers represent which viruses would have to be do by a specialty group/Study Group
Bunyaviridae[1]	*Bunyavirus*	Bunyamwera LaCrosse Oriboca California encephalitis BeAn 84381	Many species have place names
	Nairovirus	Crimean hemorrhagic fever Congo } Nairobi sheep disease Dera Ghazi Khan	Some species identified by isolate number Identical viruses; names kept for epidemiologic reasons
	Uukuvirus	Uukuneimi	Many species have place names Many species have place names
Arenaviridae	*Arenavirus*	Lymphocytic choriomeningitis } Machupo Lassa Junin	Many species have place names

a   Only taxa containing vertebrate viruses listed.

b   Examples chosen to represent various good species terms and various problem cases.

c   Comments relate primarily to clarifications, neutral opinions, and especially to problem cases where English vernacular terms do not lend themselves to reasonable formal species epithets. The absence of negative comments indicates the author's choices for acceptable species terms.

d   Subfamilies omitted.

e   Bracketed genus terms are from interim considerations by the ICTV *Herpesviridae* Study Group (B. Roizman, Chairman): species terms also derive from interim Study Group considerations.[16] This system of family, subfamily, genus, and species nomenclature suffers from repeated use of the root term "herpes". This is reasonable at the subfamily and genus levels, but is objectionable at the species level. The system also suffers from use of host terms in species epithets. What are alternatives?

f   Vernacular virus names included for information.

g   12 to 17 "subgroups" or serogroups have been defined. Hierarchial structure for the genus is presently under study by a subcommittee of the ICTV *Reoviridae* Study Group; B. M. Gorman, Chairman.

h   May be removed from genus because has 12 segments of DS RNA (D. Knudson).

i   Bracketed genus terms are taken from a paper by P. Vogt.[21] No attempt has ever been made to develop suitable terms for genera or species, and this exercise, using vernacular virus or subgroup names, only shows the difficult task that this will entail.

j   *Paramyxoviridae* Study Group of the ICTV, under Chairmanship of D. Kingsbury developed both a species name and letter/number code model for member viruses (see Section II.D.).

k   Genus structure described in detail by specialty group.[14]

l   Several "subgroups" or serogroups, which equal subgenera, have been identified.[15] For necessary precision suggested usage would be *Bunyaviridae, Bunyavirus (Simbu) buttonwillow* (family, genus (subgenus) species).

levels from the top down — that is, from the subfamily level and subgenus level down to laboratory strain level. Nevertheless, there needs to be a new review of the whole structure of the family *Retroviridae* and new effort to complete genus and species construction. In addition, there needs to be a new attempt to find terms for genera, subgenera, and species which better fit the wish of many virologists that viruses be identified by euphonic combinations of terms in set order.

Finally, the issue which is proving to be most difficult — the species nomenclature issue. The progress which has been made in the past 5 years must be credited to Matthews in his position as President of the ICTV. The issue centers on the vertebrate viruses because their taxonomy is further advanced. The issue must be considered the most important problem facing the Vertebrate Virus Subcommittee over the next few years. The exercise attempted in this chapter (Table 2) serves only to illustrate the difficulty of the overall problem and the need for both guidelines and comprehensive modeling. The product of a successful completion of an official species nomenclature for all vertebrate viruses will be more valuable than envisioned by many virologists — it will resolve ambiguities in virus identification in the literature, eliminate the vagaries of multiple vernacular synonyms in related literature (e.g., the infectious disease literature), and improve the precision of electronic data handling systems. It will provide virologists, their specialty groups, and the ICTV with a precise dictionary or atlas of the objects of their teaching and research.

REFERENCES

1. **Lwoff, A., Horne, R., and Tournier, P.,** A system of viruses, *Cold Spring Harbor Symp. Quant. Virol.,* 27, 51, 1962.
2. **Wildy, P., Ginsberg, H. S., Brandes, J., and Maurin, J.,** Virus classification, nomenclature and the International Committee on Nomenclature of Viruses, *Prog. Med. Virol.,* 9, 476, 1968.
3. **Fenner, F.,** The classification of viruses: why, when and how, *Aust. J. Exp. Biol. Med. Sci.,* 52, 223, 1974.
4. **Wildy, P.,** Classification and nomenclature of viruses (First Report of the ICNV), *Monogr. Virol.,* 5, 1, 1971.
5. **Fenner, F.,** Classification and nomenclature of viruses (Second Report of the ICTV), *Intervirology,* 7, 1, 1976.
6. **Fenner, F.,** The classification and nomenclature of viruses. Summary of results of meetings of the ICTV in Madrid, September 1975, *Virology,* 71, 371, 1976.
7. **Matthews, R. E. F.,** Classification and nomenclature of viruses (Third Report of the ICTV), *Intervirology,* 12, 132, 1979.
8. **Matthews, R. E. F.,** Classification and nomenclature of viruses (Fourth Report of the ICTV), *Intervirology,* 17, 1, 1982.
9. **Western Hemisphere Committee on Animal Virus Characterization.** An updated listing of animal reference virus recommendations, *Am. J. Vet. Res.,* 36, 861, 1975.
10. WHO/FAO Programme on Comparative Virology. WHO Report VPH/CVR, 80, 1, 1980.
11. **Kalter, S. S., Ablashi, D., Espana, C., Heberling, R. L., Hull, R. N., Lennette, E. H., Malherbe, H. D., McConnell, S., and Yohn, D. S.,** Simian virus nomenclature, 1980, *Intervirology,* 13, 317, 1980.
12. WHO Committee on Influenza. A revision of the system of nomenclature for influenza viruses, WHO Report, 1982.
13. WHO Expert Committee on Viral Hepatitis. Advances in viral hepatitis, *WHO Tech. Rep. Ser.,* 602, 1, 1977.
14. **Calisher, C. H., Shope, R. E., Brandt, W., Casals, J., Karabatsos, N., Murphy, F. A., Tesh, R. B., and Wiebe, M. D.,** Proposed antigenic classification of registered arboviruses. 1. Togaviridae, alphavirus, *Intervirology,* 14, 229, 1980.
15. **Calisher, C. H., Bishop, D. H. L., Brandt, W., Casals, J., Karabatsos, N., Shope, R. E., Tesh, R. B., and Weibe, M. E.,** Recommended antigenic classification of registered arboviruses. II. Bunyaviridae, bunyavirus, *Intervirology,* in press.

16. Roizman, B., Carmichael, L. E., Deinhardt, F., deThe, G., Nahmias, A. J. Plowright, W., Rapp, F., Sheldrick, P., Takahashi, M., and Wolf, K., Herpesviridae. Definition, provisional nomenclature and taxonomy, *Intervirology*, 16, 201, 1981.

17. Norrby, E., Bartha, A., Boulanger, P., Dreizin, R. S., Ginsberg, H. S., Kalter, S. S., Kawamura, H., Rowe, W. P., Russell, W. C., Schleisinger, R. W., and Wigand, R., Adenoviridae, *Intervirology*, 7, 117, 1976.

18. Matthews, R. E. F., The classification and nomenclature of viruses. Summary of results of meetings of the ICTV held in Strasbourg, August 1981, *Intervirology*, 16, 53, 1981.

19. Melnick, J. L., Allison, A. C., Butel, J. S., Eckhart, W., Eddy, B. E., Kit, S., Levine, A. J., Miles, J. A. R., Pagano, J. S., Sachs, L., and Vonka, V., Papovaviridae, *Intervirology*, 3, 106, 1974.

20. Bachmann, P. A., Hoggan, M. D., Kurstak, E., Melnick, J. L., Pereira, H. G., Tattersall, P., and Vago, C., Parvoviridae: second report, *Intervirology*, 11, 248, 1979.

21. Vogt, P. K., The Oncovirinae — A definition of the group in Report No. 1 of WHO Collaborating Centre for Collection and Evaluation of Data on Comparative Virology. UNI-Druck, Munich, 1976.

22. Kingsbury, D. W., Bratt, M. A., Choppin, P. W., Hanson, R. P., Hosaka, Y., terMeulen, V., Norrby, E., Plowright, W., Rott, R., and Wunner, W. H., Paramyxoviridae, *Intervirology*, 10, 137, 1978.

23. Dowdle, W. R., Davenport, F. M., Fukumi, H., Schild, G. C., Tumova, B., Webster R. G., and Zakstelskaja, X., Orthomyxoviridae, *Intervirology*, 5, 245, 1975.

24. Brown, F., Bishop, D. H. L., Crick, J., Francki, R. I. B., Holland, J. J., Hull, R., Johnson, K., Martelli, G., Murphy, F. A., Obijeski, J. F., Peters, D., Pringle, C. R., Reichmann, M. E., Schneider, L. G., Shope, R. E., Simpson, D. I. H., Summers, D. F., and Wagner, R. R., Rhabdoviridae, *Intervirology*, 12, 1, 1979.

25. Porterfield, J. S., Casals, J., Chumakov, M. P., Gaidamovich, S. Y., Hanoun, C., Holmes, I. H., Horzinek, M. C., Mussgay, M., OkerBlom, N., Russell, P. K., and Trent, D. W., Togaviridae, *Intervirology*, 9, 129, 1978.

26. Cooper, P. D., Agol, V. I., Bachrach, H. L., Brown, F., Ghendon, Y., Gibbs, A. J., Gillespie, J. H., Lonberg-Holm, K., Mandel, B., Melnick, J. L., Mohanty, S. B., Povey, R. C., Rueckert, R. R., Schaffer, F. L., and Tyrrell, D. A. J., Picornaviridae, second report, *Interviriology*, 10, 165, 1978.

27. Schaffer, F. L., Bachrach, H. L., Brown, F., Gillespie, J. H., Burroughs, J. N., Madin, S. H., Madeley, C. R., Povey, R. C., Scott, F., Smith, A. W., and Studdert, M. J., Caliciviridae, *Intervirology*, 14, 1, 1980.

28. Tyrrell, D. A. J., Alexander, D. J., Almeida, J. D., Cunningham, C. H., Easterday, B. C., Garwes, D. J., Hierholzer, J. C., Kapikian, A., Macnaughton, M. R., and McIntosh, K., Coronaviridae: second report, *Intervirology*, 10, 321, 1978.

29. Bishop, D. H. L., Calisher, C. H., Casals, J., Chumakov, M. P., Gaidamovich, S. Y., Hanoun, C., Lvov, D. K., Marshall, I. D., OkerBlom, N., Pettersson, R. F., Porterfield, J. S., Russell, P. K., Shope, R. E., and Westaway, E. G., Bunyaviridae, *Intervirology*, 14, 125, 1980.

30. Casals, J. and Tignor, G. H., The Nairovirus genus: serological relationships, *Intervirology*, 14, 144, 1980.

31. Pfau, C. J., Bergold, G. H., Casals, J., Johnson, K. M., Murphy, F. A., Pedersen, I. R., Rawls, W. E., Rowe, W. P., Webb, P. A., and Weissenbacher, M. C., Arenaviridae, *Intervirology*, 4, 207, 1974.

32. Kiley, M. P., Bowen, E. T. W., Eddy, G. A., Isaacson, M., Johnson, K. M., Murphy, F. A., Pattyn, S. R., Peters, D., Prozesky, O. W., Regnery, R. L., Simpson, D. I. H., Slenzca, W., Sureau, P., van der Groen, G., and Webb, P. A., Filoviridae: a taxonomic home for Marburg and Ebola viruses? *Intervirology*, 18, 24, 1982.

Chapter 3

CURRENT PROBLEMS IN PLANT VIRUS TAXONOMY

R. I. B. Francki

TABLE OF CONTENTS

I. INTRODUCTION

"Besides satisfying the intellectually tidy, an efficient taxonomy is useful and necessary because it emphasizes resemblances which make worthwhile generalizations apparent, and makes minor differences easier to understand. "
P. D. Cooper, 1961.[1]

The current system of plant virus classification has evolved over the past decade under the auspices of the International Committee on Taxonomy of Viruses (ICTV). It is widely used by virologists, especially those involved in diagnostic work and teaching. Since its inception, very few alternatives have been suggested by individual workers, contrary to earlier days when numerous systems appeared in the literature confusing rather than solving the problem of how best to classify viruses (see Chapter 1).

The current classification system began with the definition of sixteen groups of viruses by the first Plant Virus Subcommittee (PVS) of the ICTV[2] and their approval by the ICTV in 1971.[3] The ICTV has subsequently approved the establishment of a further eight groups; four in 1975,[4] three in 1978,[5] and one in 1981.[6] It was also recognized that a number of plant pathogenic viruses which also replicate in their insect vectors, are very similar to certain viruses of vertebrates and arthropods. These plant viruses are now members of the families *Reoviridae* and *Rhabdoviridae*. Moreover, those in the *Reoviridae* are divided into two genera, *Phytoreovirus* and *Fijivirus*.[5,6]

Although the viruses of vertebrates, bacteria, and arthropods have been classified into genera and families, plant virologists have continued to use only loosely defined "groups" for the classification of plant viruses. In some quarters there has even been very active resistance to suggestions that a uniform system of classification into families and genera be followed by all virologists. The establishment of the families *Reoviridae* and *Rhabdoviridae* was largely the result of an initiative taken by vertebrate virologists.

There is no doubt that solid foundations to plant virus taxonomy have been laid in the past decade and a half. As many as 244 plant viruses have been classified in the 24 groups and two families approved by the ICTV and a further 227 have been designated as either probable or possible members of these taxa.[6] However, much still remains to be done. There are still many viruses which remain unclassified; some because insufficient is known about them and others because they do not appear to fit into any of the existing groups or families. For the latter category, new groups will almost certainly need to be established in the future. Perhaps the greatest challenges that lie ahead are those concerned with the improvement of the current virus nomenclature. The aim of this chapter is to acquaint the reader with the current status of plant virus taxonomy, emphasizing existing problems and suggesting how they may be approached in the future.

II. CLASSIFICATION

A. Development of Methods

An essentially Adansonian approach was taken for the establishment of the first 16 groups of plant viruses in 1971.[2] All available characters of the viruses, 49 in all (Table 1), were considered without weighting. However, for most of the 99 viruses classified, the available data were incomplete. Nevertheless, the resulting clusters of viruses became the 16 groups approved by the ICTV.[3]

The approach to the eight groups established subsequently was a little different. Again, many characters were used but the approach was not strictly Adansonian. It was rather a case of recognizing viruses which obviously did not fit into the existing groups but which formed clusters by having properties in common.[4-6]

The validity of the method used in establishing the first 16 groups of plant viruses is indicated by the fact that more than a decade later they still exist.[6] Moreover, the majority of virologists find this classification useful. However, it can now be argued that further development of the classification system need not be done by an Adansonian approach. From experience gained over the past decade, we have learned to distinguish which characters of viruses are taxonomically useful. Furthermore, we have gained insight into which characters are useful for distinguishing virus groups as we know them, and which are useful for distinguishing viruses within a group. This important difference was not as evident a decade ago and hence the Adansonian approach advocated by Gibbs[7] and used by Harrison and colleagues[2] may have been a safer one to use at the time. It should be stressed here, however, that if a truly Adansonian approach is taken, the characters used must be carefully screened for those that may reflect the same property and which would thus in effect be scored more than once. Unless this is done, weighting of some characters will in fact occur. For example, two characters listed in Table 1 actually reflect the surface charge on the virus particles, their electrophoretic mobility and isoelectric point (Table 1, C7 and 8). Similarly, the percentage of nucleic acid of the virus particle by weight (Table 1, D4) is a reflection of the molecular weight and number of strands of RNA per particle (Table 1, D3 and 5) and the molecular weights of the protein subunits and their number per particle (Table 1, D9 and 10).

Table 1
A LIST OF THE CHARACTERS CONSIDERED BY
HARRISON ET AL.[2] FOR THE CLASSIFICATION OF PLANT
VIRUSES, TOGETHER WITH A CURRENT ASSESSMENT
OF THEIR IMPORTANCE IN TAXONOMY

	Useful in distinguishing		
Character	Virus groups	Viruses within groups	Of doubtful taxonomic significance
A. Behavior in hosts			
1. Host range		+	
2. Predominent symptom types		+	
3. Tissue restrictions	+		
4. Approximate concentration in sap			+
5. Type of inclusion body	+	+	
6. Intracellular location of inclusions		+	
7. Effectiveness of heat therapy			+
8. Seed transmissibility	+		
9. Geographical distribution			+
10 Cross-protecting viruses			+
B. Vector relations			
1. Taxa of vectors	+		
2. Number of vector species			+
3. Acquisition threshold feeding/access period			+
4. Inoculation threshold feeding/access period			+
5. Latent period			+
6. Persistence in vector	+		
7. Multiplication in vector	+		
8. Transmission through egg/resting spore		+	
9. Vector stage able to acquire virus			+
10. Vector stages able to inoculate virus			+
C. Particle properties			
1. Shape and symmetry	+		
2. Size	+		
3. Number and arrangement of morphological subunits	+		
4. Arrangement of nucleic acid	+		
5. Sedimentation coefficient	+		
6. Molecular weight daltons	+		
7. Electrophoretic mobility		+	
8. Isoelectric point		+	
9. Thermal inactivation point in sap			+
10. Retention of infectivity in sap at 20°C			+
11. Dilution end point in sap			+
12. Serological relationship		+	
13. Major properties of accessory particles	+		

Table 1 (continued)
A LIST OF THE CHARACTERS CONSIDERED BY HARRISON ET AL.[2] FOR THE CLASSIFICATION OF PLANT VIRUSES, TOGETHER WITH A CURRENT ASSESSMENT OF THEIR IMPORTANCE IN TAXONOMY

| | Useful in distinguishing | | |
Character	Virus groups	Viruses within groups	Of doubtful taxonomic significance
D. Particle Composition:			
Nucleic acid			
1. RNA or DNA	+		
2. Nucleotide ratio			+
3. Molecular weight	+		
4. Percentage of particle weight	+		
5. Number of strands	+		
6. Circular or linear	+		
7. Sedimentation coefficient, and of accessory nucleic acid			+
Protein			
8. Number of protein species in particle	+		
9. Molecular weight(s) of subunits	+		
10. Number of chemical subunits per particle	+		
11. Number of amino acid residues per subunit			+
12. Noteworthy amino acids included or excluded		+	
13. Enzymatic activities	+		
14. Percentage of particle weight	+		
Lipid			
15. Percentage of particle weight	+		
Other compounds			
16. Type and amount	+		

In Table 1, I have divided the virus characters used by Harrison and colleagues[2] into three categories; those that I consider useful for dividing or assigning viruses into groups, those useful for distinguishing viruses within a group, and those that appear to be of little or no use in taxonomy. This division is mine and it must be stressed that some of my views may not be shared by others. Advances over the past decade, especially in the molecular biology of nucleic acids, also indicate that a number of properties of viruses other than those used by Harrison and colleagues[2] are of considerable taxonomic significance. These are discussed later.

It is also my opinion that some virus characters are more useful for taxonomic purposes than others. Hence, I feel that an Adansonian approach to virus classification may not be as attractive as it was thought to be a decade ago. It must be stressed, however, that one would be skating on thin ice if one used only a very limited number of characters to decide the taxonomic position of a virus.

Table 2
INFORMATION USEFUL IN ASSIGNING
VIRUSES TO GROUPS

A.   Properties of virus particles
   (a)   Particle morphology
        1. Size
        2. Shape
        3. Presence or absence of envelope
        4. Capsomeric structure
   (b)   Sedimentation properties
        1. Number of components
        2. Sedimentation coefficient(s)
B.   Properties of virus coat protein
        1. Number of polypeptides
        2. Molecular weight(s) of polypeptide(s)
C.   Properties of virus nucleic acid
        1. Type (RNA or DNA)
        2. Strandedness (linear or circular)
        3. Polarity (positive or negative)
        4. Number of molecules
        5. Molecular weight(s) of molecules(s)
        6. Presence or absence of 5'-terminal M⁷Gppp
        7. Presence or absence of 5'-terminal VPg[a]
        8. Presence or absence of 3'-terminal poly (A)

  [a]   Covalently-linked polypeptide.

From a glance at Table 1 it becomes evident that more of the physical and chemical than the biological characters listed are considered useful for both grouping and differentiating viruses. The characters which have been considered as unsuitable to be used for taxonomy are rejected for one or both of the following reasons: first, because they are difficult to measure accurately without an inordinate amount of effort; and second, because even when determined with precision, they are of little consequence to taxonomy for one reason or another. For example, the so-called "physical properties in sap" including thermal inactivation point, retention of infectivity and dilution end-point (Table 1, C 9-11) have been shown to be so poorly reproducible as to be of no real value in taxonomy.[8] On the other hand, cross-protection of one virus by another (Table 1, A 10) is usually a phenomenon observed among related viruses. However, there are also many reports of cross-protection between obviously unrelated viruses and absence of cross-protection between related viruses.[9,10] Hence, this character is obviously unreliable for use in taxonomy.

Virus characters used in taxonomy should be selected with two aspects in mind. As well as being taxonomically useful, it is also important that each character can be determined precisely without an inordinate amount of effort. For example, although complete nucleotide sequences would be most informative when comparing a cluster of related viruses, the amount of effort necessary to obtain the data would usually be quite disproportionate to their taxonomic value. In most instances, the similarity or difference between the two viruses as measured by RNA-RNA or RNA-complementary DNA (cDNA) hybridization would be quite sufficient. However, one must also beware of the other extreme. That is, one must avoid generating data merely because they can be easily obtained, without regard to their discriminatory value.

B. A Useful Series of Experimental Procedures for the Allocation of a Virus to its Appropriate Group

It is now generally recognized that the physical and chemical properties of viruses are more stable than their biological properties and hence more useful for assigning the viruses to taxonomic groups. Those which I consider taxonomically most significant are listed in Table 2. Most modern laboratories are equipped with facilities for determining these properties rapidly. Thus a previously uncharacterized virus can be assigned to its correct taxonomic group after a realistic input of effort.

Most biological properties of viruses are relatively unstable and hence are less reliable for differentiation at the group level. Nevertheless, there are a number of biological characters which are typical of certain virus groups; six of these are listed in Table 1. When investigating the taxonomic position of a virus, it is probably wise to determine its physical and chemical characters without paying undue attention to its biological properties. In this way the taxonomic group to which it belongs can be established quickly. Once this has been done the biological properties can be predicted and checked if necessary. A suggested sequence of steps which can be taken and which I have found especially useful when assigning viruses to their taxonomic groups, is outlined under the following headings.

1. Electron Microscopy of Extracts from Infected Plants

The particles of many viruses can be detected and their gross morphology observed by electron microscopy in crude extracts from infected leaves.[11] This procedure is now commonly known as the "leaf dip method". Information thus gained takes only minutes to obtain and often provides invaluable preliminary taxonomic information. It may also provide useful clues as to how the virus should be purified and handled during subsequent characterization.

2. Virus Purification

Detailed characterization of a virus in physical and chemical terms needs to be done on purified preparations. At one time the purification of most viruses was a major problem and an obstacle to their characterization. However, the wide range of techniques now available allows us to prepare many viruses in adequate amounts and of sufficient purity for full characterization.[12,13] Nevertheless, there are still a number of viruses with unstable particles or which occur in very low amounts in plants making them difficult to isolate and study. These are the viruses whose taxonomy is in an unsatisfactory state as can be seen in Section IV.

3. Electron Microscopy of Purified Virus Preparations

Both the size and shape of virus particles, which can be determined from relatively low resolution electron micrographs, can provide valuable taxonomic information. The presence of an envelope is also of great significance. In fact, the gross morphology of six plant virus groups is so characteristic that their members can be identified with little or no other information (left-hand column in Table 3). Viruses of other groups can be divided into those with elongated particles (with helical symmetry) and those with small polyhedral particles (with icosahedral symmetry) about 30 nm in diameter (Table 3).

Viruses whose particles have helical symmetry are divided into seven groups which can be differentiated by their characteristic shapes and lengths. However, the determination of the particle lengths needs considerable care. It is considered that at least 100 particles should be measured from micrographs taken with a carefully calibrated microscope to obtain an accurate estimate of their modal length.[14] There are a number of potential problems that must be avoided during specimen preparation for electron microscopy of these viruses. These include prevention of particle breakage and aggregation[15] and their possible swelling or contraction in buffers containing divalent cations or chelating agents.[16] Generally, these problems can best be avoided by using, if possible, crude leaf extracts rather than concentrated or purified virus preparations.

Table 3
PARTICLE MORPHOLOGY OF THE GROUPS OF PLANT VIRUSES[a]

Viruses with highly diagnostic gross morphology	Viruses with elongated particles	Viruses with small polyhedral particles
Caulimovirus (polyhedral, ~ 50 nm)	*Tobravirus* (rigid rod, 46-114 × 22 and 180-215 × 22 nm)	*Luteovirus* (25-30 nm)
Geminivirus (geminate 30 × 18 nm)	*Hordeivirus* (rigid rod, 100-150 × 20 nm, three components)	Maize chlorotic dwarf virus group (~ 30 nm)
Reoviridae[b] (polyhedral, 65-71 nm)	*Tobamovirus* (rigid rod, ~ 300 × 18 nm)	*Tymovirus* (~ 29 nm)
Rhabdoviridae (bacilliform, enveloped, 135-380 × 45-95)	*Potexvirus* (flexuous rod, 470-580 × 13 nm)	*Sobemovirus* (~ 30 nm)
Tomato spotted wilt virus group (spherical, enveloped, ~ 82 nm)	*Carlavirus* (flexuous rod, 600-700 × 13 nm)	Tobacco necrosis virus group (~ 28 nm)
Alfalfa mosaic virus group (bacilliform, 18,36,48 and 58 × 18 nm)	*Potyvirus* (flexuous rod, 680-900 × 11 nm)	*Tombusvirus* (~ 30 nm)
	Closterovirus (flexuous rod, 600-2,000 × 12 nm)	*Comovirus* (~ 28 nm)
		Nepovirus (~ 28 nm)
		Pea enation mosaic virus group (~ 28 nm)
		Dianthovirus (31-34 nm)
		Cucumovirus (~ 29 nm)
		Bromovirus (~ 26 nm)
		Ilarvirus (26-35 nm, three components)

[a] Particle dimensions taken from Matthews.[6]

[b] This family is subdivided into two genera, *Phytoreovirus* and *Fijivirus*.[6]

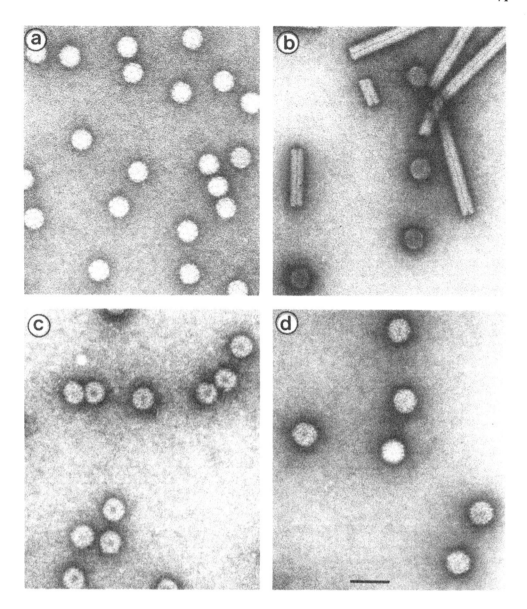

FIGURE 1. Preparations of four different viruses with T = 3 icosahedral symmetry stained in uranyl acetate showing differences in their morphological appearance: (a) turnip yellow mosaic virus *(Tymovirus)*; (b) southern bean mosaic virus *(Sobemovirus)* mixed with rod-shaped particles of tobacco mosaic virus; (c) cucumber mosaic virus *(Cucumovirus)*; and (d) tomato bushy stunt virus *(Tombusvirus)*. Bar represents 50 nm.

Electron microscopy is less informative for the assignment of viruses with small isometric particles to their appropriate groups. This is because the particles of viruses belonging to most of these thirteen groups (Table 3) are similar in size and appearance. However, even here electron microscopy can be of some use. All available data indicate that the morphology of viruses belonging to the same taxonomic group are indistinguishable. It is also apparent that some of the groups have characteristic particle differences which can be distinguished by high resolution microscopy.[17] For example, the tymoviruses can be distinguished from viruses belonging to all the other groups by their characteristic rounded particles with relatively easily resolved capsomeres arranged with T = 3 icosahedral symmetry (Figure 1a). It is interesting that particles of tymoviruses are readily distinguished from those of sobe-

Table 4
SOME PROPERTIES OF GROUPS OF PLANT VIRUSES WITH SMALL POLYHEDRAL PARTICLES[a]

Virus group	Number of sedimenting particles and S_{20w}	Coat protein polypeptide(s) and size $(M_w \times 10^3)$	Linear ss-RNA genome segments and size(s) $(M_w \times 10^6)$
Luteovirus	One, 155-127	One, ~ 24	One, ~ 2.0
Maize chlorotic dwarf virus group	One, ~ 183	?	One, ~ 3.2
Sobemovirus	One, ~ 115	One, ~ 30	One , ~ 1.4
Tobacco necrosis virus group	One, ~ 118	One, ~ 23	One, ~ 1.4
Tombusvirus	One, ~ 140	One, ~ 40	One, ~ 1.5
Dianthovirus	One, ~ 135	One, ~ 40	Two, ~ 1.5 + ~ 0.5
Cucumovirus	One, ~ 99	One, ~ 24	Three, ~ 1.3, ~ 1.1 & ~ 0.8 (& ~ 0.3)[b]
Bromovirus	One, ~ 85	One, ~ 20	Three, ~ 1.1 & ~ 1.0 & ~ 0.7 (+ ~ 0.3)[b]
Tymovirus	Two, ~ 54 and ~ 115[c]	One, ~ 20	One, ~ 2.0
Pea enation mosaic virus group	Two, ~ 99 and ~ 112	One, ~ 22	Two, ~ 1.7 & ~ 1.3
Comovirus	Three, ~ 58, ~ 98 and ~ 118	Two, ~ 42 & ~ 22	Two, ~ 2.4 & ~ 1.4
Nepovirus	Three, 45-56 86-128 115-134	One, 56-60[d]	Two, ~ 2.8 & 1.3-2.4
Ilarvirus	Three, 80-120	One, ~ 25	Three, ~ 1.1, ~ 0.9 & ~ 0.7, (& 0.3)[b]

[a] Data taken from Matthews.[6]
[b] Component in brackets is not essential for infectivity but is encapsidated. The RNA is the messenger for coat protein whose sequences are also in the smallest genomic RNA.
[c] Purified preparations of turnip yellow mosaic virus, the type member of the group, also contain at least eight minor nucleoprotein components sedimenting between 54 and 115 S.[13]
[d] This may represent an oligomer or contain repeated amino acid sequences.[6]

moviruses, tombusviruses and cucumoviruses (Figures 1b-d) all of which have also been shown to have T = 3 symmetry.[18-20] Thus, high resolution electron microscopy of virus preparations stained under standard conditions can be of some help in recognizing taxonomically significant differences in the morphology of small polyhedral particles and may be used more widely for taxonomic purposes in the future.

4. Sedimentation Properties of Virus Particles

Valuable taxonomic information can sometimes be obtained from the sedimentation properties of virus particles. This is especially so when dealing with viruses which have small polyhedral particles and which are difficult to place in their correct group on the basis of morphology. Particles of these viruses sediment at various rates as one, two, or three components with a range of sedimentation coefficients (Table 4).

The number of distinguishable sedimenting components of a virus is easily determined by centrifugation in an analytical ultracentrifuge or by density-gradient centrifugation in a preparative machine.[12] Of the thirteen groups of plant viruses with small polyhedral particles, eight sediment as a single component, two as two components and three as three components (Table 4). Although the sedimentation coefficients of these viruses differ markedly they can be of doubtful use for taxonomy because their accurate determination is technically difficult.

This can be seen from the wide range of values reported in the literature for the same virus. Furthermore, there appears to be considerable variation in the values reported for different viruses within a group. This is especially marked within some groups such as the nepoviruses (Table 4). An approximate value for the sedimentation coefficient of an unknown virus can, however, be determined very quickly by cosedimenting a preparation of it with that of another virus whose sedimentation properties are well established, as marker.[21] Similarly, in crude extracts from infected leaves, the approximate sedimentation coefficient of the virus can sometimes be determined by its sedimentation relative to the leaf ribosomes.[12] Data from such simple experiments can often provide taxonomically valuable information quickly.

5. Properties of the Viral Coat Proteins and Nucleic Acids

The numbers and sizes of coat protein polypeptides and nucleic acid segments of viruses are characteristic of the taxonomic group to which the virus belongs.[6] The 13 groups of plant viruses which have small polyhedral particles are no exception and all can be distinguished readily by a combination of their coat protein and nucleic acid characteristics (Table 4). Viruses of some of the groups have particles devoid of RNA, whereas others which contain two or four RNA segments, sediment as single components. Thus it is important to remember that the number of detectable sedimentating components in virus preparation does not necessarily reflect the number of nucleic acid segments that it contains.

There are several methods for the analysis of proteins and nucleic acids of viruses. However, because of their simplicity, polyacrylamide gel electrophoretic methods have been most commonly used for separating both protein and nucleic acids into components and estimating their molecular weights. The precision of these methods has generally been sufficient for taxonomic purposes.

A number of procedures have been described for the polyacrylamide gel electrophoretic analysis of viral protein. One of the simplest, which is also quite satisfactory, is to dissociate preparations of intact virus particles by boiling in the presence of denaturing agents such as sodium dodecyl sulfate (SDS) and 2-mercaptoethanol and subjecting the digest to electrophoresis in either a continuous[22] or discontinuous gel system.[23] Accurate estimates of molecular weight are largely dependent on the use of stable and well-characterized marker polypeptides. However, the molecular weights of some commonly used marker proteins have been significantly changed recently on the basis of amino sequence analysis.

Care is essential in interpreting the results when more than one band is observed on the gel. Although some components may represent distinct polypeptides, others may be the results of partial degradation of protein during virus purification or storage.[24,25] Extra bands can also be present due to incomplete dissociation of polypeptides,[25,26] and probably also to conformational variants.[27]

It is essential to establish if multiple bands observed in polyacrylamide gels are due to the presence of distinct polypeptides or are the results of the above mentioned artifacts. This can be done by peptide mapping of each component. Relatively simple and rapid methods for this are now available.[28,29]

Analyses of viral nucleic acids should be done on preparations devoid of protein. Numerous methods have been described for the isolation of viral nucleic acids and their subsequent analysis by polyacrylamide gel electrophoresis.[13] The accuracy of molecular weight determinations in polyacrylamide gel electrophoresis is dependent not only on use of the appropriate markers but in the case of single-stranded nucleic acids, the complete denaturation of their secondary structure. Numerous methods have been described for achieving this but it is doubtful if any of them are completely satisfactory. However, recently developed methods using materials such as methyl-mercury (II) hydroxide and glyoxal appear to be some of the best available.[30,31] Methyl-mercury (II) hydroxide has the disadvantage of being very toxic and should be handled with extreme care.

Table 5

GUIDELINES FOR DECIDING IF A VIRUS BEING INVESTIGATED IS A NEW VIRUS OR A STRAIN OF ONE ALREADY EXISTING[a]

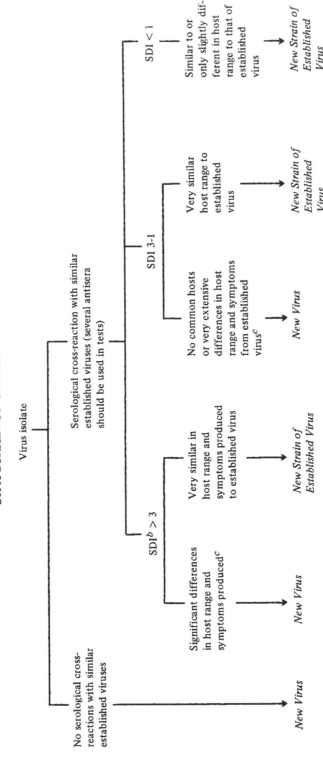

a   Scheme adapted from Hamilton and colleagues.[14]

b   Serological differentiation index (SDI) is defined as the number of twofold dilution steps separating the homologous from the heterologous titers.

c   Hamilton and colleagues[14] suggest that differences in some other properties of the virus could also be taken into account. They suggest that particle morphology, cytopathic effects, vector species, base composition of nucleic acid and electrophoretic properties of the particle as examples. Opinions expressed in the text of this chapter would not suggest the use of some of these properties.

Problems of aggregation are not as serious with nucleic acids as with proteins. Also the presence of artifactual bands due to partial digestion of nucleic acids is usually not important because when it does occur, the molecules rarely undergo cleavage at specific sites and random cleavage produces smears rather than distinct bands in the gels. However, nicking of circular single-stranded molecules can produce an unexpected band under some conditions because a single nonspecific cleavage per molecule will yield polynucleotides of the same length and hence with a single electrophoretic component but different from that of the circular molecules.[32]

It may be relevant to mention here that in the past, base ratios of viral nucleic acids were considered to be useful taxonomic characters.[2],[7] However, they are difficult to determine accurately, their estimation requires large amounts of material and they do not necessarily reflect differences in base sequences. There are examples of completely unrelated viruses such as cucumber mosaic virus *(Cucumovirus* group) and tobacco ringspot virus *(Nepovirus* group) with base ratios indistinguishable from each other.[33,34] There appears no good reason for continuing to use this property in virus taxonomy, except perhaps for a few groups with very distinctive base composition such as the tymoviruses.[6]

C. Characters Useful for Distinguishing Viruses Within Groups

Once a virus has been assigned to a taxonomic group it is necessary to determine if it is one that has already been described or if it is new and needs to be named. Unless this is done correctly the virus in question may be described as a new one whereas in fact it may be identical to one already named or sufficiently similar to it so as to be considered as one of its strains. There are examples where inadequate virus characterization has resulted in descriptions of the same virus under two, or sometimes even more different names. This only confuses the literature and should be avoided at all costs.

Even where a virus has been characterized fully, there may be difficulties in deciding if it should be considered as new. This is because there are no clear rules as to how different two virus isolates must be to warrant distinct names. Recently the PVS of the ICTV asked a number of virologists to form a study group to produce some guidelines on how the problem should be approached. The conclusion was that because virus reproduction is clonal and not sexual, the distinction between viruses and strains of the same virus has to be arbitrary.[14] Nevertheless, it was recommended that antigenic and symptomatological data are the most suitable for delineating viruses within groups. The recommended guidelines are summarized in Table 5. It is difficult to quarrel with these guidelines but it should be mentioned that other criteria can, and have been used to establish differences between isolates of related viruses. These include the amino acid composition and sequence of virus coat proteins, the nucleic acid base sequences of viral genomes and virus vector specificities. The virtues and limitations of these criteria are briefly discussed below.

1. Antigenic Properties

Serological tests have been used extensively for tracing relationships among viruses. Theoretically one could have reservations about the validity of the results because the coat protein cistron of most viruses represents only a small proportion of the total viral genome. Furthermore, only a part of that cistron is reflected in the antigenic determinants of the virus particle.[35] However, it is well established that in general, viruses which are antigenically related also share most of their other properties. Not all viruses within established taxonomic groups are antigenically related but related viruses are almost invariably included in the same group. Probably the most puzzling anomaly is the serological cross-reaction between the rod-shaped particles of tobacco mosaic virus and the small polyhedral particles of cocksfoot mild mosaic virus.[36]

When a collection of similar viruses is examined serologically they can usually be arranged into a network of interrelated individuals. The closeness of the relationship between each pair of viruses can be quantified as the serological differentiation index (SDI); defined as the number of twofold dilution steps separating the homologous from the heterologous titers. It must be remembered, however, that an accurate estimate of SDI is difficult because of several sources or error which have been discussed recently in some detail by Matthews[13] and van Regenmortel.[35] The most serious source of error is probably that individual antisera, obtained from different animals immunized with the same antigen, or from the same animal collected at different times, can vary significantly in their content of cross-reacting antibodies. Usually, there is an increase in antibody cross-reactivity in an animal with increase in the time of immunization. However, this is not a universal phenomenon. In practice, if the SDI is to have any real meaning, it must be the result of a number of determinations with antisera from several animals taken at various times during the immunization schedule. It must be stressed that the distinction between close and distant serological relationships must be an arbitrary one.

A wide variety of serological assays have been used for tracing relationships among plant viruses but liquid precipitin and agar-gel double-diffusion tests have been especially popular in the past.[35] The SDIs between viruses with small polyhedral particles have usually been determined in gel diffusion tests whereas tests with large or elongated particles have been done in liquid because of their slow diffusion rates in gels. Both types of assays are capable of detecting relatively distant relationships between viruses. Agar-gel double-diffusion tests are especially suitable for detecting differences between closely related viruses by the appropriate placement of antigen and antibody wells or intragel cross-absorption tests.[13,35]

More recently, immunoelectron microscopic and enzyme-linked immunosorbent assays (ELISA) have gained popularity among plant virologists.[35] The tests are very sensitive in detecting antibody-antigen reactions and ELISA especially, has potential as a powerful method for tracing relationships among viruses for taxonomic purposes. In its standard direct form, where heterologous antibodies are used for both coating on plates and coupling to enzyme, ELISA is extremely specific. Hence, it can be used for detecting minor antigenic differences between closely related viruses.[37-39] On the other hand, using heterologous antibodies for coupling with enzyme, ELISA is efficient at detecting remote antigenic relationships.[39] The indirect ELISA method has also been used and shown to be very efficient at detecting remote antigenic relationships among viruses.[35,40] This method utilizes antibodies to the homologous virus from two species of animals which can be a disadvantage, especially with viruses which are difficult to purify in quantity. However, it has the advantage that only one antiglobulin-enzyme conjugation is necessary for testing unlimited viruses as long as their antisera are prepared in the same animal species.

A word of warning may not go amiss here about a problem associated with the interpretation of observed minor antigenic differences between some virus isolates. It is a problem which does not appear to be widely recognized. There are viruses whose coat proteins are susceptible to partial proteolysis during purification and storage.[24,25] Du Plessis and Wechmar,[41] have demonstrated that such proteolysis can change the antigenic reactivity of an isolate of cauliflower mosaic virus to a similar degree as those observed between distinct isolates of the virus. This points to the possible misinterpretation of the significance of minor serological differences; are they due to intrinsic differences in the antigenic structure of coat proteins, or are they artifactual in origin?

2. Host Range and Symptom Expression

The infection and symptom expression in a plant inoculated with a virus is the result of a complex interaction of the host plant genome with the virus genome and the interaction is modulated by the environment. The host range and symptom expression is of little value for the characterization of viruses because there are numerous examples of similar symptoms produced on the same plant species by obviously unrelated viruses. However, minor chemical differences between isolates of related viruses can be responsible for large differences in host range and symptom expression. Because of the involvement of the host genome and environment on symptom expression it is essential that if differences are sought between two virus isolates, both must be inoculated to host plants from the same group of seedlings and maintained under the same environmental conditions. It is well known that environmental conditions, especially light and temperature, can have overriding influences on symptom expression, as can the strains or lines of test plants that are used.[42]

Most differences in symptom expression cannot be quantified and hence must be interpreted subjectively. However, they can be valuable in assessing differences between related viruses (Table 5).

3. Amino Acid Composition and Sequences

The amino acid composition of viral coat proteins has been used successfully to detect differences between related viruses. Moreover, the data have been analyzed in the computer to obtain dendrograms expressing degrees of relationship between the viruses. The most extensively studied have been the tobamoviruses and tymoviruses.[43,44] In the *Tobamovirus* group there is good agreement between relationships indicated by amino acid and serological data whereas in the *Tymovirus* group the correlation is rather poor.[44] This difference is as yet unexplained.

Determination of the amino acid composition of proteins is laborious and requires sophisticated equipment and hence is not an attractive approach to routine taxonomic investigations. Furthermore, similar compositions may arise from quite different amino acid sequences. Amino acid sequences of coat proteins would be much more informative but their determination involves an amount of work disproportionate to their taxonomic value. However, differences in amino acid sequences can be detected by peptide mapping techniques. Some of the available methods, involving proteolysis or partial proteolysis of proteins with enzymes or chemical agents such as cyanogen bromide followed by gel electrophoresis, are very rapid. They are capable of detecting some minor differences in amino acid sequences and hence variation among viruses.[29]

4. Nucleotide Base Sequences

Using currently available techniques, it is possible to determine the base sequences of complete nucleic acid molecules. Such data could be used to compare related viruses to detect differences. However, the effort required to obtain such data for taxonomic purposes would be hard to justify. The view often expressed in earlier years, that an ultimate taxonomy of viruses could be based purely on the comparisons of the nucleotide sequences of their nucleic acids, is not tenable. The significance of the nucleic acid sequence can only be assessed by also considering the organization of the genome, its functions and its products.

In taxonomy, there are now a number of useful methods for purposes of distinguishing related viruses by differences in the nucleotide sequences of their nucleic acids. Differences between the ds-DNA genomes of caulimoviruses have been detected by restriction enzyme mapping such as those reported by Hull.[45] In principle, there is also no reason why this method should not be used on double-stranded cDNA copies of ss-RNA viral genomes. Several other methods have been used for comparing the ss-RNAs of viruses; these include oligonucleotide fingerprinting and either RNA:RNA or RNA:cDNA hybridization tests.

Oligonucleotide fingerprinting can now be used with very small amounts of RNA if it is labeled in vitro with either ^{125}I or ^{32}P.[46] Fingerprinting methods have been more popular with small RNAs such as those of viroids;[47] with larger RNAs the maps tend to become so complex as to be more difficult to interpret. Although useful for detecting differences in nucleic acids, the methods are not amenable to quantitative comparisons and hence have not the potential of hybridization tests.

Viral RNAs have been compared by RNA:RNA hybridization.[48-51] The method involves the ability of various unlabeled viral RNAs to compete with a radioactive viral RNA for its homologous complementary strand in preparations of dissociated double-stranded replicative form RNA. It is laborious and has a number of other disadvantages.[50,52,53] The most important are the inability to detect base pair mismatching in the hybrids resulting in an over-estimation of sequence homology, and the inability to detect contamination of one RNA by another.

Although all hybridization methods have their problems, the RNA:cDNA method has some significant advantages over RNA:RNA tests.[52,53] Only very small amounts of ss-RNAs are required and there is no need for the preparation of ds-RNA replicative form which can be troublesome because of the low amounts usually present in virus-infected tissues. Another important advantage is that the method allows the detection and accurate estimation of any contaminants in preparations of the RNAs by $R \rightleftharpoons t$ analysis of the RNA:cDNA hybrids.

On the debit side there are some difficulties in estimating the proportion of base sequences shared by the RNAs. The hybridization assay involves annealing of labeled cDNA with the heterologous RNA to be tested for homology in a reaction mixture containing buffered salt. After incubation under appropriate conditions, S_1 nuclease is added and the mixture is further incubated to remove the unannealed cDNA. The amount of label in the nuclease-resistant fraction can then be estimated to give a measure of cDNA in hybrid form. It has been observed that the estimated homology of RNA:cDNA hybrids with extensive mismatching of base sequences, can vary with the conditions of assay. The salt concentration during the hybridization reaction has little effect on the estimated base sequence homology. During S_1 nuclease assay, however, low salt conditions result in a much lower apparent homology than at higher concentrations.[52] The concentration of the nuclease in the assay medium has also a significant effect on the estimated homology.[52] In spite of these problems the RNA:cDNA method appears to be a powerful method for comparing the RNAs of similar viruses for taxonomic purposes. If assays are done under standard conditions, the relative base sequence homologies of the RNAs can be estimated semiquantitatively.[52-56]

From the limited amount of data so far available in the literature, it is difficult to draw any definite conclusion about the merit of base sequence homologies in virus taxonomy. However, it seems that homologies can be detected only between very closely related viruses. Such viruses form clusters whose RNAs appear to have complete or almost complete homology. Relatives with partial homology appear to be rather rare.[51,53,56] Although there seems to be some correlation between relationships among viruses revealed by hybridization and serological analyses, there are some apparent exceptions. For example, cucumber green mottle mosaic virus and cucumber virus 4 are serologically only remotely related (~6 SDI units)[35] and yet their RNAs appear to have a high degree of base sequence homology.[53] In general, however, it appears that distant relationships between viruses are more readily detected by serology than by RNA homology.[35,51,56]

Assay of base sequence homologies by the RNA-cDNA procedure is relatively simple and requires no sophisticated equipment other than that required for assay of radioactivity. Once a cDNA to a particular RNA has been prepared with an isotope such as ^3H or ^{14}C, it can be stored indefinitely rather like an antiserum. In some ways and for some viruses, preparation of a highly specific cDNA presents fewer problems than the preparation of an antiserum. Only microgram amounts of RNA are required for transcribing cDNA needed for numerous assays whereas relatively large amounts of highly purified virus are required for antiserum production, especially if the virus is poorly immunogenic. Also, viral RNAs are easier to obtain in a high degree of purity than virus. The RNAs can be isolated from relatively crude virus preparations and then separated from any possible contaminants by polyacrylamide gel electrophoresis. Tracing relationships among viruses such as those of the *Luteovirus* group which occur in plants in very low amounts, may technically be much easier by RNA-cDNA hybridization than by serological methods.

At present it is difficult to assess the relative taxonomic merits of results from serological and base sequence comparisons of related viruses. Undoubtedly both methods have their advantages and disadvantages and only time will tell which will be more widely used. It is probably safe to say, however, that both approaches will find a permanent place in plant virus taxonomy.

5. Vector Specificity

Known differences in vector specificity of viruses can be used for distinguishing members of a group. However, where no difference is known, it is probably imprudent to search for it solely for taxonomic purposes. The search may be long and arduous without guarantee of success.

D. Comments on Other Characters of Taxonomic Significance

There are at least two virus properties of taxonomic significance which have not been mentioned above: the strategy of replication and the cytopathological consequences of infection. Because it was considered inappropriate to discuss them under the previous section headings, they will be discussed separately here.

1. Strategy of Virus Replication

If the descriptions of plant virus groups are compared to those of vertebrate virus families and genera in publications such as the ICTV reports,[6] the most striking difference is the disproportionate amount of information available about their modes of replication. Whereas the replication of most types of vertebrate viruses is fairly well understood, very little is known about that of most plant viruses. This imbalance is not only a reflection of the much greater investment of effort going into work on vertebrate viruses but also of the much greater technical difficulties associated with plant virus replication studies.[13] Nevertheless, some useful information is now starting to accumulate about the genome organization and mode of replication of some plant viruses, especially viruses such as the tobamoviruses,[57] tymoviruses,[13] and comoviruses.[58-59]

Undoubtedly, the strategy of virus replication is of taxonomic significance, especially at the group level. However, because of the lack of data, this character cannot be used widely in plant virus taxonomy, at least at present.

2. Virus Cytopathology

Virus infected plant cells may be recognized as such by one or more structural changes: the location and distribution of virus particles, the development of virus-induced inclusions, or the modification of normal cell organelles. These changes can and have been used for the identification of viruses by both light and electron microscopy.[60-62] Nevertheless, observations by electron microscopy are much more informative.

From the vast literature on the subject, it is becoming clear that cytopathic effects induced by plant viruses can be used for taxonomic purposes for all categories: for example, all caulimoviruses induce very characteristic viroplasms in the cytoplasm of infected cells,[61] as do the *Reoviridae*.[63] Similarly, all tymoviruses induce typical peripheral vesicles on the chloroplasts of infected cells.[64] However, the taxonomic significance of the vesiculation is not clear because similar structures have been observed in cells infected by galinsoga mosaic virus which does not appear to be related to the tymoviruses.[65]

The tobamoviruses are a good example of a group in which individual viruses induce quite different cytopathic effects.[65] Virus particle distributions and virus-induced inclusions are quite different in cells infected with distinct but serologically related viruses. Again, the significance of the cell inclusions is puzzling. Two of the tobamoviruses, cucumber virus 4 and cucumber green mottle mosaic virus induce enlarged and vesiculated mitochondria

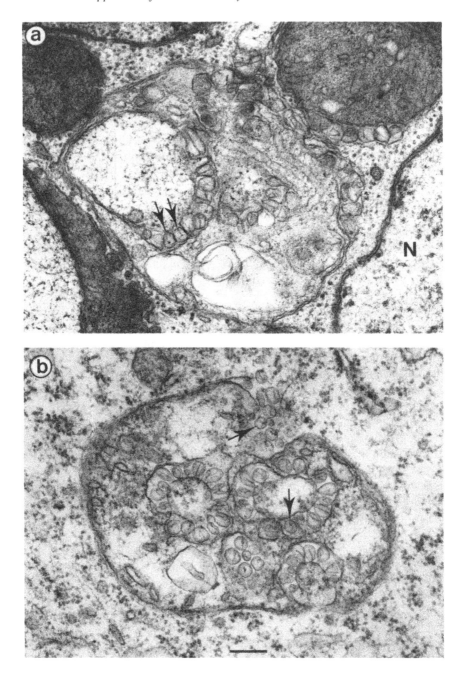

FIGURE 2.   Vesiculation of mitochondria in virus-infected leaf cells: (a) French bean infected with galinsoga mosaic virus (unclassified virus with small polyhedral particles); and (b) cucumber infected with cucumber green mottle mosaic virus *(Tobamovirus)*. Arrows point to vesicles containing densely stained strands reminiscent of nucleic acid. Bar represents 200 nm.

similar to those found in cells infected by viruses with small polyhedral particles such as turnip crinkle virus[66] and galinsoga mosaic virus[65] (Figure 2). The other eight tobamoviruses induce vesiculation of the tonoplasts, a property considered characteristic of cucumovirus infection.[67] Only four of these viruses induce, in addition, the classical x-bodies usually associated with tobamovirus infection.[68]

Significantly different cytopathic effects have also been induced by serologically closely related strains of some viruses. For example, electron microscopic examination of leaf cells of plants infected with four strains of bean yellow mosaic virus revealed that the cytopathic effects induced by three of them were very similar. However, the fourth differed from the others in that it also induced nucleolar inclusions.[69,70] The numerous examples of disagreement in the literature about the cytopathic effects of some viruses could well be explained by the authors having used different strains of the virus.

All interpretations of virus-induced cytopathic effects observed by electron microscopy should be approached critically. It is essential to examine a truly representative sample of cells. This is time-consuming but of the utmost importance if valid conclusions are to be reached. Not only a large number of cells must be examined, but they should come from different parts of more than one plant, and preferably several.

To determine the cytopathic effects of a virus it is usual to compare the cells of infected plants to those of healthy ones. However, even this may not give a true picture. Virus-infection produces a general stress on cells and hence some of the ultrastructural differences between healthy and infected cells may represent general stress rather than effects specific to a particular virus. Some of the nonspecific effects may account for the ultrastructural changes of chloroplasts often reported to be associated with virus infection. Some others involving mitochondria, golgi bodies and the development of myelin or multivesicular structures could also result from virus-induced physiological stress.

In spite of the problems of using cytopathic effects for taxonomic purposes, examination of thin sections from plant tissues infected with viruses to be identified and classified can be most informative. It is now done in many laboratories more or less as a routine procedure.

III. NOMENCLATURE

A. The Current Position of Plant Virus Nomenclature

There are few subjects which are likely to stir more heated debate among plant virologists than a suggestion that the 24 groups of plant viruses should be rearranged into families and genera. However, an even more explosive reaction could perhaps be generated if one was courageous enough seriously to propose the introduction of a species concept for plant viruses. This is surprising because the rules of the ICTV since its inception, envisaged the establishment of virus families, genera and species.[3] Moreover, many vertebrate viruses have been assigned to approved families and genera since 1970[3] and more recently, some species have also been established.[6] There seems to be support for this approach to taxonomy in all branches of virology except plant virology.

The rather primitive approach to virus nomenclature of using only one type of taxon (the group), favored by many plant virologists, is probably an over-reaction to the rather sad history of plant virus taxonomy. Since the first suggestion of a classification system for plant viruses in 1927,[71] numerous others were proposed. None were very satisfactory whereas some were disastrous, mainly because too much was attempted too soon as previously discussed by Francki[72] and in Chapter 1 of this volume. Most of the systems were based solely or predominantly on biological characters and hence were doomed to failure from the start. This was predicted by Bawden as early as 1939.[73]

In Chapter 9 of this book, persuasive arguments have been made for a nomenclature in which virus species are grouped into genera which in turn are assembled into families. It would seem that the time is now ripe for plant virologists to forget their past failures associated with virus taxonomy. The present classification of plant viruses deserves a better system of nomenclature than the "group" system and there seems to be no good reasons why we should not follow our colleagues from other branches of virology. Of course, there is no need for undue haste. A satisfactory and above all, stable taxonomy is much more likely to emerge if it is allowed to develop slowly ensuring that it is based on well-considered decisions.

B. Families and Genera

If the existing classification of plant viruses into groups is compared to the classification of vertebrate viruses, it seems that the groups correspond to either genera or families of vertebrate viruses. Groups such as the caulimoviruses and geminiviruses appear to be so distinct as to warrant family status. Some of the other groups are more likely to warrant generic rank which in turn could, in many instances, be grouped as genera of one family. Some suggestions along these lines have already been made. For example, Veerisetty[74] suggested how the viruses with elongated particles currently distributed among seven groups could be classified into nine genera within four families. Similarly, van Vloten-Doting and colleagues[75] proposed the inclusion of the bromoviruses, cucumoviruses, ilaviruses, and alfalfa mosaic virus in a single family for which they suggested the name Tricornaviridae. By glancing at the other groups of plant viruses, some further suggestions come to mind. For example, persuasive arguments could be made for considering the comoviruses and nepoviruses as two genera belonging to the same family. On the other hand, there are also a number of the existing groups which will be difficult to classify into genera or families. This will be inevitable with some of the groups of viruses with small polyhedral particles because insufficient data are available at present on their properties. Genera and families should be established from the groups only when adequate data become available to support such moves.

One may well ask about the advantages of classifying the viruses as genera and families. For the virologist interested solely in diagnosis, there are probably few, if any. The current system has proved very satisfactory. However, for the teacher or one interested in comparative virology there are definite advantages. Some of the existing groups which could well be considered as genera have many similarities and hence a teacher would find it convenient to be able to refer to them collectively; what better than to be able to refer to the viruses as a family? For example, I have often wished to refer to all the viruses with tripartite ss-RNA genomes and particles with icosahedral symmetry. These viruses are now distributed among four groups. It would be convenient to refer to them as a single family for which the name Tricornaviridae has already been suggested.[75] If one wished to write a book about these viruses, a short and informative title would be difficult to formulate unless one could use the word "Tricornaviridae" or some other approved family name.

C. The Species Question and Binomial Nomenclature

In biology, the species concept has always gone hand-in-hand with the binomial system of nomenclature. Although this system has many undoubted virtues, it also has one serious disadvantage in that any change in the classification of an organism is invariably accompanied by a change in its name. This can lead to undesirable consequences, especially with organisms which are initially poorly characterized. Later their names are changed repeatedly, and with time, they are referred to in the literature by a number of names, resulting in confusion.

Early attempts to classify viruses followed a binomial-type nomenclature. Had they received a general following, the results would have been disastrous. For example, although numerous names were proposed for viruses such as tobacco mosaic virus (Table 6), they were largely ignored and to this day the virus continues to be universally identified by its humble vernacular name.

Although most virologists approve the vernacular system of nomenclature in use, many would concede that it does have serious shortcomings. It provides little, if any, information of importance about the viruses. It highlights some biological properties of the viruses which we now consider of minor taxonomic importance. Moreover, it only refers to the disease it

Table 6
SOME NAMES PROPOSED FOR TOBACCO
MOSAIC VIRUS IN THE PAST[a]

Year	Name	Author
1922	*Stangyloplasma iwanowskii*	B. T. Palm
1927	Tobacco virus 1	J. Johnson
1937	Nicotiana virus 1	K. M. Smith
1939	*Marmor tabaci*	F. O. Holmes
1940	*Musivum tabaci*	W. D. Valleau
1941	*Phytovirus nicomosaicum*	H. H. Thornberry
1942	*Nicotianavir communae*	H. S. Fawcett
1957	*Minchorda nicotianae*	H. P. Hansen
1966	*Protovirus tabaci*	A. Lwoff & P. Tournier
1968	*Vironicotum maculans*	H. H. Thornberry
1970	*Virothrix iwanowskii*	A. E. Procenko

[a] Data adapted from Francki.[72]

causes in a single host although it may produce a number of other diseases in a wide range of different plant species. For example, the name arabis mosaic virus tells us only that the virus produces mosaic symptoms on arabis plants; a disease of little economic consequence. It does not tell us that strains of the virus can cause diseases such as raspberry yellow dwarf, rhubarb mosaic, grapevine vein banding, hop line pattern, and several others; many of which are of considerable agricultural importance. The name tells us nothing about the virus' basic properties or about its immediate relatives. It does not indicate that the virus belongs to the *Nepovirus* group which would at once inform us of many of its basic characters as well as some of its important biological properties such as that it is transmitted by nematodes and through seeds.[76] Some currently used vernacular names of viruses are even misleading. For example, lily symptomless virus can produce relatively severe symptoms in some species of *Lilium*.[77]

There is a persuasive argument that the binomial system of nomenclature has distinct advantage over any vernacular system. The generic and specific names have opposite functions and both provide important information about the organism. The generic name indicates relationship with other organisms and hence provides information about its general characteristics. The specific name expresses distinctness from its relatives. However, for this advantage to be utilized, the binomial nomenclature must be based on a sound classification. In turn, a sound classification can only be achieved through a thorough knowledge of the entities to be classified.[78] It was the lack of such knowledge about viruses that spelt disaster for earlier taxonomic systems based on a binomial nomenclature, such as those proposed by Holmes.[79]

Just prior to the establishment of the ICTV, Gibbs and colleagues[80,81] argued forcibly against the premature introduction of a binomial nomenclature system into virus taxonomy. They proposed an alternative which would alleviate the shortcomings of the vernacular system without prejudicing the development of a better nomenclature later. They suggested that vernacular names should continue to be used but that they be supplemented where precision was needed, with a cryptogram in code form indicating some of the salient features of the virus. These cryptograms were used to advantage for some time and made a positive contribution to virus taxonomy. However, their growing complexity due to discoveries such as that some viruses have divided genomes, and their inadequate taxonomic information content have made their general use less attractive than when they were first introduced.[80] In 1978, the ICTV concluded that cryptograms had outworn their usefulness and resolved not to use them in future ICTV reports.[5]

It can now be argued that our current classification system of plant viruses is soundly based and hence, that we are in a position to adopt a binomial nomenclature. If our current groups of plant viruses are reorganized into families and genera, we will have a generic name for every virus that has been classified. It will remain only to decide on a satisfactory method of delineating and naming species. This may not be as difficult as some think because much thought has already gone into how to decide if two virus isolates are distinct viruses or strains of the same virus. This has already been discussed in detail in Section II.C. as well as in Chapter 9, and it can be argued that plant virologists have, in fact, already been considering how to delineate virus species.

Even if we accept the above-mentioned ideas of how plant viruses could be classified into genera and species, we must not fall into the trap which snared virologists in the past. Generally, they classified all known viruses notwithstanding how well they had been characterized. This practice was bound to result in mistakes which, when corrected, would lead to the attendant name changes which are considered so undesirable. Thus, if any binomial system is contemplated, only well-characterized viruses should be named. Of course, the inevitable consequence of this would be a "rag bag" of viruses with vernacular or provisional names. Although by no means ideal, this could have a positive effect of encouraging workers to characterize these viruses more fully so that they can be named and classified.

It does not seem unreasonable to accept that many fully characterized viruses which now carry vernacular names are equivalent to virus species. The argument in favor of this conclusion has been discussed at length in Chapter 9 and need not be repeated here. However, it would be very cumbersome to use current vernacular names as species names because of their length. One solution would be to devise species names in such a way that they consisted of one word but would also reflect its current vernacular name. For example, the *Cucumovirus* group currently contains three definite members, cucumber mosaic (type), tomato aspermy, and peanut stunt viruses.[6] It has already been suggested that perhaps this group should be considered as one of the genera in the family Tricornaviridae.[75] The three species of the genus could then be named *Cucumovirus cucumosaicum, C. tomatospermum,* and *C. pestuntum,* respectively. One potential problem with such a move should be mentioned. It is well established that the cucumoviruses and bromoviruses have many properties in common. It can be argued that they should be combined into a single group or genus.[75] If this should be accepted by the ICTV later, some name changes would have to follow. This illustrates the main disadvantage already mentioned, of a strictly binomial system of nomenclature.

D. Some Comments on the Species Concept

Rule 11 of the ICTV rules of nomenclature states that "a virus species is a concept that will normally be represented by a cluster of strains from a variety of sources, or a population of strains from a particular source, which have in common a set of or pattern of correlating stable properties that separate the cluster from other clusters of strains".[6] Unfortunately this definition is not very helpful in recognizing virus species in practice. However, virologists are not alone in this as has been discussed in Chapter 9.

Many plant virologists are not in favor, at least for the present, of using either a species concept or a binomial nomenclature system in virus taxonomy. Because of their preconditioning by commonly accepted ideas regarding the taxonomy of eukaryotes, they see two major obstacles. First, the emphasis that is placed on phylogeny as an integral part of classification and second, the emphasis on the ability of two organisms to be capable of interbreeding as the ultimate test of conspecificity. For example, Mayr[78] has defined taxonomy as "the scientific classification of the different kinds of living organisms according to the proved or inferred phylogenetic relationships". The same author[82] defined species as "groups of interbreeding natural populations that are reproductively isolated from other such groups".

One of the major objections of many plant virologists to the adoption of a species concept and a binomial nomenclature is that very little is known about the origin and evolution of viruses. Furthermore, ideas which are expounded, are highly speculative and hence it would be ill advised to use them as any integral part of virus classification. In Chapter 9 of this book and in a previous publication,[5] Matthews has discussed at length the point that both the species concept and the Linnaean binomial nomenclature system based on it, originated many years before the theory of evolution was first proposed. Hence, there should not be undue concern about adopting them in virology.

Another common objection to the species concept is that because virus reproduction is exclusively asexual, any species concept cannot take advantage of the most convincing test of conspecificity, that of the ability to interbreed. Relevant to this, however, is the fact that it is possible to test the ability of some viruses, those with divided genomes, to interchange their genetic material. It is well established that the nucleic acid segments of some of these viruses can be reassorted to form viable pseudorecombinants or reassortants whereas others cannot. There are no known instances where viruses from different taxonomic groups have been shown to form pseudorecombinants. On the other hand, within the same groups, some members are capable of exchanging genetic material whereas others are not. At one extreme are virus strains such as those of the bipartite *Nepovirus*, raspberry ringspot virus,[83] or of the tripartite *Cucumovirus*, cucumber mosaic virus[84,85] which readily exchange their RNA segments to form viable pseudorecombinants. Strains of these viruses are serologically closely related. At the other extreme are viruses such as the serologically unrelated nepoviruses, raspberry ringspot virus and tobacco ringspot virus,[86] or the very distantly serologically related CAM and PRN strains of tobacco rattle virus[87] which do not produce viable pseudorecombinants. Thus the CAM and PRN strains of tobacco rattle virus should perhaps be considered as two distinct viruses or virus species.

If we adopt the concept that the ability of viruses to pseudorecombine is the ultimate test of conspecificity, then unfortunately, the results of some tests will be difficult to interpret. Already, there are reported instances where viruses such as the serologically related yet distinct Brazilian strain and the American yellow strain of tobacco rattle virus produce a pseudorecombinant from one heterologous combination of the two RNAs but not from the other.[88] Similar results have been reported with the A and G_{12} strains of the *Nepovirus*, tomato black ring virus.[89] There is also a situation with the tripartite genomes of viruses such as the cucumoviruses, cucumber mosaic virus, and tomato aspermy virus where only the smallest genomic RNA is interchangeable between the viruses but the two larger ones are not.[84,85] These viruses are only remotely related serologically[35,39,40] and have little if any base sequence homology.[52]

Of the 26 families and groups of plant viruses which have so far been approved by the ICTV, 12 appear to have divided genomes. This means that even if in the future some form of pseudocombination test is used for establishing conspecificity, viruses from less than half the taxonomic groups will be amenable to such tests. However, by applying this approach to distinguishing species of viruses with divided genomes, it will be possible to evaluate the importance of other virus properties for their taxonomic significance, which can then be extrapolated to groups with monopartite genomes.

E. Taxa Above the Level of Family

Before the advent of the ICTV, many of the proposed systems of virus classification were hierarchial. For example, Lwoff and colleagues[90,91] suggested that all viruses can be included in the phylum Vira which should then be subdivided into subphyla, classes, orders, and suborders based on the type of genetic material, symmetry of the nucleocapsid, presence or absence of an envelope, and rigidity or flexuousness of the nucleocapsid. Only then were the viruses to be divided into families. The majority of virologists probably agree that the

Table 7
GROUPS OF PLANT VIRUSES DIVIDED INTO
CLUSTERS BASED ON TYPE OF GENOME
AND PARTICLE MORPHOLOGY

Cluster	Properties	Virus Groups
I	ds-DNA, icosahedral, no envelope	*Caulimovirus*
II	ss-DNA, icosahedral, no envelope	*Geminivirus*
III	ds-RNA, icosahedral, no envelope	*Reoviridae* *(Phytoreovirus & Fijivirus)*
IV	ss-RNA, helical, envelope	*Rhabdoviridae* Tomato spotted wilt virus
V	ss-RNA, helical, no envelope	*Tobravirus* *Hordeivirus* *Tobamovirus* *Potexvirus* *Carlavirus* *Potyvirus* *Closterovirus*
VI	ss-RNA, icosahedral, no envelope	*Luteovirus* Maize chlorotic dwarf virus *Tymovirus* *Sobemovirus* Tobacco necrosis virus *Tombusvirus* *Comovirus* *Nepovirus* Pea enation mosaic virus *Dianthovirus* *Cucumovirus* *Bromovirus* *Ilarvirus* Alfalfa mosaic virus

ICTV was correct in starting virus classification from the group or genus and family categories. In fact, even now there appears to be no urgent need for the establishment of taxa above the family level. However, it seems that such taxa might prove useful, especially to those engaged on comparative virological work and teaching. For example, a teacher may well wish to refer to all the plant viruses with ss-RNA genomes and particles with helical symmetry. At present, these viruses are assigned to seven groups each of which has an approved name but there is no commonly used single name for the viruses included in all seven groups.

Although there is no obvious urgency for the establishment of taxa above the family level, some thought should perhaps be given as to how this could be accomplished eventually. An approach to this problem has been discussed at some length in Chapter 9. Here it will be considered only very briefly.

Because there is no general agreement about the relative importance of virus characters for the construction of a hierarchical classification system,[92] it may be advisable initially to consider only the establishment of orders as soon as all the plant virus groups are arranged into families and genera. Few virologists would dispute that the type of nucleic acid, its strandedness, the presence or absence of an envelope, and the helical or icosahedral symmetry of the nucleocapsid are all important virus characters. If all these characters are used to divide the existing families and groups of plant viruses into clusters, they will fall into six such clusters as shown in Table 7. Each of these could be considered as an order and named.

Plant virologists, especially teachers, would find it useful to be able to refer to all the viruses in clusters V and VI, each by a single distinct name. On the other hand, little would be achieved by naming clusters I-III because each would contain only one family. However, these orders, as well as all the others, would also include families of viruses infecting other organisms.

IV. COMMENTS CONCERNING INDIVIDUAL VIRUS GROUPS

A. Groups Approved by the ICTV

Some of the approved virus groups include well-characterized viruses thus constituting taxonomically sound units. On the other hand, the taxonomy of other groups is in a less satisfactory condition which varies from group to group. Complete current lists of approved members, probable and possible members of each group are included in the Fourth Report of the ICTV.[6] They are not duplicated here. However, the taxonomic status of each of the approved groups is discussed briefly.

1. Caulimovirus Group

The six members of the group are very similar in their particle structure, ds-DNA genome, and cytopathic effects. However, the serological relationships of only four of the members, cauliflower mosaic, dahlia mosaic, carnation etched ring, and strawberry vein clearing viruses have been tested and the viruses found to be interrelated.[93-95] The four possible members of the group await complete characterization but are almost certain to become true members. Comprehensive serological comparison of all members and possible members would be very informative.

2. Geminivirus Group

The five members of the group are well characterized and have remarkably similar geminate particles, circular ss-DNA genomes, and coat protein polypeptides.[6] None of them appear to be serologically related.[35] However, more critical tests with high-titered antisera could reveal distant relationships.

Two of the members, bean golden mosaic and tomato golden mosaic viruses, are transmitted by whiteflies whereas two others, maize streak and *chloris* striate mosaic viruses, are transmitted by leafhoppers. This difference in vector specificity does not appear to be correlated with any physical or chemical characteristics.[96,97] The vector of cassava latent virus is unknown.[96]

All nine probable members of the group have been shown to have typical geminate particles but they require further characterization. Again, some are transmitted by whiteflies and others by leafhoppers. Many of these viruses were initially described on the basis of their biological properties and hence some may be synonymous. It is already suspected that tobacco leaf curl and tomato yellow dwarf viruses are so closely related as to be considered the same virus or strains of a single virus.[98] Similarly, bean summer death and tobacco yellow dwarf viruses may be the same virus.[99] It will not be surprising if future work will reveal further close relationships among the probable members of the group.

3. Reoviridae

This family has been subdivided into six genera two of which, *Phytoreovirus* and *Fijivirus*, infect plants. Viruses belonging to these genera can be distinguished readily by their detailed particle structure and ds-RNA genomes. *Phytoreovirus* has particles without spikes and a genome consisting of 12 segments of ds-RNA. Matthews[6] lists wound tumor and rice dwarf viruses as the only two members but there seems little doubt that the less well characterized possible member, rice gall dwarf virus, will also be accepted as a member.[63] No antigenic relationships have been detected between the viral proteins. However, serotyping of *Reoviridae* should be done with the utmost care because several of the viruses have been shown to elicit antibodies specific to ds-RNA which cross-react with a wide range of unrelated ds-polyribonucleotides.[63]

Members of the genus *Fijivirus* differ from those of *Phytoreovirus* in that the particles have twelve spikes projecting from their icosahedral vertices and the genome consists of 10 segments of ds-RNA. The nine approved members fall into three apparently unrelated serologic groups.[6,63] Unfortunately, *Fijivirus* particles are unstable and readily lose their outer shells to produce spiked subviral particles or cores which are fairly immunogenic. Hence, serological studies are based on tests with these cores and antisera raised against them. It is conceivable that a slightly different picture of relationships could emerge if the antigens of the outer shell coat proteins were also compared. However, at present Fiji disease virus, the type member, appears to be unrelated to any other member. Maize rough dwarf, cereal tillering disease, pangola stunt, and rice black streaked dwarf viruses form a group of serologically related viruses which also have ds-RNA segments of similar sizes.[63] The remaining three viruses, those of oat sterile dwarf, *Lolium* enation and *Arrhenatherum* blue dwarf appear to be serologically indistinguishable and similar in other respects including vector specificity.[63] It seems that they should never have been described as three distinct viruses, no doubt a consequence of having been described on only their biological properties by different workers in different places.

Rice ragged stunt virus has reovirus-like particles but the structure, and possession of only eight ds-RNA segments preclude it from being assigned to any of the six genera of *Reoviridae*.[63] Although the virus is almost certainly a member of the family, its exact taxonomic status needs to be determined.

4. Rhabdoviridae

Whereas the viruses of vertebrates in this family have been classified into two genera, there is no formal taxonomy of the plant virus members. At present there are only eight plant virus members of the family whose physical, chemical, and antigenic properties have been studied to any extent.[6] There is a list of 29 probable members which have been transmitted experimentally, either by vectors or mechanical inoculation, and whose particles have been observed in the electron microscope. A further 39 possible members have been recognized only on the basis of virus-like particles observed by electron microscopy in various plant species. A herculean task awaits virologists to ascertain how many of the viruses listed in the various categories are indeed distinct, and to establish relationships among them. In the meantime, it is to be hoped that those workers detecting rhabdovirus-like particles in yet other plant species will ensure that they are not dealing with a virus which has already been described.

It has been suggested recently that the plant rhabdoviruses could be subdivided into two groups, based on the cellular location of the virus particles, their protein composition and the kinetics of transcriptase activity.[100,101] Viruses of one of the groups which includes lettuce necrotic yellows virus, have properties similar to those of the vertebrate viruses of the genus *Vesiculovirus*. The other group including potato yellow dwarf virus, has properties more like the members of *Lyssavirus*. With the very limited amount of data on only a few viruses, however, it may be premature to draw any definite taxonomic conclusions. The taxonomy of the plant virus members of the family needs to be developed but care is needed not to attempt too much without sufficient data. Unfortunately, data on the plant rhabdoviruses other than electron microscopic, are being acquired very slowly due to technical difficulties.[102]

There are a number of reports of diseased plants in which particles reminiscent of rhabdoviruses but devoid of envelopes, have been observed. Although the particles are very similar to those of rhabdovirus nucleocapsids, nothing else is known about their basic properties[102] and hence their true taxonomic position cannot be ascertained. The best studied of these particles are those associated with orchid fleck disease.

5. Tomato Spotted Wilt Virus Group

Tomato spotted wilt virus (TSWV) appears to be the only member of this group. Although the basic properties of the virus are still poorly known, it is so different from any other plant virus that its inclusion in a separate group is fully justified. However, its possible relationship to some viruses of vertebrates is more difficult to assess. It has recently been suggested that TSWV resembles some members of the *Bunyaviridae*.[65] However, it has also been claimed that the virus has positive sense RNA.[103] If this can be confirmed, it would preclude any relationship with the *Bunyaviridae*.

6. Tobravirus Group

The group consists of only two approved members, tobacco rattle virus (TRV) and pea early browning virus (PEBV), and one possible member, peanut clump virus.[6] However, there appear to be significant differences in the properties of TRV variants.[104] The differences have not been studied very thoroughly but variation in both particle lengths and antigenic properties of many of the isolates seem considerable. The serological differences between some strains of TRV appear as great as those between this virus and PEBV.[105,106] Furthermore, some of the very distantly related strains of TRV do not produce pseudorecombinants.[87] Therefore with hindsight, it would seem that perhaps some of the isolates of TRV should have been described as distinct viruses. However, any move to rectify this possible error should not be contemplated without a thorough reinvestigation of the distantly related virus variants.

7. Hordeivirus Group

The three members of the group are all distantly related serologically but only the type member, barley stripe mosaic virus, has been studied in any detail.[107] The viruses of the group are quite distinct from those of any other group in that they appear to be the only viruses with elongated particles and tripartite genomes.

8. Tobamovirus Group

Although tobacco mosaic virus (TMV) is by far the best known plant virus, the literature on *Tobamovirus* taxonomy is complicated and confusing. This probably stems from the fact that tobamoviruses have usually been identified by virtue of their characteristic particle morphology and described as strains of TMV notwithstanding possibly significant differences in many other properties. For example, sunn-hemp mosaic virus (SHMV) was described as a strain of TMV[108,109] and is still often referred to in the literature as the cowpea strain of TMV.[56] This is in spite of the fact that the two viruses are only remotely related serologically (SDI >5), have no detectable base sequence homology and differ significantly in their coat protein amino acid composition and some biological properties.[35,43,53,110] Similar situations exist in the literature with most of the other tobamoviruses. This creates confusion because many naturally occurring and induced *Tobamovirus* variants have been described and should be regarded as true strains; most numerous are those of TMV itself and tomato mosaic virus.[111,112]

Available data on the properties and relationships between the ten tobamoviruses listed by Matthews[6] and youcai mosaic virus isolated from brassicas in China,[54] especially the data concerning their serological relationships[35] and base sequence homologies[53,54,56] indicate that they should all be considered as distinct viruses. Even if they are correctly referred to as such in the future, we cannot undo the confusion already existing in the literature caused by their having been referred to as strains of TMV.

The possible members of the *Tobamovirus* group listed by Matthews[6] are very different from the approved members and their exact taxonomic position will have to be assessed carefully in the future. The *Chara corallina* virus (CCV) is most unusual in that its host is a green alga.[113] However, its particles are tobamovirus-like in structure although somewhat longer. Furthermore, a very distant serological relationship between CCV and two true tobamoviruses has been reported.[113] The remaining five viruses infect angiosperms and have a number of properties in common which, however, differ from those of the approved members of the group. They have not been studied in great detail because of difficulties in their purification. Their particle lengths are polydisperse and there is an indication that soil-borne wheat mosaic virus (SBWMV) has a multipartite genome.[114] Three of the viruses, SBWMV, beet necrotic yellow vein virus, and potato mop top virus (PMTV) have fungal vectors.[43] The most persuasive argument for inclusion of these viruses in the *Tobamovirus* group are the reports of distant serological relationships of SBWMV and PMTV to TMV.[115-117]

9. Potexvirus Group

This is a large group with some 18 members and a further 19 possible members.[6] Many of the viruses are not well characterized. Serological relationships have been reported between a dozen or more of the viruses but the degree of the relationships is difficult to assess from the available data.[35,118] The precise determination of serological relationships among the potexviruses may well present a formidable problem in that their coat proteins are known to be susceptible to degradation by host enzymes.[24,119,120] This is likely to modify the antigenic determinants.

10. Carlavirus Group

This is another large group with 23 members and a further 12 possible members.[6] Most of the members are serologically related to at least one other of the viruses[35,121] but much of the serology has not been done thoroughly. Decisions as to what constitutes a separate virus in this group are sometimes difficult to make because many of the viruses appear to be serologically closely related yet differ markedly in their host ranges.[121] Little is known about their other characters and some of the serological data appear to be contradictory. For example, Veerisetty and Brakke[122] found that alfalfa latent virus was serologically unrelated to pea streak virus whereas Hampton[123] found the two viruses to be serologically indistinguishable.

Most of the possible members of the group have been described only on the basis of particle morphology and some biological properties. Hence we can expect a number of the descriptions to be of the same or very closely related viruses. Obviously, much remains to be done to bring the taxonomy of the group into a satisfactory condition.

11. Potyvirus Group

With a list of 48 members and 67 possible members, it would appear that this is the largest group of plant viruses.[6] However, the flimsy data on which a very large proportion of the viruses have been described may well mean that many are synonymous. Most of the viruses have been identified as potyviruses by their characteristic particles and the presence of typical pin-wheel or cylindrical inclusions in the cytoplasm of infected cells. Some of the potyviruses have been compared serologically and shown to be interrelated.[35] However, serology may be a dangerous approach to tracing relationships among the potyviruses because their particles are difficult to purify without partial degradation of their coat proteins, which is likely to change the antigenic structure.[124]

At present there is some confusion about the significance of vector specificity in the taxonomy of the potyviruses. The majority of viruses which have flexuous particles between

680 and 900 nm long and which induce pin-wheel inclusions in their host cells, are transmitted by aphids in a nonpersistent manner. However, there are a number of viruses with particle morphology and cytopathic effects similar to potyviruses but which have vectors other than aphids; five are transmitted by fungi, five by eriophyid mites, and one by a whitefly.[6] None of these viruses appears to be serologically related to any of the aphid-borne potyviruses and some authors do not regard them as true potyviruses.[125] At present the physical and chemical properties of these viruses are not well known, and so it is difficult to compare some of their more basic characters to those of the better characterized potyviruses. However, it may well turn out that vector specificity is not such an important taxonomic characteristic for use with the group category as some may think. After all, there are a number of other plant virus groups where very similar viruses are transmitted by different types of vectors. The rhabdoviruses and geminiviruses are good examples.

At present, the taxonomy of the *Potyvirus* group is in a very unsatisfactory state. This stems mainly from an apparently vast variation among viruses of this type and from our inability so far, to understand the taxonomic significance of this variation. The development of a satisfactory taxonomy of potyviruses presents an immense problem which must be faced. The most urgent need is to characterize fully those viruses which have already been described. This will help to determine which of them are synonymous. It will also be necessary to decide if the potyviruses should be subdivided into lower taxa. For example, should the fungal-borne and/or mite-borne viruses be considered as distinct subgroups of the potyviruses or perhaps even completely different groups? These decisions can only be made satisfactorily when more data become available. Nevertheless, Edwardson and colleagues[62,126] have suggested that potyviruses could be divided into three subgroups based on the morphology of their cytoplasmic inclusions. However, Moghal and Francki[70] considered that two subgroups based on these characteristics would be more appropriate.

Preliminary studies on the antigenic and amino acid properties of several potyviruses indicate that they could aid classification.[124] However, difficulties associated with the preparation of undegraded virus coat proteins makes this approach somewhat hazardous.[124] This problem may, at least in part, explain the preliminary results by Abu-Samah and Randles[55] which indicate a poor correlation between relationships among some potyviruses based on serological and base sequence homology studies.

Although considerable variation in the mean particle length of potyviruses has been reported,[126] this property has been shown not to be of any potential use for subdividing viruses within the group.[70] This is due firstly to high errors in their measurement, and secondly to significant differences in the particle lengths of viruses which are similar in most other respects.[126] The ability of potyviruses to swell and contract depending on the composition of the suspending medium would be another complicating factor.[16]

12. Closterovirus Group

Eleven viruses have been assigned as members and four as possible members of the groups.[6] However, as more data become available, it is becoming doubtful whether the viruses form a coherent group that suggestions have been made that the viruses could be divided into two[127] or even three subgroups.[65] Whether these should remain as subgroups of the closteroviruses or warrant new group names, is difficult to assess at present. Most of the viruses in the group have not been extensively studied and more data are needed to make sound taxonomic decisions.

13. Luteovirus Group

At first sight this appears to be another large group of viruses with 16 members, 3 probable members and 16 possible members.[6] All the viruses are difficult to work with because they are confined to the phloem tissues of infected plants and cannot be transmitted experimentally by mechanical inoculation.[128] All transmission experiments must be done with their aphid vectors and it is difficult to obtain sufficient purified virus for adequate physical, chemical, and serological studies.

Most of the luteoviruses were initially described on their biological properties and later serological comparisons revealed that many are related,[128] but most of the data are not amenable to quantitative assessment. Some more recent serological studies using quantitative serological techniques indicate that many of the viruses are very closely related, if not identical.[129] For example, it seems that pairs of viruses such as subterranean clover red leaf and soybean dwarf or potato leafroll and tomato yellow top may be identical.[129] These conclusions are supported by more extensive studies on the biological properties of the viruses.[129] There may be many other similar cases and hence, with time, the list of luteoviruses may shrink rather than expand.

At the other end of the spectrum there is evidence that there are plant diseases which were thought to be caused by only one virus or its strains, but may actually be caused by distinct luteoviruses. For example, it now appears that the leafroll disease of potatoes can be caused either by potato leafroll virus or by beet western yellows virus.[130] Also, serological,[131] vector specificity,[132] and cytopathological differences[133] among what were referred to as isolates of barley yellow dwarf virus indicate that yellow dwarf disease of barley may be caused by what could be considered strains of at least two distinct viruses.

These brief comments give some idea of the chaotic state of *Luteovirus* taxonomy which can only be corrected by painstaking research on the biological, antigenic, chemical, and physical properties of what are now considered as definite, probable, and possible members of the group.

14. Maize Chlorotic Dwarf Virus Group

Maize chlorotic dwarf virus is the sole member of this group. It differs from viruses in all other groups in that it is transmitted by leafhoppers in a semipersistent manner. Its small polyhedral particles are also unique in that they contain large ss-RNA with a MW of about 3.2×10^6.[6] The rice tungro disease virus described by Galvez[134] has been listed as a possible member of the group.[6] However, recent work by Hibino and colleagues[135] casts some doubt on the exact etiologic agent of rice tungro disease.

15. Tymovirus Group

This is a very homogeneous group of 17 viruses and Matthews[6] lists only one possible member. Serological relationships among 14 of the tymoviruses have been traced and quantified (Figure 3) by Koenig.[136] The remaining three group members have also been shown to be related to other tymoviruses.[137-139]

Relationships between 11 of the viruses included in the serological analysis summarized in Figure 3 were also assessed from their coat protein amino acid composition data.[44] However, in general, the correlation between similarities in protein composition and serological relatedness was poor.

An interesting feature of the tymoviruses is the close serological relationship among some of the group members which are considered as distinct viruses (Figure 3) by virtue of significant differences in their host ranges. Some may perhaps question the validity of decisions to consider such serologically closely related viruses as distinct. However, some of these decsisions may be validated by the results of base sequence homology studies. So far, very little has been done to investigate the relationships among the tymoviruses based on nucleic acid homology; but it is interesting to note that no homology was detected between the RNAs of Andean potato latent and eggplant mosaic viruses[240] which are serologically very closely related (Figure 3).

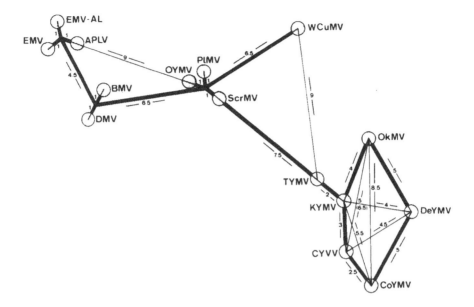

FIGURE 3. Serological relationships among tymoviruses based on serological differentiation indices (SDI) in reciprocal tests.[136] The SDI is defined as the number of twofold dilution steps separating homologous and heterologous titers. Numbers indicate SDI values between the following viruses; EMV = eggplant mosaic virus; EMV-AL = the AL isolate of EMV; APLV = Andean potato latent virus; BMV = belladonna mottle virus; DMV = *Dulcamara* mottle virus; OYMV = *Ononis* yellow mosaic virus; PLMV = *Plantago* mottle virus; ScrMV = *Scrophularia* mottle virus; WCuMV = wild cucumber mosaic virus; TYMV = turnip yellow mosaic virus (type member); OkMV = okra mosaic virus; KYMV = kennedya yellow mosaic virus; DeYMV = *Desmodium* yellow mosaic virus; CYVV = *Clitoria* yellow vein virus; and CoYMV = Cacao yellow mosaic virus.

16. Sobemovirus Group

At present this is a small group with only two members and four possible members.[6] No serological relationships have been detected between any of these viruses.[35] However, one of the possible members, cocksfoot mottle virus, has been shown to be serologically related to *Cynosorus* mottle virus which has not yet been assigned to any virus group.[141] It would seem that the *Sobemovirus* group may be a group of viruses, few of which are serologically related. More data on the physical and chemical properties of its members and possible members are needed to consolidate the taxonomy of the group.

17. Tobacco Necrosis Virus Group

Although tobacco necrosis virus (TNV) is the only group member,[6] its strains show considerable antigenic variability.[142,143] TNV is unique among plant viruses in that it can act as a helper to a satellite virus. It also stands apart from other viruses with small polyhedral particles in that it has a fungus as vector. Its physical and chemical properties also distinguish it from viruses in other groups.[142]

Cucumber necrosis virus (CNV) has been listed as a possible member of the group.[6] It is similar to TNV in its biological properties but the two viruses are not serologically related. Even from the limited available data on its physical and chemical properties, CNV seems to differ significantly from TNV[144-146] and hence its inclusion in the group should be seriously questioned.

18. Tombusvirus Group

The group consists of seven members and two possible members.[6] Five of the viruses, including the type member tomato bushy stunt virus (TBSV), are serologically closely related.[35] In some publications, artichoke mottled crinkle, carnation italian ringspot, *Pelargonium* leaf curl and *Petunia* asteroid mosaic viruses are simply referred to as strains of TBSV.[147] Thus, once again, we have the problem where, initially, the viruses were described as distinct on superficial characterization and were later shown to be closely related.

One of the two possible members of the group, turnip crinkle virus (TCV), has a remarkably similar structure to that of TBSV[148] providing a strong argument for its inclusion in the group. Nevertheless, Russo and Martelli[66] have argued for exclusion of TCV and the other possible member, saguaro cactus virus (SCV), on the grounds that they do not induce cytopathic effects characteristics of tombusviruses. This seems a somewhat tenuous argument if we consider the variety of cytopathic effects produced by some of the tobamoviruses which are generally agreed to constitute a single homogeneous virus group.

There are several unclassified viruses with small polyhedral particles that could possibly be considered for inclusion in the *Tombusvirus* group.[65] However, many of them will need to be characterized more thoroughly. Furthermore, the characters essential for inclusion in the group may also need to be better defined.

19. Comovirus Group

Matthews[6] lists 12 members of the group but two additional members are also included by Bruening.[149] Serological data collated from many laboratories indicate that all the 14 comoviruses are interrelated so that each is related to at least one other virus.[149] The degrees of the relationships are more difficult to assess because the data have originated from many laboratories using different methods. However, it seems clear that the comoviruses are a homogeneous group.

Only one possible member of the group is listed by Matthews,[6] broad bean wilt virus (BBWV). Variants of this virus have sometimes been referred to in the literature as nasturtium ringspot and *Petunia* ringspot viruses.[150] Although BBWV has very similar physical and chemical properties to those of true comoviruses,[151] no serological relationship has been reported to any of them. Furthermore, BBWV has aphid vectors[150] and not beetles like the true comoviruses[149] and differs in its cytopathic effects.[65] The virus poses an interesting taxonomic dilemma.

20. Nepovirus Group

Of the 22 members listed by Matthews,[6] 12 can be arranged into subgroups of serologically related viruses. No serological relationships have been detected between members of these subgroups or the remaining 10 viruses. However, all 22 members have very characteristic physical, chemical, and biological properties which leave little doubt that they are members of the same group.

The five possible members of the group[6] have certain properties in common with nepoviruses. However, due either to incomplete characterization or certain unusual properties, they cannot be included in the group at present. The best known of these is strawberry latent ringspot virus which poses a difficult taxonomic question. In most of its properties it appears to be a typical nepovirus but its coat protein is not characteristic of the group.[152]

21. Pea Enation Mosaic Virus Group

This is a monotypic group.[6] Pea enation mosaic virus has been thoroughly characterized and shown to have unique properties[153] which fully justify its current taxonomic position.

22. Dianthovirus Group

This is the most recently established group containing three members.[6] Carnation ringspot virus, the type member, is serologically unrelated to the other two, red clover necrotic mosaic and sweet clover necrotic mosaic viruses, which appear to be close relatives. Apart from this, all three viruses have very similar properties which differ significantly from those belonging to all other groups.

23. Cucumovirus Group

The group consists of only three members, all of which are serologically distantly related.[35,39,40] However, all three, and in particular cucumber mosaic virus (CMV), occur as numerous strains. Many of these have been studied, some in considerable detail.[39,84,154] Kaper and Waterworth[154] list 59 isolates which have been reported in the literature by various authors as strains of CMV. The most confusing and unfortunate thing about CMV is the fact that it has been referred to in the literature by at least 43 different names, coined after the diseases it causes in various plant species.[154] Some of the reports have been published in the last 5 years, one as late as 1979.

Cowpea ringspot virus[155] is listed as the only probable member.[6] Its membership is in doubt because of insufficient data on its basic properties and uncertain serological relationship with the three members.

24. Bromovirus Group

Three members and one possible member have been listed by Matthews.[6] These viruses share some properties with the cucumoviruses and a case could possibly be made for an amalgamation of the two groups. However, there are no serological relationships between viruses in the two existing groups and some of the minor differences in their biological, physical, and chemical properties can be used as arguments against any amalgamation.[75]

Recent serological work by Rybicki and von Wechmar[156] have demonstrated serological relationships among the three bromoviruses. Recent data reported on *Pelandrium* yellow fleck virus, the only possible member, argue for its inclusion in the group.[157]

25. Ilarvirus Group

Eleven distinct viruses are at present considered as members of the group.[6] A number of them appear in the literature under different names because the isolates were described and named before the viruses were fully characterized. For example, it is now considered that Danish plum line pattern, hop A and rose mosaic viruses are isolates of apple mosaic virus.[6]

Serological studies have revealed that nine viruses of the group can be divided into two subgroups.[35] However, the data are not sufficiently detailed to be able to assess the degree of the relationship between the individual viruses of each subgroup. The remaining two members of the group, lilac ring mottle and spinach latent viruses, do not appear to be serologically related to any other ilarviruses, but both qualify for inclusion in the group by virtue of their physical and chemical properties.

26. Alfalfa Mosaic Virus Group

The sole member of the group, alfalfa mosaic virus (AMV) is one of the most thoroughly studied plant viruses.[158] In its physical and chemical properties it has much in common with the ilarviruses and it has been suggested that AMV should be included in a single group with them.[75] Although there is much to be said for such amalgamation, some virologists may find the suggestion unacceptable because AMV is transmitted by aphids, whereas none of the ilarviruses appear to have insect vectors. On the other hand, others may agree with the opinion that vector specificity is a poor criterion for classification at the group level.

B. Unclassified Viruses

Some 15 years ago, Martyn[159] published a list of 630 plant viruses, and a significant number of new viruses have been described subsequently. However, Martyn's list has also been depleted by discoveries that many of the diseases assumed to have been caused by viruses actually have other etiologic agents such as mycoplasmas, rickettsia-like organisms and other agents.[13] There is also little doubt that others on the list will be shown not to be viruses and still others to be synonymous. Thus it is difficult to estimate how many recognized plant diseases are caused by distinct viruses. Even among the 471 viruses listed as definite, probable, and possible members of the established virus families and groups listed in the Fourth Report of the ICTV,[6] there must be a significant number of synonyms, as already discussed.

The "Description of Plant Viruses" published by the Commonwealth Myological Institute and the Association of Applied Biologists (CMI/AAB) included 239 viruses up to 1981. Very few of these are likely to be synonymous. Of the described viruses, 222 are members of established families and groups and 17, or about 8%, are unclassified. The proportion of all unclassified viruses is probably higher than this because viruses which are described in the CMI/AAB publication are usually those that are reasonably well characterized and hence can be placed in the existing taxa.

From the above considerations it seems that there are probably something like 600 viruses which cause recognized plant diseases of which at least 10% cannot be classified at present as definite, probable, or possible members of approved taxa. Many cannot be classified because they have not been adequately studied but there are also some that although sufficiently characterized, do not fit into any of the existing groups and families. It seems obvious that some new taxa will have to be established by the ICTV in the future to accomodate these viruses. I believe that viruses around which new groups could possibly be established in the near future are as follows:

1. Velvet Tobacco Mottle Virus

Recently, an unusual virus, causing velvet tobacco mottle (VTMoV), was discovered. In many respects it resembles viruses of the *Sobemovirus* group but it is unique in that it also encapsidates a small viroid-like circular ss-RNA which is essential for infectivity.[32,160,161] Its unusual RNA complement indicates that the virus should not be classified within the sobemoviruses or any other existing group. *Solanum nodiflorum* mottle,[162] lucerne transient streak,[163] and subterranean clover mottle viruses[164] are similar to VTMoV and have been included as a possible virus group in the most recent ICTV report.[6]

2. Carnation Mottle Virus

It has been realized for some time that carnation mottle virus (CMoV) differs in some of its basic properties from viruses in all the established groups.[45] Although the virus has small polyhedral particles about 30 nm in diameter and a monopartite ss-RNA genome, its properties are distinct from those of the luteoviruses, maize chlorotic dwarf virus, sobemoviruses, tobacco necrosis virus, tombusviruses, and tymoviruses (Table 4). Properties of viruses such as those of *Pelargonium* flower break, elderberry latent and *Narcissus* tip necrosis are also reminiscent of CMoV.[165] The possibility of establishing a new taxon to accommodate them should be considered.

3. Phleum Mottle Virus

Phleum mottle virus (PhMV) is yet another virus with small polyhedral particles about 30 nm in diameter and a monopartite ss-RNA genome whose properties are such as to apparently preclude its inclusion in any of the established groups.[165] PhMV has been shown to be serologically related to cocksfoot mild mosaic (CMMV), brome stem leaf mottle (BSLM) and *Holcus* transitory mottle (HTMV) viruses.[166] Hull[165] also suggested that *Molinia* streak virus (MSV) has certain affinities with PhMV. However, no clear serological relationship between the two viruses could be demonstrated.[166] On the other hand, MSV was shown to be serologically related to *Panicum* mosaic virus (PaMV).[167]

Although it appears that a number of viruses similar to PhMV may constitute a new group of viruses, it may be wise to compare these properties carefully to those of tobacco necrosis virus (TNV). It seems that in spite of obvious differences in vector specificity, viruses like PhMV may have many basic particle properties similar to those of TNV.[144,166] It may also be relevant to mention here that the small particles found to be associated with those of PaMV may have the same relationship to the virus as the satellite virus of TNV has with its helper.[144,167] This could be another reason why it may be worthwhile to consider the possibility that viruses discussed in this section may be sufficiently similar to TNV as to be included in the same taxonomic group.

4. Maize Rayado Fino Virus

Recently it has been suggested that maize rayado fino virus (MRFV) and oat blue dwarf virus (OBDV) have affinities, and that their properties are different from those of other viruses with small polyhedral particles and ss-RNA genomes in existing taxa.[168] However, their propagative cycles in leafhopper vectors were the characters highlighted in the comparison. The physical and chemical properties of the two viruses may be quite different and need to be compared in more detail.[168,169] Also, possible affinities of the viruses with the tymoviruses (especially in the case of MRFV) and luteoviruses (especially in the case of OBDV) should be considered before the establishment of a new virus group is contemplated.

5. Cacao Swollen Shoot Virus

There are a number of viruses with bacilliform particles about 28×130 nm such as those of cacao swollen shoot, cacao mottle leaf and *Colocasia* virus 2 which do not appear to fit into any existing virus group.[170,171] Similar particles have also been isolated from plants infected with rice tungro disease.[135] Unfortunately, little is known about the basic properties of these viruses but it seems that when this deficiency is rectified, they may warrant the establishment of a new taxonomic group.

6. Maize Stripe Virus

Recently it has been reported that maize stripe virus (MSV) is unique in having the most unusual virus particles.[172] The elongated particles appear to be nucleoprotein in composition but only about 3 nm in diameter. Rice stripe virus also appears to have similar particles and to be serologically related to MSV.[172] The apparently unique particle structure of these two viruses certainly precludes their inclusion in any existing taxonomic group.

7. Cryptic Viruses

Virus-like particles about 29 nm in diameter with very unusual properties have been isolated from a number of species of apparently healthy plants. The particles occur in the plant in very low concentrations and all attempts to transmit them experimentally, including grafting, have failed.[173] They have been referred to as cryptic viruses but only one has been characterized in any detail: that isolated from carnations and named carnation cryptic virus (CCV). The particles contain ds-RNA.[174] These unusual properties preclude the inclusion of CCV in any existing taxon of plant viruses although they could have affinities with some of the virus-like particles associated with fungi which have been designated as members of possible families in the most recent ICTV report.[6]

8. Viroids, Satellite RNAs and Satellite Viruses

It is debatable whether the taxonomy of viroids should be considered with that of viruses. Although they induce virus-like diseases, their basic properties differ very significantly[175] and hence their possible taxonomy will not be discussed here in any detail. However, viroid-like RNAs have recently been identified as functional parts of conventional viruses.[162] Hence, some sort of relationship between viruses and viroids may exist and therefore the taxonomic position of viroids in relation to viruses may need to be considered in the future.

It has been demonstrated that a number of viruses are able to support the synthesis of satellite RNAs which are capable of modulating plant disease symptoms.[13,154,176] They are encapsidated by their helper viruses and hence there seems little reason to consider their taxonomy per se. Nevertheless, the ability of the helper viruses to support their synthesis could be used in taxonomic consideration of the helpers.

There is only one definitely established example of a plant virus supporting the synthesis of a satellite virus coding for its own coat protein, that of tobacco necrosis and its satellite virus.[144] With only one such virus there seems to be no urgency for considering its taxonomy.

V. CONCLUDING REMARKS

As the numbers of any kind of entities accumulate, whether they be organisms or any type of inanimate objects, a time comes when they need to be classified. In this respect, viruses are no exception and there is no doubt that if we are to have order in the discipline of virology, an efficient taxonomy is essential. With the steady flow of information about the basic properties of viruses in the past two decades and the work of the ICTV since its inception and publication of its first report in 1971,[3] much progress in the taxonomy of viruses has been made. The widespread use of this taxonomy reported by the ICTV at regular intervals[4-6] is the best endorsement of its success. A significant proportion of adequately characterized viruses have now been classified into groups, most of which promise to remain stable taxa. While assigning viruses to these groups, we have learned much about which virus characters are taxonomically useful. Furthermore, we can better distinguish characters which are suitable for assembling viruses into groups from those which are better for distinguishing variants within the groups.

In spite of the progress that has been made, it is obvious that there is still much to be done to improve the taxonomy of viruses. There are a number of problems that have to be faced in the near future. Some of these which concern plant viruses, include the following:

1. There are still a significant number of viruses which obviously do not fit into the already existing groups of viruses. New groups will have to be established to accommodate these viruses.
2. It seems that the loose "group approach" to the classification of plant viruses is not altogether satisfactory. Plant virologists should seriously consider falling into step with their colleagues from all the other branches of virology, in redefining the existing approved groups of plant viruses as families and genera.
3. The current use of the vernacular nomenclature in virology leaves much to be desired. It lacks discipline and has encouraged the proliferation of synonyms for many viruses. A more rigorous approach to virus nomenclature and hence to taxonomy, would have many advantages. Steps need to be taken towards the development of a better, soundly based virus nomenclature including a species concept.
4. The establishment of taxa above the family level needs to be considered. Although there is no obvious urgency for this, it would be an advantage to be able to refer to clusters of virus families with some common properties by name.

Continued efforts to develop an improved virus taxonomy will not only be intellectually satisfying but will also provide a more efficient system for the orderly and readily accessible storage of information useful for many practical purposes.

ACKNOWLEDGMENTS

I thank Dr. T. Hatta for kindly allowing me to use the electron micrographs in Figures 1 and 2; Dr. R. Koenig for Figure 3; and Dr. J. W. Ashby for access to his unpublished data. Many ideas expressed in this chapter originated from numerous stimulating discussions with my colleagues, especially Professor R. E. F. Matthews, with whom I served on the Executive Committee and the Plant Virus Subcommittee of ICTV; to all these friends I am deeply grateful.

REFERENCES

1. **Cooper, P. D.,** A chemical basis for the classification of animal viruses, *Nature (London),* 190, 302, 1961.
2. **Harrison, B. D., Finch, J. T., Gibbs, A. J., Hollings, M., Shepherd, R. J., Valenta, V., and Wetter, C.,** Sixteen groups of plant viruses, *Virology,* 45, 356, 1971.
3. **Wildy, P.,** Classification and nomenclature of viruses. First report of the International Committee on Nomenclature of Viruses, *Monogr. Virol.,* 5, 1, 1971.
4. **Fenner, F.,** Classification and nomenclature of viruses. Second report of the International Committee on Taxonomy of Viruses, *Intervirology,* 7, 1, 1976.
5. **Matthews, R. E. F.,** Classification and nomenclature of viruses. Third report of the International Committee on Taxonomy of Viruses, *Intervirology,* 12, 131, 1979.
6. **Matthews, R. E. F.,** Classification and nomenclature of viruses. Fourth report of the International Committee on Taxonomy of Viruses, *Intervirology,* 17, 1, 1982.
7. **Gibbs, A. J.,** Plant virus classification, *Adv. Virus Res.,* 14, 263, 1969.
8. **Francki, R. I. B.,** Limited value of the thermal inactivation point longevity in vitro and dilution endpoint as criteria for the characterization, identification and classification of plant viruses, *Intervirology,* 13, 91, 1980.
9. **Kassanis, B.,** Interactions of viruses in plants, *Adv. Virus Res.,* 10, 219, 1963.
10. **Hamilton, R. I.,** Defences triggered by previous invaders: viruses, in *Plant Disease, An Advanced Treatise,* Vol. 5, Horsfall, J. G. and Cowling, E. B., Eds. Academic Press, New York, 1980, 279.
11. **Hitchborn, J. H. and Hills, G. J.,** The use of negative staining in the electron microscopic examination of plant viruses in crude extracts, *Virology,* 27, 528, 1965.
12. **Francki, R. I. B.,** Purification of viruses, in *Principles and Techniques in Plant Virology,* Kado, C. I. and Agrawal, H. O., Eds., van Nostrand Reinhold, New York, 1972, 295.
13. **Matthews, R. E. F.,** *Plant Virology,* 2nd. ed., Academic Press, New York, 1981a.
14. **Hamilton, R. I., Edwardson, J. R., Francki, R. I. B., Hsu, H. T., Hull, R., Koenig, R., and Milne, R. G.,** Guidelines for the identification and characterization of plant viruses, *J. Gen. Virol.,* 54, 223, 1981.
15. **Francki, R. I. B.,** Some factors affecting particle length distribution in tobacco mosaic virus preparations, *Virology,* 30, 388, 1966.
16. **Govier, D. A. and Woods, R. D.,** Changes induced by magnesium ions in the morphology of some plant viruses with filamentous particles, *J. Gen. Virol.,* 13, 127, 1971.
17. **Hatta, T. and Francki, R. I. B.,** unpublished results.
18. **Abad-Zapatero, C., Abdel-Meguid, S. S., Johnson, J. E., Leslie, A. G. W., Rayment, I., Rossmann, M. G., Suck, D., and Tsukihara, T.,** Structure of southern bean mosaic virus at 2.8Å resolution, *Nature (London),* 286, 33, 1980.
19. **Harrison, S. C., Olson, A. J., Schutt, C. E., Winkler, F. K. and Bricogne, G.,** Tomato bushy stunt virus at 2.9Å resolution, *Nature (London),* 276, 368, 1978.
20. **Finch, J. T., Klug, A., and van Regenmortel, M. H. V.,** The structure of cucumber mosaic virus, *J. Mol. Biol.,* 24, 303, 1967.
21. **Gould, A. R., Francki, R. I. B., Hatta, T., and Hollings, M.,** The bipartite genome of red clover necrotic mosaic virus, *Virology,* 108, 499, 1981.
22. **Weber, K. and Osborn, M.,** The reliability of molecular weight determinations by dodecyl sulphate-polyacrylamide gel electrophoresis, *J. Biol. Chem.,* 244, 4406, 1969.
23. **Laemmli, U. K.,** Cleavage of structural proteins during the assembly of the head of bacteriophage T4, *Nature (London),* 227, 680, 1970.

24. **Koenig, R., Stegemann, H., Francksen, H., and Paul, H. L.,** Protein subunits in the potato virus X group. Determination of the molecular weights by polyacrylamide electrophoresis, *Biochim. Biophys. Acta,* 207, 184, 1970.

25. **Al Ani, R., Pfeiffer, P., and Lebeurier, G.,** The structure of cauliflower mosaic virus. II. Identity and location of the viral polypeptides, *Virology,* 93, 188, 1979.

26. **Rice, R. H.,** Minor protein components in cowpea chlorotic mottle virus and satellite of tobacco necrosis virus, *Virology,* 61, 249, 1974.

27. **Matthews, R. E. F.,** Some properties of TYMV nucleoproteins isolated in cesium chloride density gradients, *Virology,* 60, 54, 1974.

28. **Cleveland, D. W., Fischer, S. G., Kirschner, M. W., and Laemmli, U. K.,** Peptide mapping by limited proteolysis in sodium dodecyl sulfate and analysis by gel electrophoresis, *J. Biol. Chem.,* 252, 1102, 1977.

29. **Koenig, R., Francksen, H., and Stegemann, H.,** Comparison of tymovirus capsid proteins in SDS-polyacrylamide-porosity gradient gels and partial cleavage with different proteases, *Phytopath. Z.,* 100, 347, 1981.

30. **Bailey, J. M. and Davidson, N.,** Methylmercury as a reversible denaturing agent for agarose gel electrophoresis, *Anal. Biochem.,* 70, 75, 1976.

31. **Murant, A. F., Taylor, M., Duncan, G. H., and Raschké, J. H.,** Improved estimates of molecular weight of plant virus RNA by agarose gel electrophoresis and electron microscopy after denaturation with glyoxal, *J. Gen. Virol.,* 53, 321, 1981.

32. **Randles, J. W., Davies, C., Hatta, T., Gould, A. R., and Francki, R. I. B.,** Studies on encapsidated viroid-like RNA. I. Characterization of velvet tobacco mottle virus, *Virology,* 108, 111, 1981.

33. **Francki, R. I. B., Randles, J. W., Chambers, T. C., and Wilson, S. B.,** Some properties of purified cucumber mosaic virus (Q strain), *Virology,* 28, 729, 1966.

34. **Randles, J. W. and Francki, R. I. B.,** Some properties of a tobacco ringspot virus isolate from South Australia, *Aust. J. Biol. Sci.,* 18, 979, 1965.

35. **van Regenmortel, M. H. V.,** *Serology and Immunochemistry of Plant Viruses,* Academic Press, New York, 1982.

36. **Querfurth, G. and Bercks, R.,** Relative importance of IgG- and IgM-antibodies in distant serological cross-reactivities of isometric molinia streak- and cocksfoot mild mosaic viruses and of the tobacco mosaic virus, *Phytopathol. Z.,* 85, 193, 1976.

37. **Koenig, R.,** ELISA in the study of homologous and heterologous reactions of plant viruses, *J. Gen. Virol.,* 40, 309, 1978.

38. **Bar-Joseph, M. and Salomon, R.,** Heterologous reactivity of tobacco mosaic virus strains in enzyme-linked immunosorbent assays, *J. Gen. Virol.,* 47, 509, 1980.

39. **Rao, A. L. N., Hatta, T., and Francki, R. I. B.,** Comparative studies on tomato aspermy and cucumber mosaic viruses. VII. Serological relationships reinvestigated, *Virology,* 116, 318, 1982.

40. **Devergne, J. C., Cardin, L., Burckard, J., and van Regenmortel, M. H. V.,** Comparison of direct and indirect ELISA for detecting antigenically related cucumoviruses, *J. Virol. Methods,* 3, 193, 1981.

41. **Du Plessis, D. H. and Wechmar, M. B.,** The effect of limited proteolysis on the antigenic stability of cauliflower mosaic virus, *Virology,* 107, 298, 1980.

42. **van der Want, J. P. H., Boerjan, M. L., and Peters, D.,** Variability of some plant species from different origins and their suitability for virus work, *Neth. J. Plant Pathol.,* 81, 205, 1975.

43. **Gibbs, A. J.,** Tobomovirus group, C.M.I./A.A.B. *Descriptions of Plant Viruses,* No. 184, 1977.

44. **Paul, H. L., Gibbs, A., and Wittman-Liebold, B.,** The relationships of certain tymoviruses assessed from the amino acid composition of their coat proteins, *Intervirology,* 13, 99, 1980.

45. **Hull, R.,** Structure of the cauliflower mosaic virus genome. III. Restriction endonuclease mapping of thirty-three isolates, *Virology,* 100, 76, 1980.

46. **Lee, Y. F. and Fowlks, E. R.,** Rapid *in vitro* labeling procedures for two-dimensional gel fingerprinting, *Anal. Biochem.,* 119, 224, 1982.

47. **Dickson, E.,** Viroids: infectious RNA in plants, in *Nucleic Acids in Plants,* Vol. 2, Hall, T. C. and Davies, J. W., Eds., CRC Press, Boca Raton, Fla., 1979, 153.

48. **Bol, J. F., Brederode, F. T., Janze, G. C., and Rauh, D. K.,** Studies on sequence homology between the RNAs of alfalfa mosaic virus, *Virology,* 65, 1, 1975.

49. **Vandewalle, M. J. and Siegel, A.** A study of nucleotide sequence homology between strains of tobacco mosaic virus, *Virology,* 73, 413, 1976.

50. **Palomar, M. K., Brakke, M. K., and Jackson, A. O.,** Base sequence homology in the RNAs of barley stripe mosaic virus, *Virology,* 77, 471, 1977.

51. **Piazzolla, P., Diaz-Ruiz, J. R., and Kaper, J. M.,** Nucleic acid homologies of eighteen cucumber mosaic virus isolates determined by competition hybridization, *J. Gen. Virol.,* 45, 361, 1979.

52. **Gonda, T. J. and Symons, R. H.,** The use of hybridization analysis with complementary DNA to determine the RNA sequence homology between strains of plant viruses: its application to several strains of cucumoviruses, *Virology,* 88, 361, 1978.

53. **Palukaitis, P. and Symons, R. H.,** Nucleotide sequence homology of thirteen tobamovirus RNAs as determined by hybridization analysis with complementary DNA, *Virology,* 107, 354, 1980.
54. **Palukaitis, P., Randles, J. W., Tian, Y-t, Kang, L-y, and Tien-P.,** Taxonomy of several tobamoviruses from China as determined by molecular hybridization analysis with complementary DNA, *Intervirology,* 16, 136, 1981.
55. **Abu-Samah, N. and Randles, J. W.,** A comparison of the nucleotide sequence homologies of three isolates of bean yellow mosaic virus and their relationship to other potyviruses, *Virology,* 110, 436, 1981.
56. **van de Walle, M. J. and Siegel, A.,** Relationships between strains of tobacco mosaic virus and other selected plant viruses, *Phytopathology,* 72, 390, 1982.
57. **Hirth, L. and Richards, K. E.,** Tobacco mosaic virus: model for structure and function of a simple virus, *Adv. Virus. Res.,* 26, 145, 1981.
58. **Franssen, H., Goldbach, R., Broekhuijsen, M., Moerman, M., and van Kammen, A.,** Expression of middle-component RNA of cowpea mosaic virus: *in vitro* generation of a precursor to both capsid proteins by a bottom-component RNA-encoded protease from infected cells, *J. Virol.,* 41, 8, 1982.
59. **Goldbach, R., Rezelman, G., Zabel, P., and van Kammen, A.,** Expression of the bottom-component RNA of cowpea mosaic virus: evidence that the 60 kilodalton VPg precursor is cleaved into single VPg and a 58-kilodalton polypeptide, *J. Virol.,* 42, 630, 1982.
60. **Christie, R. G. and Edwardson, J. R.,** Light and electron microscopy of plant virus inclusions, *Florida Agric. Exptl. Stn. Monogr.,* No. 9, 1977.
61. **Martelli, G. P. and Russo, M.** Plant virus inclusion bodies, *Adv. Virus Res.,* 21, 175, 1977.
62. **Edwardson, J. R. and Christie, R. G.,** Use of virus-induced inclusions in classification and diagnosis, *Ann. Rev. Phytopathol.,* 16, 31, 1978.
63. **Francki, R. I. B. and Boccardo, G.,** The plant Reoviridae, *The Viruses,* Plenum Press, New York, in press.
64. **Lesemann, D. -E.,** Virus group-specific and virus-specific cytological alterations induced by members of the tymovirus group, *Phytopathol. Z.,* 90, 315, 1977.
65. **Francki, R. I. B., Hatta, T., and Milne, R. G.,** *An Atlas of Plant Viruses,* CRC Press, Boca Raton, Fla., in press.
66. **Russo, M. and Martelli, G. P.,** Ultrastructure of turnip crinkle and saguaro cactus virus-infected tissues, *Virology,* 118, 109, 1982.
67. **Hatta, T. and Francki, R. I. B.,** Cytopathic structures associated with tonoplasts of plant cells infected with cucumber mosaic and tomato aspermy viruses, *J. Gen. Virol.,* 53, 343, 1981.
68. **Esau, K.,** *Viruses in Plant Hosts — Distribution and Pathologic Effects,* Univ. Wisconsin Press, Madison, 1968.
69. **Randles, J. W., Davies, C., Gibbs, A. J., and Hatta, T.,** Amino acid composition of capsid protein as a taxonomic criterion for classifying the atypical S strain of bean yellow mosaic virus, *Aust. J. Biol. Sci.,* 33, 245, 1980.
70. **Moghal, S. M. and Francki, R. I. B.,** Towards a system for the identification and classification of potyviruses. II. Virus particle length, symptomatology, and cytopathology of six distinct viruses, *Virology,* 112, 210, 1981.
71. **Johnson, J.,** The classification of plant viruses, *Res. Bull. Agric. Expt. Stn. Univ. Wisconsin,* 76, 1, 1927.
72. **Francki, R. I. B.,** Plant virus taxonomy, in *Handbook of Plant Virus Infections: Comparative Diagnosis,* Kurstak, E., Ed., Elsevier/North-Holland Biomedical Press, Amsterdam, 1981, 3.
73. **Bawden, F. C.,** *Plant Viruses and Virus Diseases,* Chronica Botanica, Leiden, 1939.
74. **Veerisetty, V.,** Suggestions for the classification and nomenclature of helical plant viruses, *Intervirology,* 11, 167, 1979.
75. **van Vloten-Doting, L., Francki, R. I. B., Fulton, R. W., Kaper, J. M., and Lane, L. C.,** Tricornaviridae - a proposed family of plant viruses with tripartite, single-stranded RNA genomes, *Intervirology,* 15, 198, 1981.
76. **Murant, A. F.,** Arabis mosaic virus, C.M.I./A.A.B. *Descriptions of Plant Viruses,* No. 16, 1970.
77. **Allen, T. C.,** Lily symptomless virus, C.M.I./A.A.B. *Descriptions of Plant Viruses,* No. 96, 1972.
78. **Mayr, E.,** Concepts of classification and nomenclature in higher organisms and microorganisms, *Ann. N.Y. Acad. Sci.,* 56, 391, 1953.
79. **Holmes, F. O.,** Order Virales, the filtrable viruses, in *Bergey's Manual of Determinative Bacteriology,* Breed, R. S., Murray, E. G. D., and Hitchens, A. P., Eds., Williams & Wilkins, Baltimore, 1948, 1127.
80. **Gibbs, A. J., Harrison, B. D., Watson, D. H., and Wildy, P.,** What's in a virus name? *Nature (London),* 209, 450, 1966.
81. **Gibbs, A. J. and Harrison, B. D.,** Realistic approach to virus classification and nomenclature, *Nature (London),* 218, 927, 1968.
82. **Mayr, E.,** *Populations, Species and Evolution,* Harvard Univ. Press, Cambridge, 1970.

83. **Harrison, B. D., Murant, A. F., Mayo, M. A., and Roberts, I. M.,** Distribution of determinants for symptom production, host range and nematode transmissibility between the two RNA components of raspberry rinspot virus, *J. Gen. Virol.,* 22, 233, 1974.

84. **Rao, A. L. N. and Francki, R. I. B.,** Comparative studies on tomato aspermy and cucumber mosaic viruses. VI. Partial compatibility of genome segments from the two viruses, *Virology,* 114, 573, 1981.

85. **Rao, A. L. N. and Francki, R. I. B.,** Distribution of determinants for symptom production and host range on the three RNA components of cucumber mosaic virus, *J. Gen. Virol.,* 61, 197, 1982.

86. **Harrison, B. D., Murant, A. F., and Mayo, M. A.,** Evidence for two functional RNA species in raspberry ringspot virus, *J. Gen. Virol.,* 16, 339, 1972.

87. **Frost, R. R., Harrison, B. D., and Woods, R. D.,** Apparent symbiotic interaction between particles of tobacco rattle virus, *J. Gen. Virol.,* 1, 57, 1967.

88. **Lister, R. M.,** Tobacco rattle, NETU, viruses in relation to functional heterogeneity in plant viruses, *Fed. Proc. Fed. Am. Soc. Exp. Biol.,* 28, 1875, 1969.

89. **Randles, J. W., Harrison, B. D., Murant, A. F., and Mayo, M. A.,** Packaging and biological activity of the two essential RNA species of tomato black ring virus, *J. Gen. Virol.,* 36, 187, 1977.

90. **Lwoff, A., Horne, R., and Tournier, P.,** A system of viruses, *Cold Spring Harbor Symp. Quant. Biol.,* 27, 51, 1962.

91. **Lwoff, A. and Tournier, P.,** The classification of viruses, *Ann. Rev. Microbiol.,* 20, 45, 1966.

92. **Matthews, R. E. F.,** A classification of virus groups based on the size of the particle in relation to genome size, *J. Gen. Virol.,* 27, 135, 1975.

93. **Brunt, A. A.,** Dahlia mosaic virus, C.M.I./A.A.B. *Description of Plant Viruses,* No. 51, 1971.

94. **Lawson, R. H., Hearon, S. S., and Civerolo, E. L.,** Carnation etched ring virus, C.M.I./A.A.B. *Descriptions of Plant Viruses,* No. 182, 1977.

95. **Morris, T. J., Mullin, R. H., Schlegel, D. E., Cole, A., and Alosi, M. C.,** Isolation of a caulimovirus from strawberry tissue infected with strawberry vein banding virus, *Phytopathology,* 70, 156, 1980.

96. **Goodman, R. M.,** Geminiviruses, *J. Gen. Virol.,* 54, 9, 1981.

97. **Hamilton, W. D. O., Sanders, R. C., Coutts, R. H. A., and Buck, K. W.,** Characterization of tomato golden mosaic virus as a geminivirus, *FEMS Microbiol. Lett.,* 11, 263, 1981.

98. **Osaki, T. and Inouye, T.,** Resemblance in morphology and intranuclear appearance of viruses isolated from yellow dwarf diseased tomato and leaf curl diseased tobacco, *Ann. Phytopathol. Soc. Jpn.,* 44, 167, 1978.

99. **Thomas, J. E. and Bowyer, J. W.,** Properties of tobacco yellow dwarf and bean summer death viruses, *Phytopathology,* 70, 214, 1980.

100. **Dale, J. L. and Peters, D.,** Protein composition of the virions of five plant rhabdoviruses, *Intervirology,* 16, 86, 1981.

101. **Peters, D.,** Plant rhabdovirus group, C.M.I./A.A.B. *Descriptions of Plant Viruses,* No. 244, 1981.

102. **Francki, R. I. B., Kitajima, E. W., and Peters, D.,** Rhabdoviruses, in *Handbook of Plant Virus Infections: Comparative Diagnosis,* Kurstak, E., Ed., Elsevier/North Holland Biomedical Press, Amsterdam, 1981, 455.

103. **Verkleij, F. N., de Vries, P., and Peters, D.,** Evidence that tomato spotted wilt virus RNA is a positive strand, *J. Gen. Virol.,* 58, 329, 1982.

104. **Harrison, B. D. and Robinson, D. J.,** The Tobraviruses, *Adv. Virus. Res.,* 23, 25, 1978.

105. **Maat, D. Z.,** Pea early-browning virus and tobacco rattle virus — two different, but serologically related viruses, *Neth. J. Plant Path.,* 69, 287, 1963.

106. **Harrison, B. D. and Woods, R. D.,** Serotypes and particle dimensions of tobacco rattle viruses from Europe and America, *Virology,* 28, 610, 1966.

107. **Jackson, A. O. and Lane, L. C.,** Hordeiviruses, in *Handbook of Plant Virus Infections: Comparative Diagnosis,* Kurstak, E., Ed., Elsevier/North-Holland Biomedical Press, Amsterdam, 1981, 565.

108. **Lister, R. M. and Thresh, J. M.,** A mosaic disease of leguminous plants caused by a strain of tobacco mosaic virus, *Nature (London),* 175, 1047, 1955.

109. **Capoor, S. P.,** Southern sunn hemp mosaic virus: a strain of tobacco mosaic virus, *Phytopathology,* 52, 393, 1962.

110. **Kassanis, B. and Varma, A.,** Sunn-hemp mosaic virus, C.M.I./A.A.B. *Descriptions of Plant Viruses,* No. 153, 1975.

111. **Hennig, B. and Wittmann, H. G.,** Tobacco mosaic virus: mutants and strains, in *Principles and Techniques in Plant Virology,* Kado, C. I. and Agrawal, H. O., Eds., van Nostrand Reinhold, New York, 1972, 546.

112. **Hollings, M. and Huttinga, H.,** Tomato mosaic virus, C.M.I./A.A.B. *Descriptions of Plant Viruses,* No. 156, 1976.

113. **Gibbs, A., Skotnicki, A. H., Gardiner, J. E., Walker, E. S., and Hollings, M.,** A tobamovirus of a green alga, *Virology,* 64, 571, 1975.

114. **Tsuchizaki, T., Hibino, H., and Saito, Y.,** The biological functions of short and long particles of soil-borne wheat mosaic virus, *Phytopathology,* 65, 523, 1975.

115. **Powell, C. A.,** The relationship between soil-borne wheat mosaic virus and tobacco mosaic virus, *Virology,* 71, 453, 1976.

116. **Randles, J. W., Harrison, B. D., and Roberts I. M.,** *Nicotiana velutina* mosaic virus: purification properties and affinities with other rod-shaped viruses, *Ann. Appl. Biol.,* 84, 193, 1976.

117. **Kassanis, B., Woods, R. D., and White, R. F.,** Some properties of potato mop-top virus and its serological relationship to tobacco mosaic virus, *J. Gen. Virol.,* 14, 123, 1972.

118. **Koenig, R. and Lesemann, D. -E.,** Potexvirus group, C.M.I./A.A.B. *Descriptions of Plant Viruses,* No. 200, 1978.

119. **Koenig, R.,** Serological relations of narcissus and papaya mosaic viruses to established members of the potexvirus group, *Phytopath. Z.,* 84, 193, 1975.

120. **Koenig, R., Tremaine, J. H., and Shepard, J. F.,** *In situ* degradation of the protein chain of potato virus X at the N- and C- termini, *J. Gen. Virol.,* 38, 329, 1978.

121. **Wetter, C. and Milne, R. G.,** Carlaviruses, in *Handbook of Plant Virus Infections: Comparative Diagnosis,* Kurstak, E. Ed., Elsevier/North-Holland Biomedical Press, Amsterdam, 1981, 695.

122. **Veerisetty, V. and Brakke, M. K.,** Alfalfa latent virus, a naturally occurring carlavirus in alfalfa, *Phytopathology,* 67, 1202, 1977.

123. **Hampton, R. O.,** Evidence suggesting identity between alfalfa latent and pea streak viruses, *Phytopathology,* 71, 223, 1981.

124. **Moghal, S. M. and Francki, R. I. B.,** Towards a system for the identification and classification of potyviruses. I. Serology and amino acid composition of six distinct viruses, *Virology,* 73, 350, 1976.

125. **Hollings, M. and Brunt, A. A.,** Potyviruses, in *Handbook of Plant Virus Infections: Comparative Diagnosis,* Kurstak, E., Ed., Elsevier/North-Holland Biomedical Press, Amsterdam, 1981, 731.

126. **Edwardson, J. R.,** Some properties of the potato virus Y-group, *Florida Agric. Exp. Stn. Monogr.,* No. 4, 1974.

127. **Bar-Joseph, M., Garnsey, S. M., and Gonsalves, D.,** The closteroviruses: a distinct group of elongated plant viruses, *Adv. Virus. Res.,* 25, 93, 1979.

128. **Rochow, W. F. and Duffus, J. E.,** Luteoviruses and yellows diseases, in *Handbook of Plant Virus Infections: Comparative Diagnosis,* Kurstak, E., Ed., Elsevier/North-Holland Biomedical Press, Amsterdam, 1981, 147.

129. **Ashby, J. W.,** personal communication.

130. **Duffus, J. E.,** Beet western yellows virus — a major component of some potato leaf roll-affected plants, *Phytopathology,* 71, 193, 1981.

131. **Aapola, A. I. E. and Rochow, W. F.,** Relationships among three isolates of barley yellow dwarf virus, *Virology,* 46, 127, 1971.

132. **Rochow, W. F.,** Transmission of strains of barley yellow dwarf virus by two aphid species, *Phytopathology,* 49, 744, 1959.

133. **Gill, C. C. and Chong, J.,** Cytopathological evidence for the division of barley yellow dwarf virus isolates into two subgroups, *Virology,* 95, 59, 1979.

134. **Galvez, G. E.,** Rice tungro virus. C.M.I./A.A.B. *Descriptions of Plant Viruses,* No. 67, 1971.

135. **Hibino, H., Saleh, N., and Roechan, M.,** Transmission of two kinds of rice tungro-associated viruses by insect vectors, *Phyropathology,* 69, 1266, 1979.

136. **Koenig, R.,** A loop-structure in the serological classification system of tymoviruses, *Virology,* 72, 1, 1976.

137. **Peters, D. and Derks, A. F. L. M.,** Host range and some properties of physalis mosaic virus, a new virus of the turnip yellow mosaic virus group, *Neth. J. Plant Path.,* 80, 124, 1974.

138. **Shukla, D. D. and Gough, K. H.,** Erysimum latent virus, C.M.I./A.A.B. *Descriptions of Plant Viruses,* No. 222, 1980.

139. **Lana, A. F.,** Properties of a virus occurring in *Arachis hypogea* in Nigeria, *Phytopath. Z.,* 97, 169, 1980.

140. **Kummert, J., Lacroix, J. P., and Semal, J.,** Heterology among the RNAs of tymoviruses as revealed by RNA-RNA hybridizations, *Virology,* 89, 306, 1978.

141. **Mohamed, N. A.,** Physical and chemical properties of Cynosurus mottle virus, *J. Gen. Virol.,* 40, 379, 1978.

142. **Uyemoto, J. K. Grogan, R. G., and Wakeman, J. R.,** Selective activation of satellite virus strains by strains of tobacco necrosis virus, *Virology,* 34, 410, 1968.

143. **Kassanis, B. and Phillips, M. P.,** Serological relationship of strains of tobacco necrosis virus and their ability to activate strains of satellite virus, *J. Gen. Virol.,* 9, 119, 1970.

144. **Uyemoto, J. K.,** Tobacco necrosis and satellite viruses, in *Handbook of Plant Virus Infections: Comparative Diagnosis,* Kurstak, E. Ed., Elsevier/North-Holland Biomedical Press, Amsterdam, 1981, 123.

145. **Dias, H. F. and McKeen, C. D.,** Cucumber necrosis virus, C.M.I./A.A.B. *Descriptions of Plant Viruses,* No. 82, 1972.

146. **Tremaine, J. H.,** Purification and properties of cucumber necrosis virus and a smaller top component, *Virology,* 48, 582, 1972.

147. **Martelli, G. P., Quacquarelli, A. and Russo, M.,** Tomato bushy stunt virus, C.M.I./A.A.B. *Descriptions of Plant Viruses,* No. 69, 1971.
148. **Butler, P. J. G.,** Structures of turnip crinkle and tomato bushy stunt viruses. III. Chemical subunits, molecular weights and number of molecules per particle, *J. Mol. Biol.,* 52, 589, 1970.
149. **Bruening, G.,** Comovirus group, C.M.I./A.A.B. *Descriptions of Plant Viruses,* No. 199, 1978.
150. **Taylor, R. H. and Stubbs, L. L.,** Broad bean wilt virus, C.M.I./A.A.B. *Descriptions of Plant Viruses,* No. 81, 1972.
151. **Doel, T. R.,** Comparative properties of type, nasturtium ringspot and petunia ringspot strains of broad bean wilt virus, *J. Gen. Virol.,* 26, 95, 1975.
152. **Gallitelli, D., Savino, V., and Martelli, G. P.,** The middle component of strawberry latent ringspot virus, *J. Gen. Virol.,* 59, 169, 1982.
153. **Hull, R.,** Pea enation mosaic virus, in *Handbook of Plant Virus Infections: Comparative Diagnosis,* Kurstak, E. Ed., Elsevier/North-Holland Biomedical Press, Amsterdam, 1981, 239.
154. **Kaper, J. M. and Waterworth, H. E.,** Cucumoviruses, in *Handbook of Plant Viruses Infections: Comparative Diagnosis,* Kurstak, E. Ed., Elsevier/North-Holland Biomedical Press, Amsterdam, 1981, 257.
155. **Phatak, H. C., Diaz-Ruiz, J. R., and Hull, R.,** Cowpea ringspot virus: a seed transmitted cucumovirus, *Phytopathol. Z.,* 87, 132, 1976.
156. **Rybicki, E. P. and von Wechmar, M. B.,** The serology of the bromoviruses. I. Serological interrelationships of the bromoviruses, *Virology,* 109, 391, 1981.
157. **Hollings, M. and Horvath, J.,** Melandrium yellow fleck virus, C.M.I./A.A.B. *Description of Plant Viruses,* No. 236, 1981.
158. **Jaspars, E. M. J. and Bos, L.,** Alfalfa mosaic virus, C.M.I./A.A.B. *Descriptions of Plant Viruses,* No. 229, 1980.
159. **Martyn, E. B.,** Plant Virus Names. An Annotated List of Names and Synonyms of Plant Viruses and Diseases, *Phytopathological Papers* No. 9, Commonwealth Mycological Inst. Kew, Surrey.
160. **Gould, A. R.,** Studies on encapsidated viroid-like RNA. II. Purification and characterization of a viroidlike RNA associated with velvet tobacco mottle virus (VTMoV), *Virology,* 108, 123, 1981.
161. **Gould, A. R., Francki, R. I. B., and Randles, J. W.,** Studies on encapsidated viroid-like RNA. IV. Requirement for infectivity and specificity of two RNA components from velvet tobacco mottle virus, *Virology,* 110, 420, 1981.
162. **Gould, A. R. and Hatta, T.,** Studies on encapsidated viroid-like RNA. III. Comparative studies on RNAs isolated from velvet tobacco mottle virus and *Solanum nodiflorum* mottle virus, *Virology,* 109, 137, 1981.
163. **Po, T., Davies, C., Hatta, T., and Francki, R. I. B.,** Viroid-like RNA encapsidated in lucerne transient streak virus, *FEBS Lett.,* 132, 353, 1981.
164. **Francki, R. I. B., Randles, J. W., Hatta, T., Davies, C., Chu, P. W. G., and McLean, G. W.,** Subterranean clover mottle virus: another virus from Australia with encapsidated viroid-like RNA, *Plant Pathol.,* in press, 1983.
165. **Hull, R.,** The grouping of small spherical plant viruses with single RNA components, *J. Gen. Virol.,* 36, 289, 1977.
166. **Paul, H. L., Querfurth, G., and Huth, W.,** Serological studies on the relationships of some isometric viruses of *Gramineae, J. Gen. Virol.,* 47, 67, 1980.
167. **Niblett, C. L., Paulsen, A. Q., and Toler, R. W.,** Panicum mosaic virus, C.M.I./A.A.B. *Descriptions of Plant Viruses,* No. 177, 1977.
168. **Leon, P. and Gamez, R.,** Some physicochemical properties of maize rayado fino virus, *J. Gen. Virol.,* 56, 67, 1981.
169. **Banttari, E. E. and Zeyen, R. J.,** Oat blue dwarf virus, C.M.I./A.A.B. *Descriptions of Plant Viruses,* No. 123, 1973.
170. **Brunt, A. A.,** Cacao swollen shoot virus, C.M.I./A.A.B. *Descriptions of Plant Viruses,* No. 10, 1970.
171. **James, M., Kenten, R. H., and Woods, R. D.,** Virus-like particles associated with two diseases of *Colocasia esculenta* (L.) Schott. in the Solomon Islands, *J. Gen. Virol.,* 21, 145, 1973.
172. **Gingery, R. E., Nault, L. R., and Bradfute, O. E.,** Maize stripe virus: characteristics of a member of a new virus class, *Virology,* 112, 99, 1981.
173. **Lisa, V., Luisoni, E., and Milne, R. G.,** A possible virus cryptic in carnation, *Ann. Appl. Biol.,* 98, 431, 1981.
174. **Lisa, V., Boccardo, G., and Milne, R. G.,** Double-stranded ribonucleic acid from carnation cryptic virus, *Virology,* 115, 410, 1981.
175. **Diener, T. O.,** *Viroids and Viroid Diseases,* John Wiley & Sons, New York, 1979.
176. **Schneider, I. R.,** Defective plant viruses, in *Beltsville Symp. Agr. Res. 1. Virology in Agriculture,* Romberger, J. A., Anderson, J. D., and Powell, R. L., Eds., Abacus Press, Tunbridge Wells, 1977, 201.

Chapter 4

CURRENT PROBLEMS IN BACTERIAL VIRUS TAXONOMY

Hans-Wolfgang Ackermann

TABLE OF CONTENTS

I. INTRODUCTION

After the discovery of bacteriophages in 1915 to 1917 by Twort[1] and d'Hérelle,[2] phages were found in many bacteria. By 1955, at least 5655 papers, books, or monographs on phages had been published.[3] Research was centered on the nature of phages, host range, ecology, phage typing, and therapy. The first phages were studied in the electron microscope by Ruska[4] and Kottmann[5] shortly after 1940, but electron microscopes were scarce and morphological data were slow to accumulate.

In the 1950s, reliable methods for the study of nucleic acids were developed. This, the introduction of negative staining by Brenner and Horne in 1959,[6] and the spread of electron microscopes greatly stimulated phage research. Interest now shifted to morphology, biochemistry, and genetics, especially of T-even phages. In addition, many phage typing schemes were developed during this period. By 1965, no less than 11,405 papers on bacteriophages had been published.[7] Publication volume had thus quadrupled in 10 years. At least 1,114 papers or 22.4% were on coliphages.[7]

Let us consider descriptive papers which are the raw material of taxonomy. Phages of enterobacteria, bacilli, pseudomonads, and staphylococci were the first to be studied in significant numbers, largely through the pioneering efforts of Bradley.[8-10] Subsequently, there were definitive peaks of interest in phages of enterobacteria (1965 to 1972), *Corynebacterium* and *Lactobacillus* (1969), *Mycobacterium* (1969 to 1972), *Rhizobium* (1972 to 1973), *Staphylococcus* (1960 to 1969 and 1980), and *Streptococcus* (1974 to 1977). Many phage descriptions were triggered by the realization that the identity of typing phages was easily controlled by electron microscopy. Between 1969 and 1974, much effort went into the investigation of nonmedical and sometimes arcane bacteria, for example *Actinoplanes, Ancalomicrobium,* or *Desulfovibrio.* Most of the known cubic and filamentous phages were described by 1974. In a general way, it can be said that morphological phage descriptions were published since 1960 at a rate of 110 year year and peaked in 1969, 1972, and 1977. This is shown in Figure 1. Presently, about 2400 bacterial viruses have been studied by electron microscopy and an immense body of knowledge on their physical and chemical properties has accumulated. Thus, the problems of phage taxonomy are largely numerical.

The data for Figure 1 and the following Tables 1 to 5 have been compiled from the author's literature files, which are necessarily incomplete. For reasons of space, neither the figure nor the tables can be referenced. They should be taken as no more than a basis for discussion.

II. DEVELOPMENT AND PRESENT STATE OF PHAGE TAXONOMY

A. Distribution of Phages Among the Host Groups

Table 1 shows the frequency of phage observations in various host groups. Almost all bacterial subdivisions of Bergey's Manual[11] are represented, as well as cyanobacteria and a green alga, *Euglena pyrenoidosa.* The occurrence of phages in *Euglena* is not certain and urgently needs confirmation.[12] A detailed tabulation of phage observations in over 90 host genera has been published elsewhere.[12] Table 1 also shows that tailed phages are far more numerous than cubic, filamentous, and pleomorphic phages.

B. Evolution of Phage Taxonomy

For d'Hérelle there was only one phage with many races, the *Bacteriophagum intestinale.*[13] Around 1930, it was shown that phages could be categorized by serology and resistance to physicochemical agents. Several phage strains were characterized in this way; some of them, for example C16 or S13, are still available. In 1943, Ruska used electron microscopy for virus classification and distinguished several types of bacteriophages.[14] These early attempts

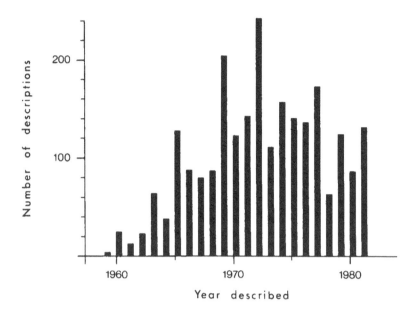

FIGURE 1. Frequency of published phage descriptions per year. The numbers exclude shadowed phages, known defective phages, and particulate bacteriocins. Completed January 1, 1982.

were clearly premature, but prompted Holmes in 1948 to classify phages into the order, *Phagineae*, a single genus, and 46 species.[15] This system was never used or accepted. In 1953, Adams stressed serological properties and proposed morphology as a criterion.[16] In his famous book published in 1959, he suggested classifying phages by a variety of criteria including serology, morphology and size, and chemical composition.[17] Adams was very interested in taxonomy and believed that phages could be classified into species. It is certain that if he had lived longer, phage taxonomy would be in a more advanced state.

When Lwoff, Horne, and Tournier published their system of viruses in 1962, it included only φX174 and two tailed phages.[18] This system was expanded in 1965 and 1966 to include cubic RNA and filamentous DNA phages.[19,20] Inexplicably, the filamentous phages were grouped among the cubic viruses. Tailed phages were given order rank and were named *Urovirales*.

Several independent attempts at phage classification proposed roughly the same groupings but no latinized nomenclature. In 1965 and 1967, Bradley published two important review papers.[21,22] Using the nature of nucleic acid and gross morphology, phages were categorized into six fundamental groups, labeled A through F. Groups A-C included phages with ds-DNA and tails that were contractile, long and noncontractile, or very short. The others were for cubic phages of the φX174-type (D), cubic phages with ss-RNA (E), and filamentous phages (F). Tikhonenko[23] recognized essentially the same categories, but grouped small cubic phages together on morphological grounds. In addition, she published a list of 472 viruses. Bradley's and Tikhonenko's proposals are compared in Table 2.

Bradley's groups A-F were immediately accepted and are still used. In 1971, to make room for the newly discovered phage PM2, viruses with long and short noncontractile tails were grouped together.[26] This proposal went more or less unnoticed, but it is found in the proceedings of the Mexico meetings of the ICTV and the six groups of bacterial viruses defined then: T-even type phages, the lambda group, PM2, φX174-type, fd-type, and RNA phages.[27] Another proposal was to define phage genera by using nucleic acid nature and phage species by morphology. It resulted in three "genera", one of them grouping PM2 and all tailed phages.[27] This is not far away from the positions of d'Hérelle or Holmes. Now

Table 1
FREQUENCY OF PHAGE OBSERVATIONS BY HOST GROUP[a]

Bergey's Manual[11]		Tailed phages				
Part	Host group	*Myoviridae*	*Styloviridae*[b]	*Podoviridae*	Other phages	Total
1	Phototrophic bacteria	1	3	2		6
2	Gliding bacteria	38	2	2		43
3	Sheathed bacteria		1			1
4	Budding and/or appendaged bacteria	4	68	6	8	86
5	Spirochetes	3	1			4
6	Spiral and curved bacteria	6	10		9	25
7	Gram-negative aerobic rods and cocci	113	148	129	10	400
8	Gram-negative facultatively anaerobic rods	251	184	164	66	665
9	Gram-negative anaerobic bacteria	2	10	1		13
10	Gram-negative cocci and coccobacilli	13	4	3		20
11	Gram-negative anaerobic cocci		2	2		4
12	Gram-negative, chemolithotrophic bacteria	1				1
14	Gram-positive cocci	33	315	15		364
15	Endospore-forming rods and cocci	182	196	38	3	419
16	Gram-positive, asporogenous rods	24	63			87
17	Actinomycetes and related organisms	4	197	6		207
18	Rickettsias	1		1		1
19	Mycoplasmas	1	2	2	16	21
—	Cyanobacteria and green algae (*Chlorella*)★	5	2	11		18
Total		681	1,208	384	112	2,385
			2,273			

[a] Excluding shadowed phages, known defective phages, and particulate bacteriocins. Completed January 1, 1982.

[b] Provisional name.

[c] Not in Bergey's Manual.[11]

Table 2
COMPARISON OF BASIC PHAGE GROUPS PROPOSED IN THE LITERATURE

		Group designation			
Nucleic acid	Morphological characteristic	Bradley, 1965 and 1967[21,22]	Tikhonenko, 1968[23]	Ackermann and Eisenstark, 1974[24]	Family (ICTV approved)[25]
ds-DNA	Tail contractile	A	V	A1, A2, A3	*Myoviridae*
	Tail long, noncontractile	B	IV	B1, B2, B3	*Styloviridae*[a]
	Tail short	C	III	C1, C2, C3	*Podoviridae*
ss-DNA	Cubic	D		D1	*Microviridae*
ss-RNA	Cubic	E	II	E1	*Leviviridae*
ss-DNA	Filamentous	F	I	F1	*Inoviridae*

[a] Provisional name.

Table 3
NUMERICAL IMPORTANCE OF PHAGE
FAMILIES AND GENERA[a]

Group	Family[25]	Genus[25]	Number of observations
A	*Myoviridae*		681
B	*Styloviridae*[b]		1,208
C	*Podoviridae*		384
D1	*Microviridae*	*Microvirus*	26
D2	Not classified		2
D3	*Corticoviridae*	*Corticovirus*	2?
D4	*Tectiviridae*	*Tectivirus*	8
E1	*Leviviridae*	*Levivirus*	36?
E2	*Cystoviridae*	*Cystovirus*	1
F1		{*Inovirus*	21
	Inoviridae	{*Plectrovirus*	13
G	*Plasmaviridae*	*Plasmavirus*	3

[a] See footnote a of Table 1.
[b] Provisional name.

that there are about 2300 known tailed phages, it is fortunate that this proposal was not accepted.

In 1974, Ackermann and Eisenstark[24] reviewed 1061 phages and expanded Bradley's original scheme to include newly found phage groups, namely cubic phages with ds-DNA and double capsids, cubic phages with ds-RNA, and rod-shaped and pleomorphic phages of mycoplasmas. In addition, tailed phages were subdivided according to head length. This resulted in 18 groups or types, most of which later became families or genera (Table 3). Simultaneously, phages of particular host groups or shape were surveyed and categorized into species (Table 4).

C. The Present State

The phages of known morphology fall into 10 clear-cut families, one of which has two genera (Table 3).[25] The size of these groupings is very variable. Tailed phages number about 2300 and constitute over 95% of observations. Many groups of cubic or filamentous phages are very small and one of them has a single member.[25]

Table 4
SUCCESS RATE OF ATTEMPTED CLASSIFICATIONS

| Phage group | Host group | Number of phages | | Number of species | Ref. |
		Surveyed	Classified		
Tailed phages	Actinomycetes	166	78	20	28
	Agrobacterium-Rhizobium	128	68	16	29
	Bacillus-Clostridium	301	113	21	30, 31
	Brucella	45	45	1	32
	Enterobacteria	375	250	24	33
	Gram-positive cocci	116	75	14	34
	Pseudomonads	146	67	17	35
	Cyanobacteria	12	12		36
	Mycoplasmas	4	3	3	37
Cubic, filamentous, and pleomorphic phages		98	98	18	37, 38
	Total	1,391	809	134	

Out of some 1400 phages surveyed by host or morphological groups, about 800 or 58% were classified into 134 species, and type viruses were proposed (Table 4). The success rate for placing the viruses in taxa was highest in cubic, filamentous, and pleomorphic phages and was lowest in viruses with long, noncontractile tails. The latter included many phages of very simple morphology.

III. PROBLEMS CAUSED BY PHAGES

A. Numerical Imbalance Between Tailed and Other Phages

Tailed phages are much more numerous than all other morphological types (Tables 1 and 3). They occur throughout the bacterial world, being found in aerobes and anaerobes, budding, gliding, spiral-shaped, or photosynthetic bacteria, in exospore and endospore formers, mycoplasmas and cyanobacteria.[12,24,39] Cubic, filamentous, and pleomorphic phages have much more restricted host ranges,[12,24,39] for example:

1.   *Microviridae* are limited to enterobacteria.
2.   Short rod-shaped and pleomorphic phages occur in mycoplasmas only.
3.   *Leviviridae* occur in enterobacteria, pseudomonads, *Caulobacter,* and possibly *Bdellovibrio.*
4.   Inoviruses of the fd-type occur in enterobacteria, pseudomonads, xanthomonads, and vibrios.

Some levi- and inoviruses even multiply in hosts belonging to different genera, for example enterobacteria and pseudomonads, which harbor certain drug-resistance plasmids and may have phylogenetic relationships.[12] It is likely that the distribution of phages reflects the course of bacterial evolution and that cubic and filamentous phages came into being long after tailed phages.[12] Whatever the cause, the imbalance in numbers between tailed and other phages is basic and unlikely to change. Phage taxonomy will have to deal with one large and many small groups.

B. Problems of Tailed Phages
1. Similarity and Diversity
Tailed phages appear as a basically homogeneous but at the same time an enormously

diversified group. They contain a single molecule of linear ds-DNA which is about 50% of particle weight, and consist of cubic capsids with helical tails and no envelopes. Many of them have base plates, tail fibers, or spikes. Morphogenesis is complex and includes several pathways. Progeny DNA is packaged into preformed heads and many tailed phages are temperate, being able to undergo a special relationship with their host. It may be added that there is a single phage, Rϕ6P of *Rhodopseudomonas sphaeroides*, that contains circular DNA.[40] This is the exception that confirms the rule; the phage may be a derivative of a plasmid that linked up with phage genes.[12]

On the other hand, tailed phages are extremely heterogeneous in fine structure and size (Figure 2), particle weight, and DNA content and composition. DNA molecular weight, for example, varies between 12 and 490 × 10⁶.[25,30] Can their taxonomy be built on these criteria? The answer is partly no.

Figures 3 to 5 illustrate head size, particle buoyant density, particle weight, and nucleic acid molecular weight in tailed phages. Data are taken from species descriptions in fully referenced review papers. Some phage ''species'', for example T-even phages, include large numbers of viruses which, taken individually, would bias the figures. It is hoped that this will give a more meaningful picture of bacteriophage characters than data from every individual virus or isolate. Figures 2 and 3 show that data are distributed over a wide range and in an approximately Gaussian way. No groups are apparent, but some individual phage species stand out because of their unusual size or DNA content. There is no correlation between physical properties and tail structure, but a parallel can be drawn between head size, particle weight, and DNA content (Figure 5).[33,39,42]

This allows predictions and control of experimental results. However, data that are extremely useful in general virology, namely strandedness and molecular weight of DNA and size and mass of the virion, fail to subdivide tailed phages or appear to be characteristic of a few individual species. We shall see in Section IV.A.1. that the base composition of phage nucleic acids has other, equally severe limitations. In cubic, filamentous, and pleomorphic phages, however, physicochemical properties are extremely valuable for classification.

2. Genus Definition

No ready way for definition of genera is apparent. It is relatively easy to categorize tailed phages into families or species, but our criteria seem to be either high-level, that is applicable to tailed phages as a whole, or species level. A possible way may be the study of the strategy of infection. Korsten et al.[43] compared T7 with other short-tailed phages and concluded that patterns of phage-directed protein synthesis, as determined by gel electrophoresis and assay of phage-coded enzymes such as RNA polymerase and SAMase, might provide evidence for phylogenetic relationships between phages. This could be the basis for ''natural'' phage genera. Another possibility is immuno-electron microscopy, since it makes possible investigation of relationships between capsids and not just the tail tips as with neutralization of infectivity.

3. Phages with Uncharacteristic Morphology

Many phages have isometric heads and long, noncontractile tails. They are particularly common in actinophages. Most of them cannot be differentiated by gross morphology and this obviously limits the value of electron microscopy. Their identification depends on precise measurements under stringent conditions and on the presence of particular structures such as collars or tail fibers. Even so, phages of identical morphology may be unrelated.

IV. PROBLEMS CAUSED BY DATA

A. Inherent Weakness of Some Frequently Determined Properties

1. Nucleic Acid Composition

Guanine-cytosine contents of tailed phages and their hosts are markedly similar (Figure

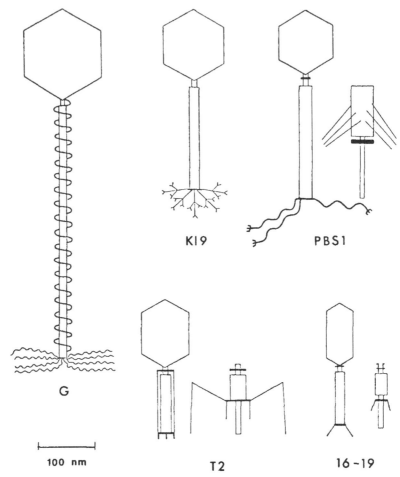

K19   **PBS1**

G

100 nm   **T2**   **16-19**

FIGURE 2.   *Morphological variation in tailed phages. Scale drawings of selected phages of Acinetobacter (545),[41] Agrobacterium (PS8),[29] Bacillus (G, PBS1, ϕ29),[30] enterobacteria (K19, T2, 16-19, T5, λ, P22, T7, 7-11),[33] Staphylococcus (3A),[34] and Streptococcus (VD13).[34]*

6). This is also evident in the pattern of nearest-neighbor base sequence frequency[44] and seems to be a general feature of bacteriophages, even those containing RNA.[33,35,39,42,45] This parallelism doubtless reflects the dependence of the phage on the metabolic machinery of the host and indicates that phages and bacteria evolved together.[12] In phage taxonomy, it strongly limits the use of base ratios as a criterion. However, the presence of unusual bases is a useful feature.

2. Serology and Nucleic Acid Hybridization

Serological relationships usually are investigated by neutralization. In tailed phages this involves the tail tip only. Neutralization tests are very sensitive, very specific, and allow differentiation of closely related phages. As a rule, morphologically different phages are antigenically unrelated.[39] If the contrary is claimed, the paper invariably is suspect. Serological properties are thus suited for species definition only. A further limitation is that one always needs two reagents, a phage and an antiserum.

Nucleic acid hybridization including construction of heteroduplexes has similar limitations. In addition, several methods or modifications are in existence and the exact threshold of relatedness is unknown. The relatively few available data confirm relationships established by morphology, serology, and other criteria.[12] The technique seems to be as sensitive as

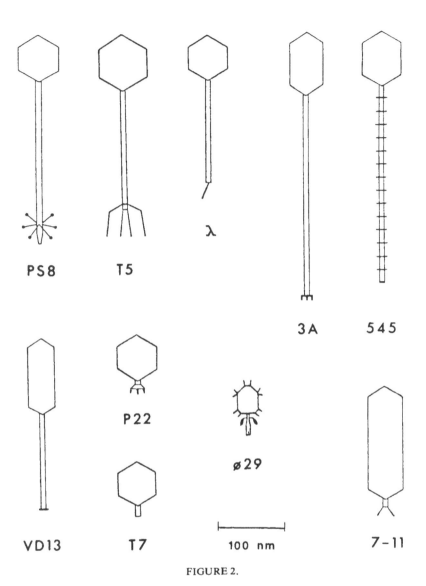

PS8   T5

λ

3A   545

P22

ø29

VD13   T7   100 nm   7-11

FIGURE 2.

serology and allows differentiation of morphologically identical phages.[47] An assessment of its taxonomic importance is needed.

3. Host- and Environment-Related Properties

Plaque morphology, adsorption kinetics, latent period, and burst size depend on the host and culture conditions as much as on the phage. Although useful for phage propagation, these properties have little or no taxonomic importance.[29,33,39,42]

B. Experimental Errors and Lack of Standardization

The most serious errors in phage descriptions occur in electron microscopy. They are due to the use of different electron microscopes and stains, and to variations of magnification within the same instrument. This is evident in Figure 7 which shows the distribution of head length in 78 phages of the T-even type. Although these phages are supposed to be of identical size, the distribution of head lengths is not even Gaussian. In principle, these problems are easily corrected by proper calibration, for example with catalase crystals.[48] The importance of size measurements lies in the very great frequency with which they are reported (Table 5).

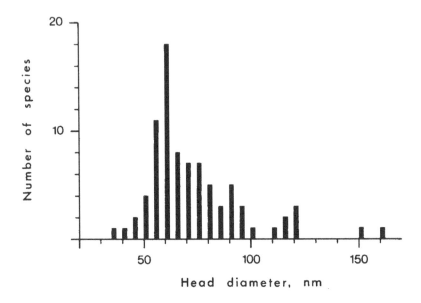

FIGURE 3.   Distribution of head size in tailed phages. Only isometric heads are represented. Based on species descriptions in References 28-35.

Other sources of error are inherent in pipetting and titration and in the absence of standardized methods. In phage serology, the error in plaque counts has been estimated as 10 to 50%.[49-51] Determination of adsorption kinetics, burst size, and resistance to the environment are probably subject to similar errors. In addition, resistance tests have been carried out under a bewildering variety of conditions. For example, chloroform sensitivity has been tested for periods anywhere between 2 min and 4 months. It is a sad consequence that most resistance data are useless for the purpose of phage taxonomy.

C. The Data Base Is Large but Insufficient

The taxonomic value of a given property depends on the frequency with which it has been reported. Table 5 shows the frequency of certain phage characters that are either frequently determined or have been considered for taxonomic use, in the author's literature files. Table 5 lists the number of phages studied and not the number of investigations. Some phages have been studied repeatedly. For example, the molecular weight of T7 DNA had been determined 17 times.[33] As far as possible, the data are complete for morphology and physicochemical properties. The table is biased in some degree because it covers few data published before 1960 and because papers describing only biological properties were not systematically recorded. For these reasons, it does not include serology and host range. Nevertheless, Table 5 shows:

1.   Morphological data predominate heavily and are the largest common denominator.
2.   Cubic, filamentous, and pleomorphic phages are over-represented with respect to physicochemical properties.
3.   Data of limited value are uncomfortably frequent, namely: (a) Properties of phages of unknown morphology; (b) GC content, adsorption velocity, latent period, and burst size; and (c) Results of resistance tests.
4.   Several kinds of data which have been considered as a possible and perhaps the only basis for phage classification, are in short supply. This is so for nucleic acid hybridization, functional genetic maps, and base or amino acid sequences. These data are published at a very slow pace or are available for only a few cubic and filamentous

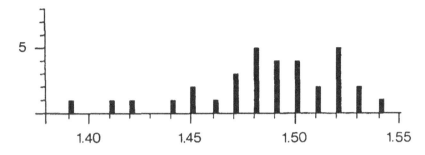

Particle buoyant density, g/ml, CsCl

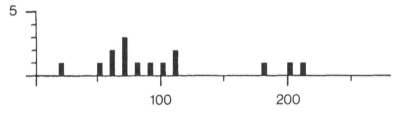

Particle weight, $\times 10^6$

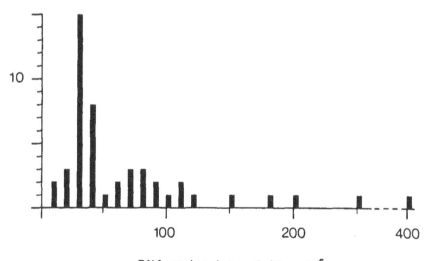

DNA molecular weight, $\times 10^6$

FIGURE 4. Distribution of major physical properties in tailed phages. Based on species descriptions in References 28-35.

phages. It is unlikely that data in these categories will ever be reported for all the 2400 phages of Table 1.

Serological data are insufficient too. They have been invaluable in classifying phages of common morphological types, for example of *Staphylococcus*,[34] which otherwise could not have been classified for lack of specific properties. Unfortunately, descriptions of "new" phages rarely include neutralization tests with antisera prepared against known phages and many serological data are for phages of unknown morphology.

In conclusion, the data base for phage classification is at first glance impressive, but is

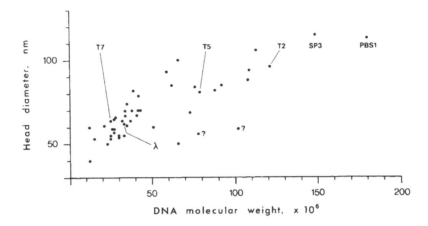

FIGURE 5.   Correlation between head size and DNA content in tailed phages. The largest DNA molecular weight known, 490×10^6, is not represented. The corresponding virus, *Bacillus* phage G, is shown in Figure 2 and has a head diameter of 160 nm.[30] In phages with elongated heads, the capsid size was taken as half the sum of length and width of phage heads. Based on species descriptions in References 28-35.

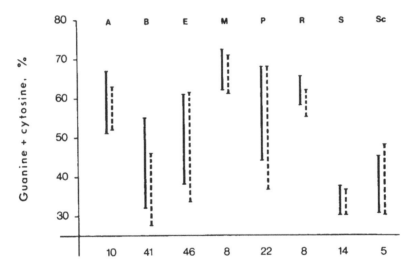

FIGURE 6.   Comparison of guanine-cytosine ranges in selected tailed phages and bacteria. Full lines, bacterial GC, dotted lines, phage GC. A, *Azotobacter;* B, *Bacillus;* E, entero-bacteria; M, *Mycobacterium;* P, *Pseudomonas;* R, *Rhizobium;* S, *Staphylococcus;* Sc, *Strep-tococcus.* The numerals below the abscissa indicate the number of phages. Bacterial GC's are from Normore.[46] Phage data are from Guay et al.[45]

in fact rather small. Phage taxonomy will have to rely heavily on morphology as the most useful common denominator. It is unrealistic to base taxonomy on properties that are determined with great difficulty and are known for a few phages only. The problem is to improve the quality of morphological phage descriptions.

V. PROBLEMS CAUSED BY MAN

The problems inherent in phages and data are compounded by man. Some manmade problems are linked to the development of phage research. Others, perhaps the most consequential of them, are due to poor habits. Herein lies a certain hope, because the elimination

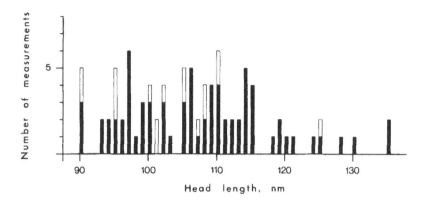

FIGURE 7. Distribution of head lengths in T-even type phages (90 determinations on 78 phages). Adapted from a graph published in 1976.[33] Empty columns represent later data.

of these habits will greatly improve phage taxonomy. Manmade problems may be identified as follows:

1. Phage research developed without a standardized nomenclature, standardized methods, and reference culture collections. As a result, synonyms are plentiful; many known phages may be identical or have been studied several times; resistance data are chaotic; and comparative serological data are insufficient. Political divisions have further hindered the exchange of information and phages.

2. Phages are easily and safely manipulated, making them targets for beginners and under-equipped laboratories.

3. Many scientists working with phages are over-specialized in their interests. There are pure morphologists, pure biochemists, or pure geneticists, working more or less in isolation and studying some properties with total disregard for the others. This has generated unbalanced papers and the dogmatic view that phages can be classified by this or that single criterion.

4. Poor habits are causing the most trouble. First, it is commonplace to call every new isolate a new phage without proof of novelty. This would be unacceptable in other branches of virology, for example with a herpesvirus. The second problem is poor electron microscopy, characterized by lack of magnification control and incomplete description. This has been criticized elsewhere.[39] As a rule of thumb, a description is complete if one can build a model. Electron microscopy is crucial for phage taxonomy, especially for the tailed phages.

5. Fraud seems to be a minor problem, although certain people published a group of 60 phages up to five times under different names.

6. Scientific journals have to share a great part of the responsibility. It is surprising how many poor micrographs are allowed to appear.

A combination of the above factors results in poor and quite useless papers. They typically describe three or four phages of a common bacterial species and contain a nondescript micrograph with few dimensions, a host range, and a wealth of data on adsorption and growth kinetics. There may be serological data, but phages are rarely compared with similar viruses of related bacteria. For added refinement, such a paper may be translated and published again in a foreign journal.

Table 5
FREQUENCY WITH WHICH PHAGE CHARACTERS HAVE BEEN REPORTED IN THE LITERATURE[a]

Parameter		Tailed	CFP[b]	Unknown	Total
			Type of phage		
Morphology		2,273	112		2,385
Nucleic acid:	Conformation	79	21	8	108
	Percentage	63	23	6	92
	Molecular weight	250	30	41	321
	Base composition	94	37	12	143
	GC only	246	11	128	385
	Base sequences		5		5
	Hybridization	68	12	12	92
	Functional genetic map	13	14	1	18
Virion:	Buoyant density in CsCl	211	42	9	262
	Sedimentation velocity	75	35	2	112
	Particle weight	53	15	1	69
	Presence of lipids	11	12		23
	Coat protein:				
	Subunit molecular weight	41	18		59
	Amino acid composition	18	23	2	43
	Amino acid sequence		9		9
	Isoelectric point	4	5		9
	Diffusion velocity	9	4		13
	Electrophoretic mobility	6	5		11
Resistance to:	Chloroform	201	27	17	245
	Citrate	24	1	12	37
	Ether	47	23	4	74
	Osmotic shock	30?	1	12	43?
	pH	156	22	49	227
	Sodium dodecyl sulphate	16	3	8	27
	Sonication	11	13	2	26
	Temperature	371	26	156	553
	Trypsin	24	8	6	38
	Ultraviolet light	103	48	107	258
	Urea	13	1	34	48
Biological:	Adsorption velocity	220	23	66	309
	Latent period and burst size	393	46	103	542

[a] Excluding known defective phages and particulate bacteriocins. Morphology excludes shadowed phages as well. Completed January 1, 1982.

[b] Cubic, filamentous, and pleomorphic.

VI. DESCRIPTION OF NEW PHAGES

A. When Is a "New" Phage New and What Should Be Done With It?

Most phage workers will some day face this question or have already done so. As an answer, let us consider host range and serological properties. First, phages usually respect host genus boundaries. Exceptions are phages of enterobacteria, cubic and filamentous phages specific for certain drug resistance plasmids (Section III.A.), and possibly some actinophages.[39] Tailed phages of enterobacteria apparently do not accept our bacteriological subdivisions and may be considered as specific for the Enterobacteriaceae family. Second, there is a strong correlation between morphology and serology, and morphologically distinct phages usually have no antigenic relationships (Section IV.A.2.). With respect to the above exceptions in host genus specificity, it may be concluded that a new phage is new if (1) no phage at all or (2) no phage of identical morphology is known in a given host genus.

The term "identical" should be taken literally. In many cases, proof of novelty can be established by the presence of particular morphological features or the dimensions of the phage. Conversely, in phages with highly characteristic fine structure, for example *Bacillus* phage PBS1 (Figure 2), electron microscopy may indicate relationships.

The situation is more problematic in phages without distinguishing morphology and in bacteria with many poorly characterized phages. Here, novelty must be proven by comparative serology or nucleic acid hybridization. If no relationships are detected, the isolate may be called a new phage and possibly a new species. Partial serological relationships and differences in resistance tests or biological properties would be insufficient. Physicochemical properties, at least in tailed phages, would be a proof, if the phage is clearly distinguished from others, for example, by the presence of unusual bases.

The subsequent investigation depends on the phage. Tailed phages all contain ds-DNA and thus cannot be differentiated by this criterion. It even seems that physicochemical properties are of limited taxonomical value in these viruses (Sections III.B.1. and IV.A.1.) and their determination is therefore not a must here. This does not diminish the descriptive value of physicochemical data. Identification of tailed phages thus depends much on careful electron microscopy, serological studies, and nucleic acid hybridization.[39] In cubic, filamentous, and pleomorphic phages, however, physicochemical properties have a high taxonomic value and should be thoroughly investigated.

B. Naming a Phage

No established system of nomenclature exists and after 65 years of phage research, it is too late to construct one. In view of the growing number of phages, the question is rather to limit confusion by eliminating sources of errors and diversifying phage names. For this purpose, we suggest:

1. Avoiding single letters and numerals, the overused letters P and φ, and dashes and superscript characters.
2. Designating new phages by the initials of the host genus and species followed by other suitable sigla.
3. Designating members of known species by the species name and the initial of the place of isolation. For example, a T2-type phage isolated in Quebec might be named "T2Q".

In conclusion, the description of new phages will be considerably improved by the adoption of simple, more or less self-evident principles.[39]

1. Morphology and host range should be studied for all new isolates.
2. New phages are new if this is proven or extremely likely.
3. In tailed phages, classification depends greatly on morphology, serology, and nucleic acid hybridization.
4. In cubic, filamentous, and pleomorphic phages, physicochemical properties should always be determined.
5. Phage names should preclude confusion.

VII. OUTLOOK

What are the prospects for the future? New phages will be described, old phages will be reinvestigated, and methods will improve. A phage reference center is being set up by the author as an instrument of the ICTV, which will concentrate on the collection of type phages.[52] Thus, phage classification will continue, hopefully with better results. Most new phages will be tailed. The discovery of entirely new phage families is not very likely, but could happen among phages of bacteria which are not of medical importance and therefore have received less attention. In archaebacteria for example, only *Halobacterium* is known to have phages at the present time.[53]

Will all phages ever be classified? This is impossible, because many phages of the literature are no longer extant and only imperfect descriptions remain from them. Phage taxonomy should improve with better electron microscopy and if more data on comparative serology, nucleic acid hybridization, and patterns of phage-directed protein synthesis become available.

REFERENCES

1. **Twort, F. W.**, An investigation on the nature of ultramicroscopic viruses, *Lancet*, 2, 1241, 1915.
2. **D'Hérelle, F.**, Sur un microbe invisible antagoniste des bacilles dysentériques, *C. R. Acad. Sci. Paris*, 165, 373, 1917.
3. **Raettig, H.**, *Bakteriophagie 1917 bis 1956*, Part 1, Gustav Fischer-Verlag, Stuttgart, 1957, 4.
4. **Ruska, H.**, Morphologische Befunde bei der bakteriophagen Lyse, *Arch. Ges. Virusforsch.*, 2, 345, 1942.
5. **Kottmann, U.**, Morphologische Befunde aus taches vierges von Colikulturen, *Arch. Ges. Virusforsch.*, 2, 388, 1942.
6. **Brenner, S. and Horne, R. W.**, A negative staining method for high resolution electron microscopy of viruses, *Biochim. Biophys. Acta*, 34, 103, 1959.
7. **Raettig, H.**, *Bakteriophagie 1957-1965*, Part 1, Gustav Fischer-Verlag, Stuttgart, 1967, 7.
8. **Bradley, D. E.**, The structure of some *Staphylococcus* and *Pseudomonas* phages, *J. Ultrastruct. Res.*, 8, 552, 1963.
9. **Bradley, D. E.**, The structure of coliphages, *J. Gen. Microbiol.*, 31, 435, 1963.
10. **Bradley, D. E.**, The isolation and morphology of some new bacteriophages specific for *Bacillus* and *Acetobacter* species, *J. Gen. Microbiol.*, 41, 233, 1965.
11. *Bergey's Manual of Determinative Bacteriology*, 8th ed., Buchanan, R. E., and Gibbons, N. E., Eds., Williams & Wilkins Co., Baltimore, 1974.
12. **Reanney, D., and Ackermann, H.-W.**, Comparative biology and evolution of bacteriophages, *Advan. Virus Res.*, in press.
13. **D'Hérelle, F.**, *Le bactériophage: son rôle dans l'immunité*, Masson & Cie, Paris, 1921, 215.
14. **Ruska, H.**, Versuch zu einer Ordnung der Virusarten, *Arch. Ges. Virusforsch.*, 2, 480, 1943.
15. **Holmes, F. O.**, Order Virales: The filterable viruses, in *Bergey's Manual of Determinative Bacteriology*, 6th ed., Breed, R. S., Murray, E. G. D., and Hitchens, A. P., Eds., Williams & Wilkins Co., Baltimore, 1948, 1126.
16. **Adams, M. H.**, Criteria for a biological classification of bacterial viruses, *Ann. N.Y. Acad. Sci.*, 56, 442, 1953.
17. **Adams, M. H.**, *Bacteriophages*, Interscience, New York, 1959.
18. **Lwoff, A., Horne, R. W., and Tournier, P.**, A system of viruses, *Cold Spring Harbor Symp. Quant. Biol.*, 27, 51, 1962.
19. **Lwoff, A., Tournier, P.**, The classification of viruses, *Annu. Rev. Microbiol.*, 20, 45, 1966.
20. Provisional Committee for Nomenclature of Viruses, Proposals and recommendations, *Ann. Inst. Pasteur*, 109, 625, 1965.
21. **Bradley, D. E.**, The morphology and physiology of bacteriophages as revealed by the electron microscope, *J. R. Microscop. Soc.*, 84, 257, 1965.
22. **Bradley, D. E.**, Ultrastructure of bacteriophages and bacteriocins, *Bacteriol. Rev.*, 31, 230, 1967.
23. **Tikhonenko, A. S.**, *Ultrastructure of Bacterial Viruses*, Izdadelstvo "Nauka", Moscow, 1968, 30, and Plenum Press, New York, 1970, 30.
24. **Ackermann, H.-W., and Eisenstark, A.**, The present state of phage taxonomy, *Intervirology*, 3, 201, 1974.
25. **Matthews, R. E. F.**, Classification and nomenclature of viruses. Fourth report of the International Committee on Taxonomy of Viruses, *Intervirology*, in press.

26. **Bradley, D. E.**, A comparative study of the structure and biological properties of bacteriophages, in *Comparative Virology*, Maramorosch, K. and Kurstak, E., Eds., Academic Press, New York, 1971, 207.
27. **Wildy, P.**, Classification and nomenclature of viruses. First report of the International Committee on Nomenclature of Viruses, in *Monographs in Virology*, Vol. 5, S. Karger, Basel, 1971, 12.
28. **Berthiaume, L. and Ackermann, H.-W.**, La classification des actinophages, *Pathol. Biol.*, 25, 195, 1977.
29. **Ackermann, H.-W.**, La classification des phages d'*Agrobacterium* et *Rhizobium*, *Pathol. Biol.*, 26, 507, 1978.
30. **Ackermann, H.-W.**, La classification des bactériophages de *Bacillus* et *Clostridium*, *Pathol. Biol.*, 22, 909, 1974.
31. **Reanney, D. C. and Ackermann, H.-W.**, An updated survey of *Bacillus* phages, *Intervirology*, 15, 190, 1981.
32. **Ackermann, H.-W., Simon, F., and Verger, J.-M.**, A survey of *Brucella* phages and morphology of new isolates, *Intervirology*, 16, 1, 1981.
33. **Ackermann, H.-W.**, La classification des phages caudés des entérobactéries, *Pathol. Biol.*, 24, 359, 1976.
34. **Ackermann, H.-W.**, La classification des bactériophages des cocci Gram-positifs: *Micrococcus, Staphylococcus* et *Streptococcus, Pathol. Biol.*, 23, 247, 1975.
35. **Liss, A., Ackermann, H.-W., Mayer, L. W., and Zierdt, C. H.**, Tailed phages of *Pseudomonas* and related bacteria, *Intervirology*, 15, 71, 1981.
36. **Safferman, R. S., Cannon, R. E., Desjardins, P. R., Gromov, B. V., Haselkorn, R., Sherman, L. A., and Shilo, M.**, Classification and nomenclature of viruses of cyanobacteria, *Intervirology*, 19, 61, 1983.
37. **Maniloff, J., Haberer, K., Gourlay, R. N., Das, J., and Cole, R.**, Mycoplasma viruses, *Intervirology*, 18, 177, 1982.
38. **Ackermann, H.-W.**, Cubic, filamentous, and pleomorphic bacteriophages, in *CRC Handbook of Microbiology*, Vol. 2, 2nd ed., Laskin, A. I. and Lechevalier, H. A., Eds., CRC Press, Boca Raton, Fla., 1978, 673.
39. **Ackermann, H.-W., Audurier, A., Berthiaume, L., Jones, L. A., Mayo, J. A., and Vidaver, A. K.**, Guidelines for bacteriophage characterization, *Adv. Virus Res.*, 23, 1, 1978.
40. **Pemberton, J. M., and Tucker, W. T.**, Naturally occurring viral R plasmid with a circular extracellular genome in the extracellular state, *Nature (London)*, 266, 50, 1977.
41. **Ackermann, H.-W., Brochu, G., and Cherchel, G.**, Structure de trois nouveaux phages de Bacterium anitratum (groupe B5W), *J. Microsc. (Paris)*, 16, 215, 1973.
42. **Ackermann, H.-W.**, Bactériophages — propriétés et premières étapes d'une classification, *Pathol. Biol.*, 17, 1003, 1969.
43. **Korsten, K. H., Tomkiewicz, C., and Hausmann, R.**, The strategy of infection as a criterion for phylogenetic relationships of non-coli phages morphologically similar to phage T7, *J. Gen. Virol.*, 43, 57, 1979.
44. **Bellett, A. J. D.**, Numerical classification of some viruses, bacteria and animals according to nearest-neighbour base sequence frequency. *J. Mol. Biol.*, 27, 107, 1967.
45. **Guay, R., Ackermann, H.-W., and Murthy, M. R. V.**, Base composition of bacteriophage nucleic acids, in *CRC Handbook of Microbiology*, Vol. 3, 2nd ed., Laskin, A. I. and Lechevalier, H. A., Eds., CRC Press, Boca Raton, Fla., 1981, 731.
46. **Normore, W. M.**, Guanine-plus-cytosine (GC) composition of the DNA of bacteria, fungi, algae, and protozoa, in *Handbook of Biochemistry and Molecular Biology, Nucleic Acids*, Vol. 2, 3rd ed., Fasman, G. D., Ed., CRC Press, Boca Raton, Fla., 1976, 65.
47. **Grimont, F. and Grimont, P.**, DNA relatedness among bacteriophages of the morphological group C3, *Curr. Microbiol.*, 6, 65, 1981.
48. **Luftig, R.**, An accurate measurement of the catalase crystal period and its use as an internal marker for electron microscopy, *J. Ultrastruct. Res.*, 20, 91, 1967.
49. **Hiatt, W. C.**, Kinetics of the inactivation of viruses, *Bacteriol. Rev.*, 28, 163, 1964.
50. **Hershey, A. D., Kalmanson, G., and Bronfenbrenner, J.**, Quantitative methods in the study of the phage-antiphage reaction, *J. Immunol.*, 46, 267, 1943.
51. **Lanni, F., and Lanni, Y. T.**, Antigenic structure of bacteriophage, *Cold Spring Harbor Symp. Quant. Biol.*, 18, 159, 1953.
52. **Ackermann, H. -W., Martin, M., Vieu, J. -F., and Nicolle, P. Felix d'Hérelle:** his life and work and the foundation of a bacteriophage reference center *ASM NEWS*, 48, 346, 1982.
53. **Torsvik, D. and Dundas, I. D.**, Bacteriophage for *Halobacterium salinarium, Nature (London)*, 248, 680, 1974.

Chapter 5

CURRENT PROBLEMS IN INSECT VIRUS TAXONOMY

John F. Longworth

TABLE OF CONTENTS

I. TAXONOMIC MILESTONES IN INVERTEBRATE VIROLOGY

The general history of virus taxonomy is reviewed in the first chapter, and it is not my intention to repeat that account here. Nevertheless events over the past 50 years have shaped the present state of invertebrate virus taxonomy. Therefore in this section a summary of developments relating specifically to viruses affecting insects is given.

In Glaser[1] three general types of insect virus diseases were recognized (1) sacbrood of honey bees, (2) the "polyhedral affections" of certain lepidoptera, and (3) two diseases of *Pieris brassicae*, the cabbage butterfly (these were both probably granulosis viruses). In the period 1912 to 1924, a number of such 'filterable' virus diseases of insects had been described, and Glaser stated that inclusion bodies undoubtedly played an important role in the arbitrary classification he presented. However, the significance of polyhedra in virus ecology was not appreciated in that period, for Glaser described them as crystal-like degeneration products of the disease. It was not until 1948 that Bergold[2] using electron microscopy demonstrated the presence of virus in the polyhedra of a nuclear polyhedrosis virus, and this was soon followed by the demonstration that inclusion bodies of granulosis viruses contained similar rod-shaped virus particles. A distinct and separate type of polyhedrosis virus with spherical particles was first demonstrated by Smith and Wyckoff.[3] Though 'filterable' virus, which did not produce large inclusion bodies, had been recognized for some time, the first description of their appearance in the electron microscope was made by Wasser,[4] who described a small spherical virus of *Pseudaletia unipuncta,* and this was placed in the genus Morator (Holmes) with the specific name *Morator nudus (nudus =* nude, naked).

Using Holmes' binominal system[5] and the eight criteria proposed by the virus subcommittee[6] at the 1950 International Congress of Microbiology; Bergold[7] proposed four genera, as follows:

- Genus: *Borrelina* Paillot; (Nuclear polyhedrosis virus).
- Type species: *Borrelina bombycis* Paillot.
- Host: *Bombyx mori* L.

Several other species were named:

- Genus: *Bergoldia* Steinhaus; (Granulosis virus).
- Type species: *Bergoldia calypta* Steinhaus.
- Host: *Cacoecia murinana* (Hbn).

Several other species were listed.

- Genus: *Smithia* Bergold; (Cytoplasmic polyhedrosis virus).
- Type species: *Smithia rotunda* Bergold.
- Host: *Arctia villica* L.
- Genus: *Morator* Holmes; (small isometric viruses).
- Type species: *Morator nudus* Wasser.
- Host: *Cirphis (Pseudaletia) unipuncta* (Haworth).

A further genus was proposed by Weiser[8] to include the virus of the beetle *Melolontha melolontha* (an entomopoxvirus) which formed characteristic inclusion bodies, and was named *Vagoiavirus*.

These groupings and their nomenclature developed independently of other fields of virology[9] to the detriment of insect virology in that modern concepts were generally ignored.[10]

In 1957, Hughes[11] published a list and bibliography of 259 papers dealing with insects reported to have virus diseases, largely of the nuclear polyhedrosis, granulosis and cytoplasmic polyhedrosis virus types. Martignoni et al.[12] added 473 new references, including one to a newly described virus from *Tipula paludosa*[13] namely *Tipula* iridescent virus.

The viruses in this rapidly growing assemblage were most commonly named by coupling the virus vernacular name with the Latin binomial of the insect host, for example "nuclear polyhedrosis virus of *Bombyx mori*". Such a system reflected the general belief of the time that these viruses were host specific and that each new isolation from a new insect species represented a new and different virus.

The growing confusion in the literature was exacerbated by the use of Holmsian concepts for classification, and it was timely in 1966 that the International Committee on Nomenclature of Viruses (ICNV) founded subcommittees to consider the whole area of nomenclature of viruses, including those of invertebrates.

At this point, it is appropriate to recall that invertebrates are animals, and indeed comprise over 97% of the animal species on earth.[14] As Andrewes[15] said, it is remarkable how frequently arthropods are involved throughout the virological scene. There are the insect-pathogens proper, the many arthropod-transmitted plant viruses and the arboviruses of vertebrates which form, in fact, a large proportion of known viruses, and he commented upon the striking fact that among the known groups or genera of viruses there are so many instances of morphologically similar viruses affecting more than one host group.

It seems likely, as Fenner[16] has suggested, that every invertebrate species will have its own suite of viruses and while many of the viruses recognized so far are like those of vertebrates, others are distinctive. Most vertebrate viruses have been described from a very small number of species — primarily man and his domestic animals; similarly only a minute fraction of the vast number of invertebrates has been studied. Thus, most invertebrates are essentially virgin territory for virologists.[16]

The first proposals of the Invertebrate Virus Subcommittee of the ICNV were approved by the ICNV and published in its general report.[17] The approved proposals were described in detail by Vago et al.,[9] who drew attention to a most important aspect of the new nomenclature of invertebrate viruses, which was the relationships that were revealed between invertebrate and vertebrate viruses. This was a strong influence unifying the various fields of virology. Important groups of vertebrate viruses, such as *Poxvirus, Parvovirus, Rhabdovirus, Reovirus,* and *Enterovirus,* which were described as genera in the ICNV report, also included invertebrate viruses, and at that stage two new genera were constituted, one for both the invertebrate nuclear polyhedrosis and granulosis viruses (*Baculovirus*) and one for the iridescent viruses (*Iridovirus*).

Vago et al.[9] called for urgent studies on characterization of invertebrate viruses so that their classification could be developed. The decade since then has seen a satisfying tendency towards definitive studies on several invertebrate viruses. This work was materially assisted by the development of several insect cell culture systems.

Harrap and Tinsley[18] described ideas on insect virus nomenclature which they had previously submitted to the ICNV in 1967, and these worked towards a Latinized binomial system as recommended by the ICNV. Their proposed generic names have now largely been superseded, but these were combined with the specific name of the host insect, thus, a nuclear polyhedrosis virus of the insect *Aglais urticae* became "*Nupovirus urticae*". These authors recognized the problems which could arise if two unrelated viruses were isolated from the same insect, but suggested that adequate characterization would facilitate separating these, and distinct isolates of virus of the same genus, from the same host, could be identified further by means of a number as a suffix to the specific name. They considered that advantages of order in nomenclature arising from such a system would far outweigh the ethical argument against taking names from one phylum to classify another.

Table 1
FAMILIES CONTAINING VIRUSES PATHOGENIC FOR INVERTEBRATES

Family	Invertebrate virus members	Study group	Responsible ICTV committee
Nodaviridae	Nodaviruses	*Nodavirus*	Invertebrate
Nudaurelia β virus	*Nudaurelia* β-like virus	*Nudaurelia* β virus	Invertebrate
Baculoviridae	Baculoviruses	*Baculovirus*	Invertebrate
Parvoviridae	Densoviruses	*Parvovirus*	Coordination
Reoviridae	Cytoplasmic polyhedrosis virus	*Reovirus*	Coordination
Rhabdoviridae	*Drosophila* σ virus Blue-crab 1 virus	*Rhabdovirus*	Coordination
Poxviridae	*Entomopoxvirinae*	*Poxvirus*	Coordination
Iridoviridae	Iridescent viruses	*Iridovirus*	Coordination
No family name approved	Many small RNA[a] viruses	Unclassified small RNA virus	Coordination

[a]   Some may eventually be placed in the family *Picornaviridae*, and then would become the responsibility of the Vertebrate virus subcommittee

A different viewpoint was advanced by Tinsley and Kelly[19] to provide a logical method of naming the "iridescent" viruses of insects and to avoid the problems of the then current practice of defining them by using host generic names. For example, three such viruses had already been isolated from three different species of the genus *Aedes*. Tinsley and Kelly[19] proposed an arrangement based upon the adenovirus system, in which each isolate from the genus would be given a type number. Further development of a more permanent system of nomenclature would follow from detailed comparisons of physical, chemical, and biological properties of the various isolates. This arrangement is still in operation and is accepted and widely used by invertebrate virologists. Recently Kelly et al.[20] have examined the serological interrelatedness of insect iridoviruses and found that types 6 and 24 have no antigens in common with each other or any other iridovirus isolate. Types 1,2,9,10,16,18,21, 22,23,24,25, and 28 form a broad serogroup, and type 29 shares common antigens with some but not all of them. Within the broad serogroup, three clusters of types were proposed; types 1, 22, and 25; types 21 and 28; and types 9 and 18. None of these isolates had been cloned, however, prior to this study.

The second report of the ICTV[21] designated the following groups as families: poxviruses, parvoviruses, iridoviruses, reoviruses, and rhabdoviruses, and some genera were named within each of these families. This development was consolidated in the third ICTV report.[22] In the family *Poxviridae*, the former genus *Entomopoxvirus* was accorded the status of a subfamily. The ICTV Study groups concerned with these families formally come under the Coordination Subcommittee of the ICTV, though they work closely with the Vertebrate Virus Subcommittee. A new study group has been set up under the auspices of the Coordination Subcommittee to consider unclassified small RNA viruses and has already proposed two new families of invertebrate small RNA viruses which were accepted by the ICTV in 1981. These now become the responsibility of the Invertebrate Virus Subcommittee. Table 1 lists the families which contain invertebrate pathogenic viruses, the viruses concerned, and the study groups and committees responsible for them.

Research into invertebrate pathogenic viruses has so far centered largely around baculoviruses, and the need to increase basic knowledge of them as a prerequisite to their use in biological control programs. They are virulent, specific, stable, easy to produce in quantity, and their apparent restriction to the class Insecta make them strong candidates as safer alternative methods of pest control than expensive and environmentally unacceptable chemical insecticides. Unequivocal methods of identification were considered necessary both to establish that the target insects died as a result of application of the test virus, and to monitor lack of effect on nontarget hosts, including vertebrates. Consequently in the past decade, there has been a rapid growth in our understanding of baculoviruses, their biochemistry and their replication. The nuclear polyhedrosis virus of *Autographa californica,* has a particularly wide host range in Lepidoptera, and has a future world-wide potential as a biological insecticide with a relatively restricted host range within the Insecta. Several groups of workers in the U.S., Canada, England, and Europe have brought together genetic studies on *A. californica* NPV and there is now a comprehensive physical map of the genome. This map is now being extended, as gene functions are located, into a genetic map. Indeed with the intensification of research on this virus it is in the forefront of invertebrate virus genetics.

There have been several important workshop meetings of virologists engaged in baculovirus research in recent years, and the central place occupied by baculoviruses in the developing field of biological insecticides has tended to concentrate research funds onto that group of viruses. While scores of scientists may attend a meeting devoted to baculoviruses, by contrast, the ICTV study group on small RNA viruses consists of 10 people and comprises almost all scientists engaged in research on those viruses. There are thus few virologists working full-time on invertebrate viruses from families other than baculoviruses. It should be pointed out that these other families may present no less of a challenge and will be no less interesting than their vertebrate counterparts. Indeed, some enteroviruses affecting insects display a majority of features in common with vertebrate enteroviruses. They can be propagated readily in mg amounts in routinely cultured insect cell lines.

Many of the viruses pathogenic for invertebrates are not the responsibility of the Invertebrate Virus Subcommittee of the ICTV (Table 1). These are dealt with more fully in Chapter 2, which deals with viruses of vertebrates. Indeed, there is no guarantee at present that invertebrate virologists will always be represented on the appropriate study groups. This may be a result of the relatively small research effort in invertebrate pathogenic viruses other than baculoviruses. Recently, chairmen of appropriate ICTV vertebrate virus study groups have been asked to give consideration to representation on each group of a suitable number of invertebrate virologists, and this may provide an adequate balance.

II. SMALL RNA VIRUSES OF INVERTEBRATES

The classification of small RNA viruses of invertebrates was reviewed by Longworth.[23] From an assemblage of more than 30 small RNA viruses that had been described in varying detail, two groups were defined unequivocally.

The first comprised a small number of viruses with divided ss-RNA genomes similar to Nodamura virus. This group now comprises over six isolates. With the exception of Nodamura virus most are serologically related, and have either been isolated from insects, or as contaminants in insect cell lines. This group was accorded family status at the ICTV meeting at Strasbourg in 1981. This family, *Nodaviridae*, has a single genus *Nodavirus*, with Nodamura virus as type species.

The second group included serologically related viruses primarily from larvae of limacodid and saturniid moths in Africa, Asia, and Australia. The best characterized virus, from *Nudaurelia cytherea capensis*, has T = 4 symmetry. Some of the morphological properties of the viruses in this group resemble those of caliciviruses. This group too was accorded family status by the ICTV in 1981, though the proposed family name was rejected. *Nudaurelia* β virus is the type species of the single genus *Nudaureliavirus*, and the vernacular name of the family is the *Nudaurelia* β virus group.

Other groups of viruses proposed by Longworth[23] included one comprising a pair of serologically related viruses from the orthopteran *Teleogryllus commodus* (Cricket paralysis virus) and the dipteran *Drosophila melanogaster (Drosophila* C virus); it may well also include a range of apparently similar but inadequately characterized viruses. These two viruses, together with a third virus from *Gonometa podocarpi*, were at one stage included in the genus *enterovirus* by Fenner.[21] Subsequently, however, in the third and fourth ICTV reports[22,24] these were included in *Picornaviridae* under the general umbrella of unclassified small RNA viruses.

In the second report of the ICTV, Fenner[21] sets out the characteristics and membership of the family *Picornaviridae*. He considered that the many small isometric nonenveloped RNA viruses recovered from insects, which are sometimes loosely termed picornaviruses, should only be placed in this family if their RNA content and strategy for characteristic of the *Picornaviridae*. This is an admirably logical standard and should be applied equally to viruses infecting vertebrates or invertebrates. With these criteria in mind, I believe that a firm decision should be taken in the near future on the classification of cricket paralysis virus and *Drosophila* C virus. They are structurally very similar to enteroviruses, having a capsid diameter of 27 nm, an ss-RNA genome with a MW of about 3.0×10^6 and a particle density of 1.34 g/mℓ in neutral cesium chloride. However, they have a low GC content, 38.7%, and post-translational cleavage of a polyprotein translation product occurs, though processing is very rapid.[25] Thus these viruses are in fact better qualified as candidates for inclusion in the genus *enterovirus* than Hepatitis virus A, for though this is a vertebrate virus with the right physicochemical properties for inclusion, its genome strategy is not yet known. Longworth[23] proposed that they certainly warrant classification as picornaviruses even taking into account the low GC content, the unusually low S value of the RNA after formaldehyde treatment, the variability in buoyant density of the virion at alkaline pH, and especially the very rapid post-translational protein processing events with these invertebrate viruses. However, whether they represent a separate genus, or belong to the genus *Enterovirus*, has yet to be established.

Tentative groupings of the other unclassified small RNA viruses were suggested, and the properties of representatives of each of the established and tentative groups are listed in Table 2. These other viruses require definitive characterization before the groupings can be clearly established and the classification less equivocal.

Nodaviridae, and the insect enteroviruses, replicate extremely well in some established insect cell lines, and very high yields of virus are often obtained. These virus/host systems present stimulating opportunities to study the replication of viruses with ss-RNA genomes in vitro, and perhaps more safely than with vertebrate viruses. Perhaps more rapid progress will be made with other small RNA viruses of invertebrates when they too can be grown in suitable insect cell lines and virologists are less dependent on often unreliable methods of culture of the insect hosts.

New and apparently unique isolates of isometric ssRNA viruses are being made with increasing frequency. Scotti et al.[26] have described a new virus, tentatively called Kedogou virus, from the fly *Chrysomia albicans* (Diptera: Calliphoridae) from Senegal, West Africa. The virus particles are enveloped, with a diameter of 80 nm, and these contain 30 nm cores. Infectivity is destroyed by detergents, and organic solvents. This virus superficially resembles togaviruses, but has a lower S-value (140) and a higher density (1.303 g/cm³) in neutral CsCl. The genome is a single-stranded RNA and the S-value of the RNA is approximately 28S. At present the virus appears to replicate only in a clone of *Aedes albopictus* cells. No replication has been observed in any of the vertebrate cell lines tested. The virus was detected as a result of a survey by the Japan Institute of Health, of insects associated with a human disease; the extracts of the insects were screened in both insect and vertebrate cell lines. This provides just one example of the enormous range of viruses still awaiting discovery, let alone adequate description leading to a sound classification.

Table 2
PROPERTIES OF PROPOSED AND ESTABLISHED GROUPS[a] OF SMALL RNA VIRUSES OF INVERTEBRATES

Group status	Diameter (nm)	Density (g/mℓ)	$S_{20,w}$ (S)	Proteins, MW × 10³	pH stability	RNA, MW × 10⁶	RNA %	Nucleotide composition %				Poly (A)	E_{260}/E_{280}
								A	G	C	U		
Family *Nodaviridae*	29	1.34	135	40 (major) 39 and 43 (minor)	To 3.0	Two species: 1.15 and 0.46	20.5	22.4 and 24.8	27.6 and 22.8	27.0 and 28.2	23.0 and 24.8	—[b]	1.61
Family *Nudaurelia* β-like viruses	29	1.298	210	61	ND[c]	1.8	11.0	24	28	25	23	ND	1.45
Proposed enteroviruses	27	1.34	167	33 31 30 7.0	3.0	3.1	28	31.6	20.8	17.9	29.8	Present	1.72
"Group 5" (proposed)	29	1.35	180	36.5 32 29 12.0	ND	N	37	ND				ND	ND
"Mini" viruses (proposed)	13—15	ND	42	ND	6.0	+	18—22	ND				ND	ND

[a] Several unique single viruses are not included.

[b] - = absent.

[c] ND = not determined.

III. *BACULOVIRIDAE*

In the first ICTV report[17] the genus *Baculovirus* had *Bombyx mori* nuclear polyhedrosis virus as type species and the genus was divided into two subgroups comprising nuclear polyhedrosis and granulosis viruses. At that stage the basic similarity of the virions of both subgroups was well known, and it had been established that the genome consisted of high MW ds-DNA. By 1976, the second ICTV report[21] showed that it was now established that the genome was a single molecule of ds circular supercoiled DNA. The baculovirus group was given family status, and the single genus *Baculovirus* was divided into three subgroups. *B. mori* nuclear polyhedrosis virus was designated as type species of subgroup A, and *Choristoneura fumiferana* granulosis virus was the type species of subgroup B. The status of subgroup C was listed as "other probable members of the family" with *Oryctes* virus as type species; this interesting virus displays all the physicochemical and morphological features of *Baculoviridae*, without formation of proteinaceous inclusion bodies in the infection process. The Third ICTV report[22] basically continued this classification, noting that Subgroup C was becoming a diverse group, based on morphological variation of virus structure, and might require further subdivision as data became available. The type species of subgroups A and B were changed to *A. californica* nuclear polyhedrosis virus and *Trichoplusia ni* granulosis virus, reflecting the fact that over the previous decade, characterization of these viruses had proceeded at a substantially greater pace than that of the original types, for which relatively few data were available. The Fourth Report[24] expands the descriptive data on the family and indicates that for subgroup A, the nuclear polyhedrosis viruses, the type species, the *A. californica* isolate, is representative of the MNPV subtype where the virions may contain one to many nucleocapsids within each viral envelope, and many virions per inclusion body. Subgroup A contains isolates from over 200 species of insects and Crustacea. *B. mori* nuclear polyhedrosis virus is representative of the SNPV subtype which also has many virions per inclusion body but each virion only contains one nucleocapsid.

Much more information became available on some viruses from Subgroup C between publication of the third and fourth ICTV reports. The result is that the well-characterized *Oryctes* virus remains the type species of subgroup C. A new subgroup, D, is proposed for a fascinating collection of viruses with double-stranded, circular, but polydisperse DNA genomes. Up to the present time, all of these have been isolated from parasitic Hymenoptera with which the viruses have a unique obligatory association. A diverse group of virus particles with general structure similar to *Oryctes* virus has been isolated from mites, Crustacea, Coleoptera, Arachnida, and a fungus, but all await detailed study and the provision of more data before they can be classified further.

Preliminary proposals to accord family status to these viruses with polydisperse DNA genomes have been made to the Invertebrate Virus Subcommittee of the ICTV. Placing them in subgroup D of the family *Baculoviridae* at this stage distinguishes them from *Oryctes*-type viruses of subgroup C whose genome, like that of other baculoviruses, is a single large closed circle of ds-DNA. It also serves as a holding measure while the classification of these novel viruses is fully considered. There are some problems to be resolved, however, for although it has been demonstrated that the DNA is polydisperse, and that there is little or no homology between the various size classes of closed circles, there is as yet no information on the function of the various genome fragments, nor on the genome strategy, nor indeed what components of the genome constitute an infective entity. Solution of these problems will be difficult, for no susceptible invertebrate cell line has yet been developed and all females in natural populations of the hymenopterous parasite host have been found to be infected.

Thus the family *Baculoviridae* consists of well over 300 recorded isolates in Lepidoptera, Hymenoptera, Diptera, Coleoptera, Crustacea, and Arachnida. An extremely small propor-

Table 3
GENOME SIZE OF THREE BACULOVIRUS SUBGROUPS

Subgroup		Mean. Mol. wt ($\times 10^6$)	Range	Number of determinations
A	MNPV	82.5	(68-100)	9
A	SNPV	85.0	(72-100)	7
B	GV	76.0	(69-95)	4
C	*Oryctes virus*	92.0		1

tion of these has been well characterized so that we have a good understanding of the basic properties of the family but a great deal of comparative information needs to be provided before further meaningful classification of the family can be achieved. The current classification is based entirely on morphological differences in the manner in which nucleocapsids are enveloped and how enveloped virus particles are occluded within polyhedra. Few comparative physicochemical studies on the various baculovirus subgroups have been made. However, Burgess[27] measured genome size, using the electron microscope, of three subgroup A MNPV's, three subgroup A SNPV's and three subgroup B GV's and found that DNA size does not appear to be related to the conventional classification of NPV's and GV's into different subgroups. Using Burgess's data and electron microscope measurements from other workers compiled by Harrap and Payne,[28] average values of genome size were calculated for four distinctive morphological groups and these are presented in Table 3. Clearly genome size bears little relationship to these groups. Restriction endonuclease analysis has a useful role in virus identification, though most studies so far with baculoviruses center on the genome of *A. californica* nuclear polyhedrosis virus for which an increasingly detailed physical map is becoming available, and a genetic map is in the early stages of development. There have been few comparative restriction endonuclease studies on the baculovirus genome, though highly specific and distinguishable cleavage patterns have been obtained and it is apparent that mixtures of viruses can be readily resolved.[29,30] Indeed, with the increasing availability of insect cell lines it is becoming a real possibility that comparative studies should be made on cloned baculovirus isolates. Serological relationships exist within and between the subtypes and subgroups, and there is wide serological cross-reactivity between inclusion body proteins of different viruses. Although serological techniques have been shown to be useful and effective in identification tests for baculoviruses from closely related hosts,[31] there have been no comprehensive surveys to assess the value of serological methods in confirming current morphological subgroupings within *Baculoviridae* or in defining new ones.

A novel virus was recently described from the honeybee.[32] This is enveloped, measures 450×150 nm, but comprises a coiled filamentous nucleocapsid 3000×40 nm, containing apparently linear ds-DNA with a MW of 12×10^6. The DNA represents between 11 and 20% by weight of nucleocapsid. Twelve structural proteins, four of them major, were resolved by analysis of the virion on SDS gels.

The comparative morphology of ds-DNA enveloped invertebrate viruses is shown diagrammatically in Figure 1. All but the honeybee virus,[32] which has linear ds-DNA, are currently classified within the family *Baculoviridae*. Subgroup C formerly contained a diverse assemblage of viruses including the viruses associated with parasitoid Hymenoptera[33] which are now placed in Subgroup D and whose ds-DNA is closed, circular but is present in numerous pieces from MW 1.5 to 8.0×10^6. A proposal is currently being prepared to create a new family for these viruses with the proposed name 'Polydnaviridae' reflecting the polydisperse nature of the DNA. Subgroup C presently contains *Oryctes* virus with similar morphology to the "Polydnaviruses" but with a ds-DNA genome with a single closed

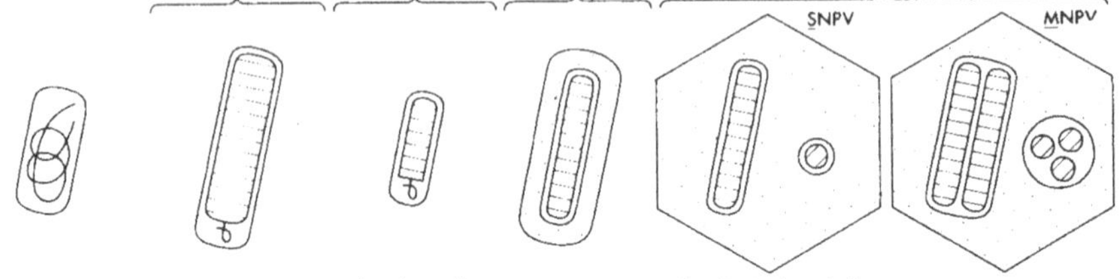

	Apis filamentous virus	Parasitoid—associated viruses subgroup D	Oryctes virus subgroup C	Granulosis viruses subgroup B	Nuclear polyhedrosis viruses subgroup A	
					SNPV	MNPV
DNA:	linear	closed, circular, polydisperse	closed, circular, single	closed, circular, single		
	12.0 × 10^6	1.5 to 8.0 × 10^6	92 × 10^6	77 × 10^6	85 × 10^6	82.5 × 10^6
Nucleocapsid:	filamentous	bullet—shaped, tailed	bullet—shaped, tailed	bacilliform		
Dimensions(nm):	3000 × 85	330 × 85	160 × 50	220—320 × 30—50		
Nucleocapsids per envelope:	1	1	1	1	1	1—25
Inclusion body:	absent	absent	absent	present	present	present
Virus particles per inclusion:	No	No	No	1	many	many

(Baculoviridae spans subgroups D, C, B, and A)

FIGURE 1. Comparative morphology of some ds-DNA enveloped invertebrate viruses.

circular high MW piece of DNA; it also contains a virus isolated from cultured *Heliothis* cells, which has a morphology more typical of Subgroup B and A virus particles. A striking feature of viruses of the latter two subgroups is the inclusion of virus particles, at the end of the infection process, in characteristic proteinaceous inclusion bodies. As has been indicated earlier, the differences between (1) subgroups B (GV) and A (NPV) and (2) within subgroup A (SNPV and MNPV) reflect (1) the number of virus particles included within each proteinaceous inclusion body and (2) the number of nucleocapsids enclosed within a common envelope.

Thus the *Baculoviridae* appear to be a large family, but the number of viruses that have been described may reflect in part the ease of detection of the inclusion bodies with the light microscope. Many other kinds of enveloped ds-DNA viruses may yet be described and I am sure that the *Oryctes* virus, the Subgroup D viruses and the *Apis* virus[32] are just the beginning of the spectrum that will be found.

IV. *REOVIRIDAE*

The cytoplasmic polyhedrosis viruses (CPV) in the family *Reoviridae* are also found in protein inclusion bodies. Although it has been suggested that the virus particles have a double protein shell, they are most likely to have a single capsid analogous to the core of *Reovirus*. In the classification of other ds-RNA viruses serological typing has traditionally been used. Early serological studies on cytoplasmic polyhedrosis virus particles[34] suggested that many of the numerous isolates were likely to be closely related serologically. More recently Payne and Rivers[35] demonstrated that CPV's could be grouped and typed by gel analysis of the RNA genome segments. Thirty-three isolates of cytoplasmic polyhedrosis viruses were compared and major differences were observed in the gel profiles of the RNA of many of the viruses, differences which were reinforced by polyacrylamide gel electrophoresis of the virus structural proteins. On the basis of these studies a provisional classification was proposed in which viruses with similar RNA gel profiles are included within the same ''type'', while isolates differing in the molecular weights of most, or all of the RNA segments are assigned to different types. Using this system 11 distinct cytoplasmic polyhedrosis virus types were recognized. These types were fully described, and in addition to the "type" isolates other similar viruses were listed. As Payne and Rivers[35] pointed out, the fractionation of ds-RNA provides a method of classifying these viruses based on properties of the virus rather than by reference to the insect host from which the virus was isolated. This has basic advantages over classification systems incorporating host names, for different cytoplasmic polyhedrosis virus isolates may replicate in the same host. Though serological typing is the conventional base for classification of ds-RNA viruses, the ease with which gel fractionation of the RNA can identify mixtures of viruses indicates that this technique should be employed prior to serological comparisons. It seems essential that these techniques should both be used since viruses which share the same RNA profile may not be identical. Differences between viruses with two RNA segments of similar size may be revealed by detailed homology studies.[36]

V. THE POLYHEDRON: THE INVERTEBRATE VIRUS INCLUSION BODY

An intriguing feature of many baculoviruses, all entomopoxviruses and, from the *Reoviridae*, all cytoplasmic polyhedrosis viruses, is that at the end of the infection process within infected cells numerous proteinaceous inclusion bodies are produced which occlude from one to many virus particles. These inclusion bodies, characteristic of these three diverse groups of viruses, have the following features in common: (1) they are predominantly composed of a single protein, 30,000 mol wt, which exists as a polymer of 300,000 mol wt; the polymers which make up the inclusion body are arranged in a face-centered cubic lattice, (2) the inclusion bodies are extremely stable, though soluble at alkaline pH, and when the insect host dies, they maintain their infectivity within the decaying insect body, either on foliage or in soil, for many months, perhaps for years, (3) when ingested subsequently by a healthy larva, the inclusion body solubilizes in the midgut lumen, which is at approximately pH 10.0, and the virus particles are released. Though inclusion bodies of a particular virus may solubilize within the lumen of the gut of many insect species, infections usually only result in the specific host and are initiated within the midgut cells. Solubilization is assisted by an alkaline protease enzyme in baculoviruses and entomopoxviruses, and in the case of baculoviruses this is located on the surface of the virus particle.

Recently Van der Beek et al.[37] isolated a virus specific mRNA of 240,000 mol wt from cultured *Spodoptera frugiperda* cells, 24 hr postinfection with *A. californica* nuclear polyhedrosis virus. The viral mRNA was isolated by means of hybridization with viral DNA coupled with cellulose, then translated in vitro in a cell-free wheat-germ embryo system to yield a protein product that was electrophoretically and serologically similar to *A. californica* nuclear polyhedrosis virus inclusion body protein. This indicates that with baculoviruses at least this protein is virus-coded.

When virus is passaged in larvae by oral infection with baculovirus inclusion bodies, each larva produces numerous enveloped virus particles in most of its cells. Many of these particles become occluded within inclusion bodies, up to 10^9 of which may be produced in a single larva. Many such virus particles, however, do not become so occluded and after appropriate purification from infected larvae can also be used experimentally to infect further larvae, usually by injection, or to infect cultured cells. It should be emphasized that the inclusion bodies are quite insoluble at neutral pH. Thus their occluded virus particles are not available to infect cultured cells. These can only be infected by virus particles which have been purified either following liberation from inclusion bodies by alkali treatment, or by extraction and purification from infected larvae.

In early passages within cultured cells, the inoculum is most often culture medium from the infected cells. Though the cells themselves may have produced normal virus-containing inclusion bodies, the infective virus in the culture medium is in the form of enveloped virus particles which had been released from cells without occlusion in inclusion bodies. Prolonged passage in vitro in this way frequently leads to diminished production of inclusion bodies, markedly reduced incorporation of enveloped virus particles into inclusion bodies, and production of many enveloped virus particles. Also, alkaline protease is not present in inclusion bodies produced in vitro. The virus particles produced in vitro in this way, however, when injected into larvae, elicit "normal" inclusion body production. Thus serial passage in vivo, with inclusion bodies as the inoculum at each passage leads to continued production of "normal" infective inclusion bodies. Passage in vitro, using as inoculum at each passage enveloped particles which had not become occluded, leads to diminished production of inclusion bodies which are "abnormal", but undiminished production of infective enveloped virus particles. There are many questions to be asked in this fascinating situation, but it appears that with baculoviruses, inclusion bodies are an effective adaptation to larva-to-larva transmission. They package a number of infective virus particles, often for a significant period outside the host, and in an epizootic enormous numbers are released into the environment. As an example, soil samples from a field in Salisbury, Maryland that had not been planted with a cole crop for 9 years contained 5.7×10^{10} inclusion bodies per hectare of *T. ni* NPV and *Pieris rapae* GV.[38]

From the evolutionary and taxonomic point of view, it is remarkable that this adaptation to efficient larva-to-larva, and particularly to generation-to-generation transmission of infection has occurred in a very similar way in such different virus families as *Baculoviridae*, *Reoviridae*, and *Poxviridae*. Further, within each family there are examples of viruses whose ecology does not call for protective inclusion bodies. These may be invertebrate viruses like the *Oryctes* virus where vertical transmission between infected adult and developing larva occurs in protected, stable niches in the environment, or viruses like the vertebrate and plant reoviruses where transmission often occurs via an insect vector, or the vertebrate poxviruses where airborne and contact transmission is common.

VI. INSECT CELL LINES

The advances that have been made in invertebrate cell culture have had a profound effect on some areas of invertebrate virology. It will be useful to provide a very brief indication of the availability and potential of these cell lines.

There are now many invertebrate cell lines. One of the earliest was developed by Grace[39] from pupal ovarial tissue of the Australian Gum Emperor-moth *Antheraea eucalypti*, and in 1976 Hink[40] listed a total of 121 cell lines from 56 species of invertebrates, from Lepidoptera, Diptera, Orthoptera, and others, and the first Coleopteran cell line has been established.[41] Commonly, invertebrate cells grow best at 25°C, with a relatively short doubling time of 12 to 24 hr and defined, commercially available culture media are often used. Insect hemolymph supplements were added to early culture media, but now fetal calf serum is routinely used.

The cultured cells can be grown by standard procedures in monolayers, allowing assays of viral infectivity, cloning of virus and studies on virus replication. Often, relatively large scale spinner production is possible, and high yields of virus can be obtained. With the nodavirus, Flock House virus and *D. melanogaster* line 2 cells for example, P. D. Scotti (personal communication) obtained 40 mg of virus from a 400 mℓ spinner cell culture at a cell density of $4.0 \times 10^7/m\ell$. Studies have been in progress for some years with the aim of producing some baculoviruses on a large scale for eventual field use, but so far economically acceptable yields have not been achieved.

Thus far, invertebrate cell lines have been readily infected with SNPV and MNPV baculoviruses (but not granulosis viruses), with iridoviruses, the Lepidopteran entomopoxviruses (but not those derived from Coleoptera) and the nodaviruses and enteroviruses. The invertebrate reoviruses — the cytoplasmic polyhedrosis viruses, have elicited only limited single-cycle infections in some cell lines and typical polyhedra have been produced, however, those infections which have been achieved could not be passaged. No invertebrate cell lines have so far been found to support replication of the *Nudaurelia* β-like virus family, nor the invertebrate members of the family *Parvoviridae* — the densoviruses. The new Coleopteran cell line may have interesting potential; the *Oryctes* Subgroup C baculovirus (isolated from a Coleopteran host) will not replicate in commonly used Lepidopteran cell lines but grows well in the new line derived from the beetle *Heteronychus arator*. The line supports growth of nodaviruses and iridoviruses and studies are in progress to determine its susceptibility to entomopoxviruses derived from Coleoptera.

Thus there is scope for detailed in vitro investigations on the biology of many invertebrate viruses and for comparative studies on their vertebrate counterparts.

VII. ADEQUATE DESCRIPTION OF NEW VIRUSES

Recently Hamilton et al.[42] and Ackermann et al.[43] have published guidelines for the characterization of plant viruses, and bacteriophages, respectively. Hamilton et al.[42] state that the (plant virus) literature contains too many examples of known viruses redescribed under new names, or of viruses so inadequately described that it is doubtful whether they really are "new". This is equally true for invertebrate viruses and it should be added that in the literature concerning invertebrate viruses there are far too many instances of putative viruses being described solely from their gross pathology in the invertebrate host, perhaps with the addition of electron micrographs of ultra-thin sections of affected tissues; the canons of Koch are often ignored and little information on the virus, apart from its diameter, is presented.

I have selected two examples from recent literature; one illustrates the worst aspects of "new" descriptions of viruses and the other provides a good example of data which facilitate useful attempts at classification.

The first example is a paper from the *Journal of Invertebrate Pathology*, describing a pathogenic "nonoccluded" virus from a moth. Very good electron micrographs were presented, the pathology was described, and infection tests were carried out. However, the only information presented on the virus was its diameter, and the fact that it contains ss-RNA. The authors then devoted a substantial part of the discussion to the problems of classification of ss-RNA viruses of invertebrates.

By contrast, a paper by D'Arcy et al.[44] described purification and characterization of a virus from the aphid *Rhopalosiphum padi* and its biological effects on the aphid were described in a later communication.[45] The virus was 27 nm in diameter, had a S_{20w} of 162, and a buoyant density of 1.379 g/ml. It contained one ss-RNA molecule of 31 S and 3 major proteins of molecular weights 31,100, 29,600, and 28,100. Its GC content was 47%. With these data, the authors were able to draw very close comparisons with the invertebrate *enterovirus* group and the speculative Group 5 (see Table 2). Definitive classification was precluded, more because of inadequate data on Group 5 than any inadequacies in the data on the aphid virus. Incidentally, this appears to be the first definitive report of a virus solely pathogenic for aphids. Its widespread occurrence in *R. padi* and other aphid species may indicate that virus infection in aphids may be a common phenomenon and that another group of viruses may be defined.

In my view, reports of "new" viruses should contain, as a basic minimum, the following information:

1. Particle dimensions, sedimentation coefficient, and density.

2.   Particle symmetry, if this can be judged from negatively stained images.
3.   Nucleic acid type, percentage by weight of particle and molecular weight. Number of segments and their size.
4.   Number and number of kinds of polypeptides in the virus particle and their molecular weight.

For the ss-RNA viruses at least this would provide the opportunity for a direct comparison with the groupings that have been approved or proposed.

I believe that journals have a responsibility to ensure that inadequate reports, especially of the "virus-like particle" type are firmly excluded or at least relegated to short communication standard. This problem was also highlighted by Matthews.[22] He indicated that new virus isolates are frequently published without detailed data that allow proper classification, and that a reduction in the number of such publications is the responsibility not only of virologists and their laboratories but also the editors of journals publishing virological papers.

The review by Harrap and Payne[28] on structural properties and identification of insect viruses gives a good overview of recent information on such features of insect viruses as morphology, physical and antigenic characteristics and the properties of their proteins and nucleic acids, and thus indicates the type of information required for effective comparative treatment of insect viruses. However, it seems appropriate to suggest that, following guidelines for plant virus and bacteriophage characterization, serious consideration should be given to the preparation of a comprehensive set of guidelines for descriptions of new viruses in the wide range of families of invertebrate pathogenic viruses. It would be of considerable value too, to outline the basic procedures that may be used in virus diagnosis and description.

VIII. CONCLUSIONS

The taxonomy of invertebrate viruses has made some progress in recent years. This is especially true of baculoviruses whose potential as biological control agents has stimulated intense activity of their molecular and cell biology in order to characterize the viruses and to assess the safety of some members of the family.

Progress has also been made with small RNA viruses where the availability of susceptible insect cell lines has opened up new and safe opportunities for study of the replication of ss-RNA virus genomes. In the future more efforts should be made to characterize ss-RNA viruses from poorly studied groups. A preliminary study of their host range in currently available cell lines may provide new opportunities for study of viruses which are poorly characterized, and which can be propagated only in intractable insect hosts. The formation of specific ICTV study groups for *Nodaviridae* and *Nudaurelia* β-like viruses will focus attention on these newly established families.

The cytoplasmic polyhedrosis viruses are quite well described. Recent information on the wide variation in the size of genome segments between isolates presents opportunities for further refinement of their classification.

Perhaps the biggest obstacle to rational classification in the past has been a paucity of reliable data on the insect viruses themselves. This is somewhat less true today, except perhaps for *Rhabdoviridae* and *Entomopoxvirinae*, but it is still a problem, and can only be overcome by a greater input from virologists — certainly from specialist invertebrate virologists, but it may be that recognition of the very close similarity for example between vertebrate and invertebrate enteroviruses will lead to a greater involvement of vertebrate virologists also.

Certainly with the increasing number of invertebrate cell lines becoming available neither the opportunity of working with cloned virus preparations nor the chance to develop a comprehensive approach to the molecular biology of invertebrate viruses should be missed.

REFERENCES

1. **Glaser, R. W.**, Virus diseases of insects, in *Filterable Viruses*, Rivers, T. W., Ed., Baillière, Tindall and Cox, London, 1928, 279.
2. **Bergold, G. H.**, Die isolierung des polyeder-virus und die natur der polyeder, *Z. Naturforsch.*, 2b, 122, 1948.
3. **Smith, K. M. and Wyckoff, R. W. G.**, Structure within polyhedra associated with virus diseases, *Nature (London)*, 166, 861, 1950.
4. **Wasser, H. B.**, Demonstration of a new insect virus not associated with inclusion bodies, *J. Bact.*, 64, 787, 1952.
5. **Holmes, F. D.**, Order Virales, the filterable viruses, in *Bergey's Manual of Determinative Bacteriology*, 6th ed., Williams & Wilkins, Baltimore, 1948, 1223.
6. **Andrewes, C. H.**, Viruses and Linnaeus, *Acta. Pathol. Microbiol. Scand.*, 28, 211, 1951.
7. **Bergold, G. H.**, Insect viruses, *Adv. Virus. Res.*, 1, 1, 1953.
8. **Weiser, J.**, *Vagoiavirus* gen. n., a virus causing disease in insects, *J. Invertebr. Pathol.*, 7, 82, 1965.
9. **Vago, C., Aizawa, K., Ignoffo, C., Martignoni, M. E., Tarasevitch, L., and Tinsley, T. W.**, Present status of the nomenclature and classification of invertebrate viruses, *J. Invertebr. Pathol.*, 23, 133, 1974.
10. **Tinsley, T. W. and Harrap, K. A.**, *Comprehensive Virology*, Vol. 12, Fraenkel-Conrat, H. and Wagner, R. P., Eds., Plenum Press, New York, 1978, 1.
11. **Hughes, K. M.**, An annotated list and bibliography of insects reported to have virus diseases, *Hilgardia*, 26, 597, 1957.
12. **Martignoni, M. E. and Langston, R. L.**, Supplement to an annotated list and bibliography of insects reported to have virus diseases, *Hilgardia*, 30, 1, 1960.
13. **Xerox, N.**, A second virus disease of the leatherjacket *Tipula paludosa*, *Nature (London)*, 174, 562, 1954.
14. **Borradaile, L. A. and Potts, F. A.**, *The Invertebrates*, Cambridge University Press, Cambridge, 1958, 795.
15. **Andrewes, C. H.**, Animal viruses, in *Viruses and Invertebrates*, Gibbs, A. J., Ed., North Holland, 1973, chap. 1.
16. **Fenner, F.**, The classification and nomenclature of viruses. Summary and results of meetings of the International Committee on Taxonomy of Viruses in Madrid, September, 1975, *Virology*, 71, 371, 1976.
17. **Wildy, P.**, Classification and nomenclature of viruses, in *Monographs in Virology*, 5th ed., S. Karger, Basel, 1971.
18. **Harrap, K. A. and Tinsley, T. W.**, A suggested latinized nomenclature of occluded insect viruses, *J. Invertebr. Pathol.*, 17, 294, 1971.
19. **Tinsley, T. W. and Kelly, D. C.**, An interim nomenclature system for the iridescent group of insect viruses, *J. Invertebr. Pathol.*, 16, 470, 1970.
20. **Kelly, D. C., Ayres, M. D., Lescott, T., Robertson, J. S., and Happ, G. M.**, A small iridescent virus (type 29) isolated from *Tenebrio molitor*: a comparison of its proteins and antigens with six other iridescent viruses, *J. Gen. Virol.*, 42, 95, 1979.
21. **Fenner, F.**, Classification and Nomenclature of viruses. Second report of the International Committee on Taxonomy of Viruses, *Intervirology*, 7, 1, 1976.
22. **Matthews, R. E. F.**, Classification and nomenclature of viruses. Third report of the International committee on Taxonomy of Viruses, *Intervirology*, 12, 132, 1979.
23. **Longworth, J. F.**, Small isometric viruses of invertebrates, *Adv. Virus. Res.*, 23, 103, 1978.
24. **Matthews, R. E. F.**, Classification and Nomenclature of viruses. Fourth report of the International Committee on Taxonomy of Viruses, *Intervirology*, 1982.
25. **Moore, N. F., Kearns, A., and Pullin, J. S. K.**, Characterisation of cricket paralysis virus-induced polypeptides in *Drosophila* cells, *J. Virol.*, 33, 1, 1980.
26. **Scotti, P. D., Yamazaki, S., Shiga, S., Natori, K., Susuki, K., Akao, Y., Kurahashi, H., and Kono, R.**, Kedogou virus, an unusual enveloped RNA virus isolated from an insect host, in press.
27. **Burgess, S.**, Molecular weights of Lepidopteran baculovirus DNAs: derivation by electron microscopy, *J. Gen. Virol.*, 37, 501, 1977.
28. **Harrap, K. A. and Payne, C. C.**, The structural properties and identification of insect viruses, *Adv. Virus. Res.*, 25, 273, 1979.
29. **Miller, L. K. and Dawes, K. P.**, Restriction endonuclease analysis for the identification of baculovirus pesticides, *Appl. Environ. Microbiol.*, 35, 411, 1978.
30. **Burgess, S.**, Eco R1 restriction endonuclease fragment patterns of eight lepidopteran baculoviruses, in press.
31. **Harrap, K. A., Payne, C. C., and Robertson, J. S.**, The properties of three baculoviruses from closely related hosts, *Virology*, 79, 14, 1977.
32. **Bailey, L., Carpenter, J. M., and Woods, R. D.**, Properties of a filamentous virus of the honey bee *(Apis mellifera)*, *Virology*, 114, 1, 1981.

33. **Stoltz, D. B. and Vinson, S. B.,** Viruses and parasitism in insects, *Adv. Virus. Res.,* 24, 125, 1979.

34. **Cunningham, J. C. and Longworth, J. F.,** The identification of some cytoplasmic polyhedrosis viruses, *J. Invertebr. Pathol.,* 11, 196, 1968.

35. **Payne, C. C. and Rivers, C. F.,** A provisional classification of cytoplasmic polyhedrosis viruses based on the size of the RNA genome segments, *J. Gen. Virol.,* 33, 71, 1976.

36. **Payne, C. C., Mertens, P. P. C., and Katagiri, K.,** A comparative study of three closely related cytoplasmic polyhedrosis viruses, *J. Invertebr. Pathol.,* 32, 310, 1978.

37. **Van der Beek, C. O., Saaijer-Riep, J. D., and Vlak, J. M.,** On the origin of the polyhedron protein of *Autographa californica* nuclear polyhedrosis virus. Isolation, characterisation and translation of viral messenger RNA, *Virology,* 100, 326, 1980.

38. **Thomas, E. D., Reichelderfer, C. F., and Heimpel, A. M.,** Accumulation and persistence of a nuclear polyhedrosis virus of the cabbage looper in the field, *J. Invertebr. Pathol.,* 20, 157, 1972.

39. **Grace, T. D. C.,** Establishment of four strains of cells from insect tissues grown *in vitro, Nature (London),* 195, 788, 1962.

40. **Hink, F. W.,** A compilation of invertebrate cell lines and culture media, in *Invertebrate Tissue Culture,* Vago, C., Ed., Academic Press, New York, 1976, 319.

41. **Crawford, A. M.,** A Coleopteran cell line derived from *Heteronychus arator* (Coleoptera: Scarabaeidae), *In Vitro,* in press.

42. **Hamilton, R. I., Edwardson, J. R., Francki, R. I. B., Hse, H. T., Hull, R., Koenig, R., and Milne, R. G.,** Guidelines for the identification and characterisation of plant viruses, *J. Gen. Virol.,* 54, 223, 1981.

43. **Ackerman, H. W., Andvrier, A., Berthiaume, L., Jones, L. A., Mayo, J. A., and Vidaver, A. K.,** Guidelines for bacteriophage classification, *Adv. Virus Res.,* 23, 1, 1978.

44. **D'Arcy, C. J., Burnett, P. A., Hewings, A. D., and Goodman, R. M.,** Purification and characterization of a virus from the aphid *Rhopalosiphum padi, Virology,* 112, 346, 1981a.

45. **D'Arcy, C. J., Burnett, P. A., and Hewings, A. D.,** Detection, biological effects of a virus of the aphid *Rhopalosiphum padi, Virology,* 114, 268, 1981b.

Chapter 6

CURRENT PROBLEMS IN FUNGAL VIRUS TAXONOMY

K. W. Buck

TABLE OF CONTENTS

I. INTRODUCTION

Two types of association between fungi and viruses can be distinguished. First, fungi can act as vectors for the transmission of a number of soil-borne plant viruses. Association between virus and vector is generally specific and such viruses can be carried either externally or internally by the fungus.[1] There is, however, no evidence that these viruses replicate in the fungus and they are properly classified as plant viruses. Second, fungi can be hosts to viruses which replicate, sometimes to high titers, within the fungal cells. These are the fungal viruses (mycoviruses) with which the present chapter is concerned.

Compared to animal and plant virology, fungal virology is a comparatively recent subject. The first fungal viruses, isolated from diseased specimens of the cultivated mushroom *Agaricus bisporus*, were described only 20 years ago.[2] A major stimulus to mycovirus research came from the discovery that the antiviral and interferon-inducing activities for animals in extracts of two fungi, *Penicillium funiculosum* and *P. stoloniferum*, were due to double-stranded RNA (dsRNA) of viral origin.[3-5] These reports led to much activity and a number of extensive screening programs were set up. As a result viruses and virus-like particles have now been described from over 100 fungal species in more than 60 genera, distributed among all classes of fungi. Several reviews have been published.[6-10] However, two problems for the taxonomy of fungal viruses have arisen from the way in which fungal virology has developed.

First, the subject has attracted only a few virologists and many of the investigators have been organic chemists, plant pathologists, or mycologists, who have been more interested in the effects of the viruses on their hosts than in the virus particles themselves. Consequently, the virus particles have been often inadequately described from the taxonomic point of view. Although the Fungal Virus Subcommittee of the ICTV was set up in 1975 only two tentative virus groups, containing a total of five viruses, were suggested, and no formal proposals were made, at the meeting of the ICTV in The Hague in 1978.[11,12]

Second, the early promise that fungal viruses may influence the production of antibiotics, mycotoxins, or other fungal metabolites, may be generally responsible for reduction (or increase) in virulence of phytopathogenic fungi, or may be generally involved in cytoplasmically transmitted properties such as senescence or vegetative death, has not been upheld by subsequent work. Most fungal viruses give rise to latent infections, in the sense that, although the viruses may replicate well, there are no obvious effects on the phenotypes of the fungi. There is, at present, only one fungal species in which viral dsRNA has been unequivocally proved to be responsible for a change in fungal phenotype, namely the production of killer toxin by killer strains of the yeasts, *Saccharomyces cerevisiae*. Such strains contain two species of dsRNA, L, which encodes the virus capsid polypeptide and M, which encodes the killer protein.[13] In a few other species in which a cytoplasmically transmitted genetic element has been proved to determine a phenotype, e.g., killer protein in *Ustilago maydis*,[14] hypovirulence in *Endothia parasitica*[15] and in *Rhizoctonia* solani,[16] and a disease of *Helminthosporium victoriae*,[17] association with dsRNA in persuasive, but not unequivocal. Even in species where the association of virus particles with a disease, e.g., in the cultivated mushroom, *Agaricus bisporus*, appeared to be well established[2] the discovery of similar particles in apparently healthy mushrooms has raised questions concerning the precise role of the virus particles in the disease.[18] A major problem has been the difficulty in infecting virus-free fungi with purified virus particles. Although methods of infecting fungal protoplasts with partially purified virus preparations have been described[107,108] most investigators have found this problem intractable.[19] As a result of such problems interest in fungal viruses has declined in the last few years, funding has become more difficult and there are today fewer than twenty laboratories around the world engaged in mycovirus research. It is therefore to be expected that future progress in fungal virus taxonomy will be made much more slowly than that in animal, plant, and bacterial virus taxonomy.

On a more positive note the handful of investigators remaining in fungal virology are carrying out detailed studies of a number of mycoviruses and useful taxonomic data are gradually accumulating. At the 1981 meeting of the ICTV at Strasbourg the report of the Fungal Virus Subcommittee included a document for discussion in which three virus families and six virus genera, including 36 viruses as members, probable members, or possible members, were suggested. These suggestions have been included in the Fourth Report of the ICTV.[20]

In this chapter the taxonomic problems associated with incomplete characterization of the large number of virus-like particles which have been reported in fungi will be outlined first and then more specific problems associated with the taxonomy of some of the better characterized mycoviruses will be considered.

II. VIRUS-LIKE PARTICLES IN FUNGI

Several different morphological types of virus-like particles (VLPs) have been described. Examples of these are given in Table 1 together with comments relevant to their taxonomy. The major problem with most of these particles is that they have been inadequately characterized. Many have been examined only by electron microscopy and some of these only in thin sections of fungi. Taxonomy of these VLPs and confirmation, or otherwise, of their viral nature must clearly await further studies.

The most numerous morphological group of VLPs comprises the small isometric particles. Although the most frequently used isolation methods may have been biased towards the selection of this type of particle there is little doubt that small isometric particles are of common occurrence in fungi and screening programs have suggested that they occur in 10 to 15% of randomly sampled fungal isolates. Particles of this type from a number of fungi have been isolated and purified, and characterized sufficiently for their classification to be considered. They all have in common a genome of dsRNA, but considerable diversity exists among them. There is therefore no question of grouping them all together in one genus or even in one family. Despite this diversity, none of these particles resemble any of the existing families or groups of dsRNA viruses, namely the *Reoviridae* family, the *Cystoviridae* family, or the group of animal viruses with bisegmented dsRNA genomes.[20] It is therefore necessary to consider creating new families, genera, or groups for these small isometric dsRNA mycoviruses.

Even with the better characterized dsRNA viruses of fungi the question has been frequently raised as to whether these particles should properly be regarded as viruses, and some workers still prefer to regard them as VLPs and their dsRNA genomes as plasmids.[38] This view appears to arise from three features of these viruses. (1) Most of the viruses do not cause disease or exert any deleterious effects on their hosts. Indeed, it has been argued that the ability to produce killer protein by *Saccharomyces cerevisiae* could be an ecological advantage to the strains which produce it.[39] Moreover, killer strains of the yeast, *Kluyveromyces lactis,* produce a killer protein determined by a linear DNA plasmid[40] and fungal killer proteins are analogous to bacteriocins which are encoded by closed circular DNA plasmids.[41] (2) Mycoviruses are transmitted only by intracellular routes.[19] There is no extracellular phase to their life cycle. Transmission, therefore, more closely resembles that of plasmids than that of most viruses. (3) The difficulty of demonstrating infectivity of the particles. There is no reported example where the infectivity of a highly purified single mycovirus has been unequivocally demonstrated.

Despite these problems, the majority of mycovirologists now use the term virus for the better characterized isometric dsRNA particles, reserving the term VLP for poorly characterized particles. Arguments in favor of this usage have been discussed at length[42] and may be summarized:

Table 1
MORPHOLOGICAL TYPES OF VIRUS-LIKE PARTICLES DETECTED IN FUNGI

Particle morphology	Particle size (nm)	Examples of fungal species in which detected	Key reference	Comments on properties relevant to taxonomy
Rigid rods	17 × 350 17 × 350 25 × 180 25 - 28 × 280 - 310	*Agaricus bisporus* *Peziza ostracoderma* *Mycogone perniciosa* *Lentinus edodes*	21 22 23 21	No data on the nucleic acid or capsid protein is available for any of the rigid rod VLPs. The particles in *A. bisporus* and *P. ostracoderma* superficially resemble plant tobamoviruses.
Flexible filaments	13 × 500 17 × 100 - 1500 15 - 17 × 100 - 1500	*Boletus edulis* *Helminthosporium sacchari* *Lentinus edodes*	24 25 21	No data on the nucleic acid of any of these VLPs is available. The VLPs of *B. edulis* superficially resemble plant potexviruses. The VLP of *H. sacchari* is serologically related to that from *L. edodes* and has a single capsid polypeptide, mol wt 23×10^3.
Bacilliform	19 × 50 19 × 48	*Agaricus bisporus* *Microsphaera mougeotii*	26 25	The *A. bisporus* particles resemble particles of alfalfa mosaic virus and contain single-stranded (ss) RNA. However, no evidence as to a possible divided genome is available and the molecular weight of the single capsid polypeptide (16×10^3) is significantly below that of alfalfa mosaic virus (24×10^3).
Club shaped, pleomorphic	100 (maximum length)	*Endothia parasitica* *Agaricus bisporus*	27 28	VLPs from *E. parasitica* sediment between 115 and 190S, are associated with dsRNA and probably contain lipid.
Pleomorphic with membrane envelope	120-170 nm (diameter)	*Neurospora crassa*	29	Particles contain ssRNA and are associated with mitochondria.
Herpes-like	280 × 130 110 (nucleocapsid diameter)	*Thraustochytrium* sp.	30	Particles contain DNA. Nucleocapsid and some aspects of replication cycle show a superficial resemblance to herpes viruses.
Head and tail	43 (head diameter) 53 (head diameter)	*Penicillium brevicompactum* PBV-1 PBV-2	31	Contain dsDNA, PBV-1 is similar to λ-phage and PBV-2 is similar to T5 phage. Can be detected only after replication in bacteria. There is no evidence that these viruses replicate in fungi.
Isometric, geminate	20 (monomer diameter) 20 × 30 (dimer)	*Neurospora crassa*	32, 33	The dimer particles show some resemblance to plant geminiviruses, but no data on the nucleic acid or capsid protein is available.

Large isometric	200 (diameter)	*Aphelidium* sp.	34	The *Aphelidium* VLPs contain DNA and show some similarities to iridoviruses, but much more information is required before assignation to that group could be made.
	130 (diameter)	*Schizophyllum commune*	35	
	130 (diameter)	*Coprinus lagopus*	36	
Small isometric	25-50 (diameter)	About 70 fungal species	8, 37	Particles from about 20 species have been reasonably well characterized. All these have dsRNA genomes, although considerable diversity exists. See text for further details.

1. Structurally the particles closely resemble those of authentic viruses. They consist of a genome of discrete molecules of completely base-paired dsRNA. This type of nucleic acid is not found in other groups of organisms, only in viruses. The RNA is surrounded by a capsid, consisting of multiple copies of polypeptide subunits arranged in an ordered fashion. The structures are consistent with icosahedral symmetry (usual for isometric viruses). In three cases, *Saccharomyces cerevisiae* virus 1,[43] *Penicillium stoloniferum* virus S,[44] and *Gaeumannomyces graminis* virus 38-4-A,[45] the dsRNA genome has been shown to encode the major capsid polypeptide. DNA plasmids are not enclosed in protein capsids.

2. For two fungi where a search has been made, namely *Saccharomyces cerevisiae* and *Gaeumannomyces graminis*, no DNA provirus copies of dsRNA could be detected. Although Vodkin[57] originally claimed that *S. cerevisiae* carried DNA copies of L dsRNA (from *S. cerevisiae* virus 1), two other independent groups of workers[60,56] have failed to confirm this result. It is now considered likely that Vodkin's "dsRNA" hybridization probe, isolated by CF 11 cellulose chromatography, was contaminated with fragments of ribosomal RNA, which are known to be present in CF 11 "dsRNA" fractions isolated from RNase-treated whole cell RNA (M. Vodkin, quoted as personal communication in Ref. 54). Hastie et al.[56] showed conclusively that there could be no more than 0.075 copies of L dsRNA per yeast nuclear genome. In the case of *G. graminis* no DNA copies of dsRNA from two viruses (3bla-B and 3bla-C) could be detected using a method capable of detecting one DNA copy per ten host genomes.[58] Thus, it can be concluded that in these two fungi, at least, dsRNA is not produced as a result of transcription of the host genome.

3. All the isometric dsRNA mycoviruses, which have been tested, have been found to possess a RNA-dependent RNA polymerase activity capable of transcribing or replicating the dsRNA genome.[44] Possession of a virion-associated RNA polymerase is also a property of dsRNA viruses in other families (*Reoviridae*, *Cystoviridae*). The ability to promote replication of genomic dsRNA, together with coding information for the virus capsid, is probably all that is require for autonomous replication of the isometric dsRNA mycoviruses. Thus, although infectivity and autonomous replication in cells are undoubtedly essential properties of viruses, the failure to demonstrate infectivity conclusively is probably the result of the technical difficulties of introducing the viruses into the fungal cells rather than any intrinsic lack of ability of the particles to replicate autonomously.

4. Ability to cause disease need not be regarded as an essential property of a virus. Many mild strains of viruses are known and in nature they may well be more common than the pathogenic varieties which have been most studied. Pathogenicity is a property of virus/host interactions and a virus which causes severe disease in one host may give rise to a symptomless infection in another. If the viruses which have evolved are those which are best able to promote their own replication and survival, mycoviruses may be considered efficient in this respect. Moreover, their intracellular mode of transmission may have resulted from their adaptation to the mycelial habit peculiar to fungi and to have allowed co-evolution of fungi with their viruses.[46] The possession of a protein coat, normally considered to be a protection for the virus genome in an extracellular environment, may help to ensure the survival of mycoviruses intracellularly during fungal sporulation when much nucleic acid turnover in the cytoplasm occurs.

III. PROBLEMS IN THE TAXONOMY OF ISOMETRIC dsRNA MYCOVIRUSES

In virus classification in the last decade increasing weight has been placed on in vitro properties of virions, on the structure and organization of the virus genome and on the strategy of virus replication and gene expression. Much less weight has been placed on biological properties and the interaction of the virus with its host. This is fortunate for mycovirus taxonomy, in view of the absence of biological effects for most viruses, since it enables groupings to be made which are in line with those made for other viruses. Major emphasis has to be placed on in vitro properties of mycovirus particles and their components because this is the main information available. In view of the absence of a method for synchronous infection of fungi with their viruses, the limited amount of information available on virus replication has come from studies in vitro of the virion-associated RNA polymerases and the structure of virus replicative intermediates isolated from infected cells. Some useful information on the organization of virus genomes has come from studies of virus variants and virus deletion mutants. Problems in evaluating the relative importance of the different taxonomic criteria available will be discussed with particular emphasis on properties considered by the Fungal Virus subcommittee of the ICTV to be most important for Fungal Virus Taxonomy.[12]

A. Properties of the Virion

1. Dimensions and Symmetry

Particle diameter has been agreed by the Fungal Virus subcommittee of the ICTV to be an important taxonomic criterion. A problem is that values reported from different laboratories for the same virus often differ significantly. For example, a diameter of 25 to 30 mm was reported for both *Penicillium stoloniferum* virus S and F by Banks et al.,[5] whereas Bozarth et al.[47] reported a diameter of 34 nm for the same two viruses. Measured particle diameters can be affected by buffer, stain, method of microscope calibration, and difficulties in defining the precise edge of particles. There may be significant differences in diameters between particles penetrated or unpenetrated by stain and between diameters of single particles and those packed together in ordered arrays. For individual icosahedral particles the diameter will vary slightly depending on the orientation of the particle. Ideally, for comparative purposes all viruses should be measured under the same conditions by one operator. For practical taxonomic purposes, the possible variations in measured particle diameters will need to be taken into consideration in weighting this particular criterion.

Although no electron microscopic or X-ray diffraction studies of the symmetry of isometric dsRNA mycoviruses have been made, analyses of the chemical compositions of a number of viruses are consistent with icosahedral capsids, constructed from 60 structure units (T = 1). The capsids of *Penicillium stoloniferum* virus S,[48,49] *Aspergillus foetidus* viruses S and F,[50] *Helminthosporium maydis* virus,[51] *Saccharomyces cerevisiae* virus 1,[54] and *Ustilago maydis* P1 virus[52] contained 120 polypeptide subunits, whereas those of *Penicillium chrysogenum* virus and *Penicillium cyaneo-fulvum* virus[53] contained 60 polypeptide subunits. Although this type of information is undoubtedly valuable, a problem is that few laboratories are equipped or have the time for this type of study. Indeed all the analyses described above were obtained in only three laboratories.

2. Sedimentation Coefficients

A problem here is that many dsRNA mycoviruses have multiple dsRNA segments, most of which are encapsidated separately, giving rise to multicomponent systems. Although S values of individual particle types can theoretically all be measured, and indeed generally there has been very good agreement between S values for the same virus determined in different laboratories, there may be uncertainty about the significance of all the dsRNA components (e.g., possible satellite and defective RNAs, see later) and hence particle types. Another difficulty is that because the viruses replicate asynchronously in a given fungal cell population and because the viruses do not lyse their host cells, nor are released from them, mycovirus preparations frequently contain intermediates of replication in addition to mature virus particles. For example, preparations of *Penicillium stoloniferum* virus S contain, in addition to particles with dsRNA (presumably the mature virions), particles with ssRNA of half the molecular weight of the dsRNA (virus messenger RNA strands) and particles with both dsRNA (of the same size as genomic dsRNA) and ssRNA (of heterogeneous size from very small up to the size of a full-length transcript).[48] The latter particles were shown to be replicative intermediates, since on incubation with nucleoside triphosphates the ssRNA was converted to full-length dsRNA in a reaction catalyzed by the virion-associated RNA polymerase.[44]

A similar situation exists for *Saccharomyces cerevisiae* virus 1 particles isolated from log phase cells. In addition to mature virions containing dsRNA, particles of lower S value contained partially ds and partially ssRNA molecules. These could be converted to full length ds molecules by the particle-associated RNA polymerase and nucleoside triphosphates in vitro and hence were regarded as replicative intermediates or subviral particles.[59] Another example is *Helminthosporium maydis* virus.[51] This virus sediments as three components with S values of 283, 212, and 152. The 283S particles were shown to contain a dsRNA of molecular weight 6.3×10^6 (later revised to 5.7×10^6),[52] while the 152S particles were shown to be protein capsids devoid of nucleic acid. The status of the 212S particles is unknown; they contain an undefined RNA, probably single-stranded. It is clear that while sedimentation coefficients are important as taxonomic criteria, they must be used with caution and allowance made for a degree of uncertainty with regard to the spread of values for a particular virus.

3. Composition

None of the isometric dsRNA mycoviruses contain lipid, so that only nucleic acid and protein composition need be considered. Although an important criterion, some methods often used to determine the percentage of nucleic acid and protein in a virion are not theoretically sound. For example, measurements of A_{260}: A_{280} ratios, based on comparisons with standard mixtures of enolase and yeast RNA,[61] make no allowances for differences in the aromatic amino acid composition of virus proteins and that of the standard or for differences in hyperchromicity between dsRNA and yeast ssRNA. Moreover, the values are critically dependent on small amounts of nucleic acid or light scattering impurities. As an example of the problems that this can cause Nash et al.[62] reported an $A_{260:280}$ ratio of 1.70 for *Penicillium chrysogenum* virus, while Wood and Bozarth[63] reported 1.50 and Buck and Girvan[53] reported 1.54 for the same virus. Colorimetric analyses for protein using bovine serum albumin as standard also make the invalid assumption that the specific absorption of the color produced by the standard is the same as that of the virus protein.

A problem with the use of buoyant densities in caesium chloride for determining particle composition is that buoyant densities depend on the amount of Cs^+ ions bound and hydration of the particles as well as chemical composition. For example, the buoyant densities of viruses in different genera of the *Picornaviridae* vary over the range 1.33 to 1.45 g/cm³ yet their chemical compositions are similar.[11] Estimates based on comparison of sedimentation coefficients of "full" and "empty" particles,[64] may be more sound provided that their capsids are the same. Even with this method assumptions are made that the diffusion coefficients of "full" and "empty" particles are the same and that the partial specific volume of the virion is an additive property of those of its protein and nucleic acid components. However, the method seems to be the most reliable of the rapid, semiempirical methods for determining particle composition. Nevertheless, direct measurements of protein (by amino acid analysis) and nucleic acid (by phosphorus and base composition analysis) are to be preferred. Some of these difficulties have led to widely divergent values for the same virus, e.g., Wood and Bozarth[63] reported 11 to 15% dsRNA in particles of *Penicillium chrysogenum* virus, whereas Buck and Girvan[53] reported 19.7% dsRNA for the same virus. Such problems are compounded for many viruses by the problems of multicomponent systems composed in many cases of particles with defective, satellite, or replicative intermediate RNAs.

4. Molecular Weight

Virion molecular weight has been determined for several dsRNA mycoviruses, e.g., *Penicillium stoloniferum* virus S (L particles),[48] *Aspergillus* foetidus virus S (Sla particles), *A. foetidus* virus F (F4 particles),[50] *Penicillium chrysogenum* virus,[63,53] *Penicillium cyaneofulvum* virus,[53] *Helminthosporium maydis* virus,[51] *Ustilago maydis* P1 virus.[32] In all cases, it has been calculated from the Svedberg equation $M = RTs/D (1 - \bar{v}\rho)$ where M = mol wt, R = gas constant, T = absolute temperature, s = sedimentation coefficient, D = diffusion coefficient, $\bar{v}$ = partial specific volume and ρ = density of solvent. Measurement of molecular weight therefore involves measurement of three virus properties, s, D, and $\bar{v}$. It is preferable to correct values to standard conditions, 20°C, water and zero concentration. Possible problems in the measurement of s due to the multicomponent nature of the virus systems have been overcome, when necessary, by careful fractionation of the particles by sucrose or caesium chloride density gradient fractionation. Measurements of D by boundary spreading have been made either in the ultracentrifuge using ultraviolet optics or in static apparatus using Schlieren optics. The former has the advantage of enabling measurements to be made at low concentrations, but the disadvantage that if the particles are not completely homogeneous, some boundary spreading due to particle separation by sedimentation during measurements may occur leading to erroneously high values for D. The latter has the advantage that slight particle heterogeneity will have little effect on the value (providing all particles are the same size), but the disadvantage that higher virus concentrations are needed, giving problems if the virus is prone to dimerization or higher orders of aggregation. Partial specific volumes of virions have not been determined directly but have been calculated from those for the virus protein and nucleic acid assuming proportional additivity. The major problem here is that a knowledge of particle composition is required with all the difficulties discussed previously. In general, however, molecular weights have been very carefully determined and values may be considered reliable. One disagreement has been reported. Wood and Bozarth[63] reported a molecular weight of 13.0×10^6 for *Penicillium chrysogenum* virus (based on s = 150S, D = 1.03×10^{-7} cm²/sec, $\bar{v}$ = 0.728 cm³/g), while Buck and Girvan[53] reported a molecular weight of 9.8×10^6 (based on s = 144.6S, D = 1.17×10^{-7} cm²/sec, $\bar{v}$ = 0.69 cm³/g). Subsequently, it was considered[37] that the most reliable value for the molecular weight of this virus (11×10^6) resulted from the combination of an average sedimentation coefficient, the value of D of the former authors and the value of $\bar{v}$ of the latter authors. Few laboratories, in which mycovirus research is carried out, are equipped for such studies, but virion molecular weights are valuable additional pieces of information.

5. Serological Reactions

There has been good agreement among members of the Fungal Viral Subcommittee of the ICTV as to the taxonomic value of positive serological realtionships between viruses. Due to difficulties in purification of some mycoviruses, there is a risk that use of antisera, which contain antibodies to normal host constituents such as proteins or polysaccharides, may give spurious indications of relationship.[65] Cross-absorption of antisera with extracts from virus-free fungi may be a prerequisite for their use.[66,67] It is also important to ensure that antisera are free from antibodies to dsRNA which could also give spurious indications of relationships. Antisera elicited by injecting rabbits with particles of two dsRNA plant viruses (rice dwarf and maize rough dwarf viruses) were shown to contain antibodies to dsRNA as well as to capsid protein[68] and cross-reactions were demonstrated to result from precipitation of dsRNA rather than of viral proteins. So far such an effect has not been reported in serological reactions among mycoviruses, and in view of the absence of cross-reactions between most viruses studied, it may be uncommon.

Two dimensional double diffusion in gels (Ouchterlony gel diffusion test) has been the most widely used for detecting relationships between mycoviruses, relationships being indicated by the formation of fusing or spurring precipitin lines. For example, *Penicillium chrysogenum* virus and *Penicillium cyaneo-fulvum* viruses gave rise to fusing precipitin lines when tested against antisera to either virus,[53] whereas *Penicillium chrysogenum* virus and *Penicillium brevicompactum* virus gave rise to fusing precipitin lines when tested against an antiserum to the former virus, but gave rise to a spur when tested against an antiserum to the latter virus.[63] Double-diffusion analysis has also been used to determine antiserum titers to a number of related mycoviruses.[55] It gave indications of close and distant serological relationships between different viruses from *Gaeumannomyces graminis*. However, some caution needs to be exercised in interpreting the degree of relationship between viruses from serological results. Some determinants may be highly immunogenic in one animal, but not in another. Two viruses from *G. graminis* appeared to be closely related when tested against a rabbit antiserum to one of them, but unrelated when tested against a mouse antiserum to the same virus.[55] Other serological tests, such as enzyme-linked immunosorbent assay (ELISA) or electron microscopy with serum-activated grids[69] have been used for detecting mycoviruses, but not so far for detecting relationships between them. It is possible that the specificity of these methods may limit their usefulness to detecting relationships between closely related viruses, but their sensitivity can be 1000-fold greater than conventional gel diffusion analysis.[65]

Absence of serological relationships between mycoviruses has been concluded from failure of one virus to react with antisera to another, but for some experiments homologous antiserum titers have not been reported, so that the possibilities of distant serological relationships could not be excluded. In other cases, absence of serological relatedness has been demonstrated by formation of crossing precipitin lines when two viruses reacted with antiserum raised to a mixture of the two viruses, e.g., *Penicillium stoloniferum* viruses S and F[47] and *Aspergillus foetidus* viruses S and F.[70] Bearing in mind that the antigenic determinants of a virus may be encoded by a small proportion of its genome and that a few mutations can markedly alter its immunological properties, the absence of serological relationships should not preclude the placement of a virus in a genus or family.

B. Properties of the Virus dsRNA Genome

1. Number of Segments

It is necessary to determine for a given virus whether the genome is undivided or divided and, if divided, into how many segments. A problem here is that isolates of dsRNA mycoviruses may contain satellite dsRNAs, defective dsRNAs, or a mixture of viruses and/or virus variants. A second problem is the absence of a reliable infection method by which the minimum number of components essential for virus replication could be tested. Even if such a method existed the sedimentation coefficients or buoyant densities of the individual virus components of some viruses are so close that sufficient separation for this purpose could not be obtained. A better prospect would be to separate individual dsRNA components by polyacrylamide gel electrophoresis. Although native dsRNA is unlikely to be infective, it should be infective after denaturation to release the positive coding strand as has been found for the replicative form dsRNAs of a number plant viruses.[71] However, until such systems are developed indirect approaches have to be used to deduce the likely number of "essential" dsRNA components. The specific problems of satellite dsRNAs, defective dsRNAs and virus mixtures will be considered in turn.

a. Satellite dsRNAs

Killer strains of *Saccharomyces cerevisiae* secrete protein toxins, which are lethal to sensitive strains of the same or related species.[72] Most sensitive yeast strains are infected with virus particles which contain a single dsRNA molecule (L dsRNA, molecular weight of 3.4×10^6) and which are able to replicate autonomously. L dsRNA encodes the major capsid polypeptide,[43] 88,000 mol wt, and there is ample coding capacity to encode also the virion RNA polymerase, although this polypeptide has not yet been identified, assuming that it differs from the major capsid polypeptide. The virus in sensitive strains therefore has an undivided genome.

Killer yeast strains are also infected with L virions, but contain, in addition, a dsRNA component (M dsRNA, molecular weight of 1.4×10^6) which encodes the protein toxin[13] and also determines immunity to toxin. L and M dsRNAs have little sequence homology.[73] M dsRNA is enclosed in virus particles with a capsid identical to that of L virions. L dsRNA encodes the polypeptide required both for its own encapsidation and that of M dsRNA. M dsRNA therefore falls within the definition of a satellite RNA proposed by Mossop and Francki.[74] It requires a helper virus (L virions) to produce its coat protein, it is not required for replication of the helper virus and it contains genetic information not present in the helper virus. Consistent with this is the fact that strains of yeast with no dsRNA or only L dsRNA are known, but strains containing M dsRNA only have not been found; M dsRNA occurs only in the presence of L dsRNA. It is interesting that M dsRNA replication requires, in addition to L dsRNA gene products, more than 20 host genes. This remarkable requirement was established by the isolation and characterization of nuclear mutants (*mak* mutants) in which M dsRNA was unable to replicate. Replication of L dsRNA was not affected in any of the *mak* mutants.[75] No DNA provirus copy of M dsRNA could be detected in yeast DNA[60] and M virions were shown to carry an RNA-dependent transcriptase activity similar to that found in L virions.[76,77] M dsRNA is therefore not transcribed from host DNA. Thirteen types of yeast killer strains have been described with distinct toxin and resistance functions.[78] The first three of these, k1 to k3, all have L and M dsRNAs; the L1 and L2 and the M1 and M2 dsRNAs from k1 and k2 killers, respectively, have been shown by fingerprint and sequence analysis to be distinct but related.[79,80]

The yeast killer system illustrates the point that the problem of lack of infectivity experiments can sometimes be solved in other ways. It is clear in the systems discussed above the the L1 and L2 virions can confidently be regarded as viruses with undivided genomes and that, in k1 and k2 killer strains, these viruses can act as helpers for the replication of the M1 and M2 satellite dsRNAs, respectively.

Killer systems have also been described in the basidiomycete fungus, *Ustilago maydis*.[91] In many ways, these are analogous to the yeast killer systems, but more complex patterns of dsRNAs are found and these create taxonomic problems which have been solved in ways different from those used for the yeast system. Killer strains of *U. maydis* secrete protein toxins, which kill or inhibit the replication of sensitive strains of the same or related species.[82] Three killer specificities (P1, P4 and P6) are known in *U. maydis*; each killer is insensitive to its own toxin, but sensitive to the other two. Ability to produce toxin is cytoplasmically transmitted and occurs only in strains carrying dsRNA virus particles.[83,14] Immunity to killer is also cytoplasmically transmitted. The dsRNA appears segmented when examined by gel electrophoresis and each killer and immune strain exhibits a distinct pattern of three to seven dsRNA components. The segments are of three size groups, designated H (heavy, molecular weight 1.8×10^6 to 4.2×10^6), M (medium, molecular weight 0.4×10^6 to 1×10^6) and L (light, molecular weight 0.25×10^6).[52] dsRNA is encapsidated in virus particles, composed of a single polypeptide species, 73,000 mol wt. The viruses of the P1, P4, and P6 killer strains were serologically related and probably identical.[84,85] The functions of individual dsRNA components have been examined by analysis of deletion mutants which have lost some of the original complement of dsRNA components.[86-88] Some of these were obtained by mutagenesis with ultraviolet irradiation and with *N*-methyl-*N*-nitrosoguanidine; others were found among sensitive wild strains and also among progeny from crosses of killer and nonkiller strains of the fungus.[86,89] Some sensitive strains carried only one H dsRNA segment and none in the M or L dsRNA size class. Nevertheless, in such strains the H dsRNA segment was still encapsidated in virus particles. It is likely, therefore, that the genetic information required for the structure and replication of the virus particles resides in a single H dsRNA segment. Virus particles containing such a single segment can therefore be regarded as an autonomously replicating virus with an undivided genome.

Other studies have shown that the M2 segment for all three killer specificities carries the killer determinant and that, in the case of a P1 killer, the L segment determined cytoplasmic immunity to P1.[90] No mutants carrying M and L segments in the absence of H segments have been obtained, indicating the dependence of M and L dsRNA replication on particles containing an H segment. Furthermore, M and L dsRNAs are encapsidated in protein coats identical to those which encapsidate H dsRNA segments. It seems reasonable to conclude that in the *Ustilago* killer systems particles containing an H segment act as a helper virus for the replication of satellite dsRNAs (M2 and L) which determine the killer and immunity functions. Consistent with this conclusion is the observation that dsRNA is common in wild populations of *U. maydis*, but killer strains are relatively rare accounting for only 1 to 2% of wild populations.[91] In a recent study of smut galls on maize in Connecticut, Day[81] found dsRNA in 48 of 49 samples tested; over half of these carried only one segment of dsRNA and this was in the H size class. It should be pointed out that most of the evidence for the helper and satellite dsRNA systems in *U. maydis* is based on genetical evidence which can indicate which dsRNA segments are associated with a particular phenotype but does not allow identification of gene products. Direct evidence by in vitro translation that an H dsRNA segment encodes the virus capsid polypeptide (and the virion RNA-dependent RNA polymerase) and that the M2 dsRNA segment encodes the killer protein awaits further investigation.

A mycovirus satellite dsRNA, not associated with a known killer system, has recently been described.[45] Again, in the absence of infectivity data, evidence for satellitism was indirect. Two closely serologically related dsRNA viruses, designated 019/6-A and 38-4-A, obtained from different isolates of *Gaeumannomyces graminis*, were compared. 019/6-A virus contained two dsRNA components, 1 and 2, with molecular weights of 1.27×10^6 and 1.19×10^6. 38-4-A virus contained three dsRNA components, 1, 2, and 3, with molecular weight of 1.27×10^6, 1.19×10^6, and 1.09×10^6; each of these three RNAs was encapsidated separately, but in identical capsids composed of one polypeptide, 55,000 mol wt. T1 oligonucleotide fingerprint analysis and nucleic acid hybridization analysis revealed that 019/6-A virus RNA 1 and 38-4-A virus RNA 1 were very similar (>95% nucleotide sequence homology), whereas 019/6-A virus RNA 2 and 38-4-A virus RNA 2 were less closely related (approximately 50% nucleotide sequence homology). RNA 3 of 38-4-A virus had less than 10% nucleotide sequence homology with RNAs 1 and 2 of both viruses. Full length ssRNA transcripts of each of the three dsRNA components of 38-4-A virus were obtained by in vitro RNA synthesis and used to program a reticulocyte lysate system for in vitro protein synthesis. The three ssRNA transcripts each directed the synthesis of a single polypeptide with molecular weights of 63,000, 55,000, and 52,000. Peptide mapping showed that these three polypeptides were unrelated in sequence and that RNA 2 encoded the virus capsid polypeptide. Since 019/6-A virus and 38-4-A virus are related and 019/6-A virus has only two RNAs it is likely, although unproven, that RNA 1 of each virus encodes the virion-associated RNA polymerases.[92] RNA 3 is therefore most likely not required for replication of 38-4-A virus, it encodes a protein unrelated to those encoded by RNAs 1 and 2, and it utilizes the polypeptide encoded by RNA 2 for its encapsidation. RNA 3 may therefore be regarded as a satellite RNA as defined by Mossop and Francki.[74]

The problem here is that, unlike the yeast and *Ustilago* virus systems in which strains were available carrying dsRNA deletions in one virus, with the *Gaeumannomyces* virus system comparison is made with a different virus, albeit closely related which has fewer dsRNA components. The possibility remains, however unlikely, that 38-4-A virus requires an additional dsRNA component for its replication, which is not required for 019/6-A virus replication. Unequivocal proof that 38-4-A virus dsRNA 3 is a true satellite would require, either infectivity data, or the production of a deletion mutant of 38-4-A virus lacking RNA 3.

b. Defective dsRNAs and Virus Variants

Suppressive sensitive strains of *Saccharomyces cerevisiae*[96] lack M dsRNA, but contain a small (S) dsRNA of molecular weight $< 1.0 \times 10^6$, as well as L dsRNA.[158,159] When such strains are mated with killer strains, the resulting diploids are frequently nonkillers and all four meiotic products are sensitive and lack M dsRNA (but contain L and S dsRNAs). S dsRNAs are packaged in virus particles with capsids identical to those which encapsidate L and M dsRNAs. Since S dsRNA appears to replace M dsRNA whenever the two are present in the same cytoplasm, particles containing S dsRNA have been considered to be analogous to the defective-interfering (DI) particles of animal viruses.[54] In making this analogy, it was presumed that S and M dsRNAs are replicated by the same proteins and that they are therefore under the same genetic control and possess similar replicase recognition sites. There is now a considerable body of evidence to support this conclusion. There are two size classes of S dsRNA: S3, S14, and S733 of molecular weight approximately 0.5 $\times 10^6$ and S1, S4, S2, and S17 of molecular weight approximately 1×10^6. There is good evidence based on oligonucleotide fingerprinting studies[73] and electron microscopic heteroduplex analysis[93] that these RNAs are derived from M dsRNA. For example, S3 was shown to arise from M by internal deletion, while S1 was shown to be a tandem, direct duplication of S3. S4 was derived from S1 by a small internal deletion. Furthermore, sequence studies have shown that the 3' termini of S1, S3, S733, S14, and M are identical for at least the first 23 nucleotides at one end[94] and at least the first 34 nucleotides at the other,[95] implying similar polymerase recognition sites. Moreover, at least one of the nuclear genes which control the replication of M dsRNA, but not that of L dsRNA (see previous section), also controls the replication of S dsRNAs.[96]

It is noteworthy that S1, S3, and S733, numerous generations after the original isolation of the suppressive parents, have diverged somewhat from their original sequences, since they possess new T_1 oligonucleotides not present in M.[73,95] This generation of new sequence variants after prolonged replication has been compared[54] to the evolution of sequence variants in vesicular stomatitis virus persistent infections.[97]

Another possible case of the occurrence of defective or variant dsRNAs occurs in a virus isolated from *Gaeumannomyces graminis* and designated 45/9-A virus.[55] This virus has four dsRNA components, designated 1, 2, 3, and 4 with molecular weights of 1.30×10^6, 1.22×10^6, 1.14×10^6, and 1.11×10^6. T1 oligonucleotide fingerprints of each of these RNAs showed that dsRNAs 1, 3, and 4 were related, but that dsRNA 2 was distinct.[98] RNA 4 appeared to be more closely related to RNA 1 (about 40% homology) than RNA 3 was to RNA 1 (about 20% homology). It is likely that RNA 3 and RNA 4 are derived from RNA 1, though they are not simple deletion products, since their fingerprints have unique T1 oligonucleotides, as well as oligonucleotides present in the fingerprint of RNA 1. The simplest hypothesis to explain these results is that RNA 3 and RNA 4 are derived from RNA 1 by deletion, followed by subsequent mutation as discussed above for the yeast S dsRNAs. Whether RNA 3 or RNA 4 encode protein products is not known, so that it cannot be assessed whether they are defective RNAs or variant RNAs. There was no indication, however, of heterogeneity in the major capsid polypeptide of this virus, although the possibility of variant RNA polymerases cannot be eliminated.

There is good evidence for the occurrence of variant dsRNAs in *Ustilago maydis*. Although some strains have four dsRNA components in the H size class, designated H1, H2, H3, and H4 with molecular weights of 4.2×10^6, 3.1×10^6, 2.1×10^6, and 1.8×10^6 variants have been found with only H1, or with only H3 and H4.[87,88] All these variants were able to produce virus particles and it may be deduced that the genetic information required to produce capsid protein and support virus replication resides on at least two or three of these RNAs, H1, and H3 and/or H4. These H dsRNAs must therefore be variants of each other. Whether there is microheterogeneity in the capsids of *U. maydis* particles or whether deletions of H1 occur only in noncoding regions is not known.

Another example of variation occurs in *Gaeumannomyces graminis*. One virus isolate, 3bla-B, previously throught to be a single virus with four dsRNA components[55] has been shown to be a mixture of two serologically related virus variants, each with two dsRNAs.[58]

In other fungi additional dsRNAs of unknown status may be present. For example, *Penicillium brevicompactum* and *P. chrysogenum* viruses, which are serologically related, each have three dsRNA components each with the same molecular weight in each virus, i.e., 2.21×10^6, 2.08×10^6, and 1.98×10^6.[63] However, *P. cyaneo-fulvum* virus, which is serologically indistinguishable from *P. chrysogenum* virus, has four dsRNA components, three of the same molecular weight as those in *P. chrysogenum* virus and an additional dsRNA of molecular weight 1.93×10^6.[53] One variant of *P. chrysogenum* virus apparently has an additional dsRNA component of molecular weight approximately 0.2×10^6.[99,100] *Aspergillus foetidus* virus S and *A. niger* virus S are serologically indistinguishable, but *A. niger* virus S contains in addition to three dsRNA components of the same molecular weight as those in *A. foetidus* virus S, an extra dsRNA component.[101]

It is quite likely that satellite RNAs, defective RNAs and variant dsRNAs and viruses will occur commonly in the isometric dsRNA mycoviruses. The intracellular mode of transmission of mycoviruses may favor the survival and accumulation of such dsRNAs. The selection pressure which would tend to eliminate nonessential RNAs when cells are infected at low multiplicity is not exerted in intracellular transmission. This situation may be comparable to the multiple passaging of animal viruses at high multiplicity which favors accumulation of defective-interfering RNAs. The evolution of sequence variants of S dsRNA in yeast has been discussed above. It is noteworthy, however, that both dsRNA and polypeptide components of mycoviruses migrate as tight bands in gel electrophoresis and that in well-characterized variants of *Gaeumannomyces graminis* viruses, there appears to be conservation of particular molecular weight species of dsRNA, although sequence variation between different dsRNAs of the same molecular weight is found.[55,45]

The association of possible satellite, defective, or variant dsRNAs with mycoviruses poses considerable problems for their taxonomy. In the examples discussed above, evidence is available to enable such dsRNAs to be distinguished with reasonable certainty, even if not completely unequivocally. However, for other viruses the situation is much less clear. For example, *Aspergillus foetidus* virus F, has five dsRNA components[70,50] but how many of these components are essential for virus replication is unknown. There are several other viruses which are reasonably well characterized, but uncertainty about the numbers of "essential" dsRNA components creates difficulties regarding their taxonomy (see Tables 3 to 9).

c. Mixtures of Unrelated Viruses

Another probable consequence of the intracellular mode of transmission of mycoviruses is the occurrence of mixed infections of unrelated viruses in fungi. Although in some cases the occurrence of mixtures of viruses in purified mycovirus preparations could be the result of heterogeneity in the original fungal isolate, in a number of instances it has been shown by immunofluorescent studies[102] or by making single uninucleate conidial isolates[103,55] that two or more viruses can coexist in the same cell. Thus, two viruses occur together in some isolates of *Penicillium stoloniferum*,[47] *Aspergillus foetidus*,[70] *A. niger*,[101] and *Helminthosporium victoriae*.[104] *Gaeumannomyces graminis* isolates have been obtained which contain one, two, three, and four viruses, respectively,[105] and evidence was obtained for five serologically distinct capsids in virus particles isolated from *Thielaviopsis basicola*.[106]

Such mixed infections create problems for taxonomy with regard to numbers of dsRNA components (1) when the mixed nature of the virus preparation has not been recognized or (2) when it is not known whether each of the viruses is able to replicate independently. For example, after infection of virus-free *Penicillium stoloniferum* protoplasts with a crude mixture of *P. stoloniferum* viruses S and F, only PsV-S could be detected in fungal cultures regenerated from infected protoplasts.[107] Further, after ultraviolet irradiation of suspensions of *P. stoloniferum* conidia, isolates lacking virus F, but harboring virus S were obtained.[109] These experiments show that virus S can replicate in the absence of virus F but not vice versa. *P. stoloniferum* viruses S and F are very similar in their biophysical properties and both have two dsRNA components of similar molecular weight. On these grounds it would be expected that virus F would be able to replicate in the absence of virus S. This hypothesis has been confirmed by the isolation of single conidial cultures of *P. stoloniferum* which contain virus F, but in which no virus S was detected.[103]

Another example is found in isolates of *Helminthosporium victoriae*, some of which contain a 190S virus, and others of which contain 190S virus, together with a serologically unrelated 145S virus.[104] Some colonies, derived from isolates containing both viruses, contained only the 190S virus following the preparation and regeneration of protoplasts.[17] No colonies containing only 145S virus were obtained and it was suggested that the 145S virus, which was associated with disease symptoms in this fungus, may only replicate in the presence of the 190S virus. However, the genome of the 145S virus which contained four dsRNA components each of molecular weight $> 2 \times 10^6$ would seem more than large enough to support autonomous replication. Other cases pose similar problems. For example, it is not known whether *Aspergillus foetidus* viruses S and F can replicate independently of each other, although the sizes of their genomes suggest that they should be able to.[50]

For taxonomic purposes it seems reasonable to classify virus particles which have been separated from mixtures and which are serologically distinct, as distinct viruses, provided that they are generally similar in their properties to other viruses which are known to be able to replicate autonomously.

d. Minimum Number of dsRNA Segments Essential for Virus Replication

Mycoviruses with undivided genomes pose few taxonomic problems when virus isolates containing only one segment of dsRNA can be obtained. For each virus it is noteworthy that only 1/3 to 1/2 of the theoretical coding capacity of the dsRNA is required to encode the capsid polypeptide, assuming no overlapping genes.[9] For example, *Helminthosporium maydis* virus dsRNA has the capacity to encode a protein of 350,000 mol wt, whereas the capsid polypeptide has a molecular weight of 121,000.[51] Similarly, *Saccharomyces cerevisiae* virus 1 L dsRNA has a coding capacity for a protein of 189,000 mol wt, whereas the capsid polypeptide has a molecular weight of 88,000.[43] This suggests that these dsRNAs probably contain other genes, e.g., those required for RNA replication, i.e., they are probably at least dicistronic. The stringent packaging requirements of small polyhedral viruses should select against the accumulation of redundant sequences in a single RNA molecule that fills the capsid and in general long noncoding sequences have not been found in such viruses. So far, however, no virus gene products, other than capsid polypeptide, have been found for any of the isometric dsRNA viruses with undivided genomes.

There are a number of mycoviruses which have more than one dsRNA component but which resemble some of the mycoviruses with undivided genomes in many of their properties (see Table 4). These pose a taxonomic problem since it is possible that only the largest dsRNA (which resembles that of the undivided genome viruses in size) is required for virus replication. However, until this can be proved by infectivity experiments with isolated components or by isolation of virus variants with only one dsRNA component these viruses cannot be classified with the undivided genome mycoviruses.

Mycoviruses which are considered to have a genome divided into two RNA segments both essential for virus replication generate more taxonomic problems than those with undivided genomes. The problems involved with virus isolates that include possible satellite, defective RNAs, etc., are the same, but even when isolates containing two dsRNA components can be obtained it is still necessary to show that both components are necessary for replication. Such viruses must be distinguished from virus systems which consist of a helper virus with an undivided genome and an extra segment of satellite dsRNA, e.g., as in the yeast killer system. This could be achieved by infection with separate components or by in vitro translation of each component to give proteins both of which can be shown to be required for structure and replication of the virus particles. Neither of these two requirements have been achieved completely for any mycovirus, so once again indirect methods have to be employed.

First it is necessary to consider the minimum number of genes required for mycovirus replication. The genome of a nondefective virus, by definition, must encode its own capsid protein, so at least one gene is required for this purpose. Transcription and replication of dsRNA require at least one RNA-dependent RNA polymerase and such enzyme activities have been found in mycoviral or subviral particles.[42,44] Whether the enzyme is encoded by mycovirus dsRNA is not known for any mycovirus. It is well established that positive-strand RNA bacteriophages, such as Qβ and MS2, encode a polypeptide which forms part of the RNA replicase holoenzyme.[110] Similarly both positive- and negative-strand RNA viruses of animals encode protein(s) required for replication of the RNA genome.[111,112] Genetic experiments have implicated virus-encoded proteins in the replication of the dsRNA viruses reovirus[113] and bacteriophage φ 6.[114] In the case of positive-strand RNA plant viruses, host polymerases able to transcribe virus RNA (but not to replicate it) have been isolated,[115] however, genetic experiments with viruses with monopartite genomes[116,117] and infection experiments with individual components of viruses with multipartite genomes[118] have shown that virus gene products are essential for virus RNA replication. Even in the case of small ssDNA or dsDNA viruses which utilize host DNA polymerases to replicate their DNA at least one virus-encoded product is required also for virus DNA replication.[119] It seems very likely, therefore, that mycovirus dsRNA encodes at least one product which is required for dsRNA replication. It has already been mentioned above that all the known dsRNA mycoviruses with monopartite genomes have ample coding capacity for this function. The possibility exists that the major capsid polypeptide could fulfil this role, but this seems unlikely since the RNA polymerase must move along the dsRNA during transcription. Furthermore, there are no analogies for such a dual role in other viruses.

Perhaps the best evidence for the requirement for two dsRNA components comes from consideration of specific mycoviruses *Penicillium stoloniferum* virus S has two dsRNA components, molecular weights of 1.10×10^6 and 0.94×10^6.[47,48] It has a capsid composed of one major polypeptide, molecular weight 42,000 which approximates the coding capacity of the smaller of the two RNAs. A minor virion polypeptide has a molecular weight of 52,000 which approximates the coding capacity of the larger of the two RNAs. This latter polypeptide is thought to be the virion RNA-dependent RNA polymerase[120,121] because (1) it is present at a level of one molecule per particle and (2) it is absent from virus capsids lacking dsRNA.[49] In another virus of similar type, *Gaeumannomyces graminis* virus 019/6-A, which has two dsRNA components of molecular weight 1.27×10^6 and 1.19×10^6,[55] it has been shown by in vitro translation that the smaller dsRNA encodes the virus capsid polypeptide and the larger dsRNA encodes an unrelated polypeptide,[98] probably the virion transcriptase.[92] Since the size of each polypeptide approximated the coding capacity of its respective dsRNA, it is likely that each dsRNA is monocistronic. After irradiation of conidia of *P. stoloniferum* originally containing both virus S and virus F, single conidial cultures containing only virus S and single conidial cultures containing no virus were isolated. Analysis of the virus from cultures containing only virus S showed that both dsRNA components were retained in these cultures.[109] This is perhaps the best direct evidence for the requirement for two dsRNA components in *P. stoloniferum* virus S. Since the irradiation procedure resulted in complete deletion of virus F, and in some isolates of both viruses, it would have been expected that, if virus S required only one dsRNA component, some isolates containing only one dsRNA would have been obtained.

It would be useful to create a taxonomic group comprising isometric dsRNA mycoviruses with two dsRNA components and sharing other properties. Since it is unlikely that rigorous proof that both dsRNA components are required for virus replication will be obtained for all these viruses in the near future, less stringent, but nevertheless useful, taxonomic criteria must be applied. It is noteworthy that the molecular weights of the two dsRNA components of viruses of this type are similar. Their size is probably limited by the requirement to be packaged in particles of a given diameter (see Section III.B.3). It is therefore possible to define a group of isometric mycoviruses with two dsRNA component of similar size, the coding capacity of one which approximates the size of the major capsid polypeptide. This simple definition (together with some additional criteria, see Table 3) allows the classification of several viruses (see Tables 6 to 8) and clearly distinguishes them from viruses with undivided genomes which may have an additional satellite or defective dsRNA component.

The problems in considering viruses which may need three dsRNA components for replication are much more difficult. For example, *Penicillium chrysogenum* virus has three dsRNA components which migrate as closely spaced bands in polyacrylamide gel electrophoresis and have molecular weight approximately 2×10^6.[53,63] In vitro translation studies have shown that each dsRNA component encodes a different polypeptide and that the smallest dsRNA encodes the capsid polypeptide (J. E. Hopper, personal communication). The molecular weight of the capsid polypeptide approximates that of the smallest RNA which is therefore likely to be monocistronic. Arguments similar to those used for the two dsRNA component viruses suggest that at least one further dsRNA component is required to encode replication functions. The three dsRNA component have withstood a wide variety of mutational and strain selection procedures, mainly in connection with penicillin production[122,123] and there is no evidence for any mutant with less than three dsRNA components (other than virus-free strains). Furthermore, the serologically related viruses from *Penicillium brevi-compactum*[124] and *P. cyaneo-fulvum*[53] have three and four dsRNA components respectively. However, the argument that two dsRNA components are required for replication functions in addition to that required for capsid polypeptide is more difficult to sustain and further investigations are required to solve this problem.

2. Molecular Weights of Segments

A wide variety of methods has been used for making size estimates of mycovirus dsRNAs, including native gel electrophoresis in agarose or polyacrylamide, denaturing gel electrophoresis, sucrose density gradient centrifugation, sedimentation equilibrium, electron microscopy, and 5′ end determination. The advantages and disadvantages of each of these methods, and factors which lead to some uncertainty in the sizes of dsRNAs, have been reviewed.[54] The molecular weight of dsRNA segments is an important criterion in the taxonomy of mycoviruses and the major problem is the likelihood of significant errors in determinations.

The most commonly used method for determining molecular weight of mycovirus dsRNAs is polyacrylamide gel electrophoresis of native dsRNA using, directly or indirectly the ten dsRNA segments of reovirus as standards. The average sizes of the three L (large), three medium (M), and four S (small) dsRNA segments had previously been determined by electron microscopy and sizes for each individual segment interpolated graphically assuming a linear relationship between log molecular weight and electrophoretic mobility in polyacrylamide gels.[125] This linear relationship has been assumed in many determinations of mycovirus dsRNAs and frequently extrapolations outside the range of the reovirus markers (0.8×10^6 to 2.5×10^6) have made. However, it was subsequently found by Bozarth and Harley[126] that extrapolated values did not agree with results obtained with another set of dsRNA standards from bacteriophage ϕ 6 whose size had also been determined by electron microscopy.[129] This led to a reevaluation of relationships between gel mobility and dsRNA molecular weight.[126] Using a set of dsRNA standards from mycoviruses, bacteriophage ϕ 6 and reovirus, the molecular weights of which had all been determined by electron microscopy, the relationship between log molecular weight and mobility was found to be a smooth curve with all points falling very close to the curve. This relationship has been confirmed by Kaper and Diaz-Ruiz[128] and by Buck and Ratti[127] who also used electrophoresis of denatured dsRNA together with ssRNA markers of well-defined molecular weight. The extent to which linear extrapolation from reovirus dsRNA standards could lead to error is illustrated by some subsequent redeterminations. For example, two of the dsRNAs in *Aspergillus foetidus* virus S originally determined to be 2.76×10^6 and 0.1×10^6 [50,70] were revised to 4.1×10^6 and 0.26×10^6, respectively.[127] A small dsRNA in *Penicillium stoloniferum* virus F was revised from 0.23×10^6 [47] to 0.46×10^6 [126] and two dsRNAs, molecular weights originally determined to be 2.9×10^6 and 0.06×10^6,[14,83] have been revised to 4.2×10^6 and 0.25×10^6, respectively.[52] Differences of a smaller magnitude can also arise from the use of various values in electron microscopy for the internucleotide distance. For example, Bozarth et al.[52] have revised the value for the molecular weight of *Helminthosporium maydis* virus dsRNA determined by electron microscopy from 6.3×10^6 (based on 239 per μm)[51] to 5.7×10^6 (based on a revised value of 215 per μm).[128] It is noteworthy that the molecular weights of reovirus and bacteriophage ϕ 6 dsRNAs were also based on 239 per μm. It is clear that before values for dsRNA molecular weight can be used with confidence in the taxonomy of mycoviruses special attention must be paid to the methods by which the values have been obtained.

3. Mode of Encapsidation

A general property of mycoviruses with more than one dsRNA segment is that each segment is encapsidated separately in identical capsids giving rise to multicomponent virus systems.[131] This is one property which distinguishes isometric dsRNA mycoviruses from other isometric dsRNA virus families such as the *Reoviridae* and the *Cystoviridae*. It applies also to satellite and defective RNAs which are encapsidated separately from the genomic RNAs; indeed this mode of encapsidation may well have evolved to allow a helper virus to support the replication of additional satellite dsRNAs, e.g., in killer strains. The size of the virus particle has probably evolved to be able to encapsidate (separately) the genomic dsRNA components and nothing much larger. It is noteworthy that in cases where satellite dsRNA or other additional dsRNA is significantly smaller than the genomic dsRNA segments, more than one dsRNA molecule may become encapsidated. For example, in *Ustilago maydis* P1 virus, the genomic dsRNA (H1 segment) has a molecular weight of 4.2×10^6 and this is encapsidated separately. However, M dsRNA segments of molecular weight 0.6×10^6 to 0.7×10^6 were encapsidated in sets of one, two, or three molecules.[52] In *Aspergillus foetidus* virus S the two larger dsRNA components, molecular weight 2.6×10^6 and 4.1×10^6 are each encapsidated separately in S1a and S2a particles, respectively, but a small dsRNA, molecular weight 0.26×10^6, is encapsidated together with each of the larger dsRNAs, in S1b and S2b particles which were slightly more dense than S1a and S2a particles, respectively.[50]

Separate encapsidation of mycovirus genomic dsRNA components is clearly an important taxonomic criterion. A problem arises with viruses for which this property has not been determined. For these it may be possible to deduce the likelihood of separate encapsidation by comparison of particle sizes and dsRNA molecular weights with those of other viruses whose dsRNA segments are known to be separately encapsidated.

4. Nucleic Acid Hybridization, Fingerprinting, and Sequence Studies

Solution hybridization has been used to study relationships between viruses isolated from *Gaeumannomyces graminis*.[45] Recently a gel transfer hybridization procedure has been developed in which dsRNA components are denatured, separated by agarose gel electrophoresis, and then transferred to nitrocellulose; the nitrocellulose is then incubated with a labeled probe prepared from ssRNA transcripts of dsRNA from another virus and hybridization is detected by autoradiography.[58] Using this procedure hybridization was detected between dsRNA of two *G. graminis* viruses, which did not show any direct serological relationship. Taxonomic problems arising from results of hybridization analysis, as exemplified by a study of strains of tobacco mosaic virus,[130] are as follows (1) a negative result may not be significant in that the features of a virus which determine its overall properties and taxonomic group may be determined by only a small proportion of its nucleotide sequence. (2) Relationships between viral nucleic acids may not parallel the serological relationships found between the virus capsids because the antigenic determinants are encoded by only a small proportion of the genome. The latter problem has been encountered with some *G. graminis* viruses within one virus group. Additionally, hybridizations should be carried out under a range of conditions of stringency and care taken to avoid spurious hybridization which may arise under conditions of low stringency.

Oligonucleotide fingerprinting studies have been used to study relationships between dsRNAs of some *Gaeumannomyces graminis* viruses[45] and dsRNAs in *Saccharomyces cerevisiae*.[73,95] Since fingerprints are very sensitive to small changes in nucleic acid sequences, the taxonomic usefulness of the method is limited to groupings of mycoviruses with closely related dsRNAs; both of the problems outlined for results of nucleic acid hybridization studies apply also to fingerprinting studies.

3′ end nucleotide sequences have been determined for L and M dsRNAs associated with *Saccharomyces cerevisiae* viruses 1 and 2 from killer strains k1 and k2 respectively. Sequences for the two sets of dsRNAs were very similar, enabling consensus sequences for transcriptase and replicase initiation sites to be proposed and confirming the close relationship between these two virus systems.[80]

It is unlikely that nucleic acid hybridization, fingerprints, or sequences will be obtained for many dsRNA mycoviruses in the foreseeable future. However, when such data are available and indicate positive relationships, they provide useful supplementary information of taxonomic value.

C. Properties of the Capsid Proteins

1. Number of Components

The capsids of many dsRNA mycoviruses are composed of one major polypeptide species.[131] The problem arises, when additional polypeptide species are present, in assessing their significance. Additional polypeptides may arise from (1) impurities, (2) additional structural polypeptides, (3) RNA polymerase bound to dsRNA, (4) degradation products, or (5) aggregates. The possibility of impurities being present in a mycovirus preparation can generally be assessed by purifying the virus by a variety of procedures, although some mycoviruses are apparently difficult to purify[66,67] while others may become insoluble during purification.[55] Genuine virion components may be recognized by their constant ratio to the major capsid polypeptide under a variety of isolation conditions. Such minor virion polypeptides have been recognized in *Penicillium stoloniferum* viruses S and F[49] and in *Aspergillus foetidus* virus F[50] and are considered to be the virion RNA polymerase. Ideally this should be confirmed by in vitro translation studies. Degradation products may sometimes be recognized by isolation of the virus under different conditions. For example, *Aspergillus foetidus* virus S has two capsid polypeptides, σ1 and σ2. Some virus preparations contained mainly σ1, while others contained mainly σ2. σ2 was considered to be a degradation product of σ1.[50] This has recently been confirmed by comparison of partial proteolysis fingerprints of these two polypeptides (H. A. Sargent and K. W. Buck, unpublished results). The formation of aggregates, insensitive to reducing agents, chelating agents, urea, guanidinium chloride or alkali, was a particular problem for a virus, F6-B, isolated from *Gaeumannoyces graminis*.[55] In other cases where more than one capsid polypeptide has been reported, e.g., *Allomyces arbuscula* virus,[132] *Helminthosporium victoriae* 190S virus,[104] *Gaeumannomyces graminis* viruses 3bla-A and F6-A,[55] the status of the additional polypeptides is uncertain, creating problems for the taxonomy of these viruses.

2. Molecular Weights of Components

The molecular weights of mycovirus capsid polypeptides have been determined by SDS-polyacrylamide gel electrophoresis using a range of marker proteins of known molecular weight. Possible sources of error with this method have recently been outlined by Hamilton et al.[133] Specific problems with mycoviruses have arisen when the molecular weight of a standard has been redetermined by amino acid sequencing, e.g., the molecular weight of *Penicillium chrysogenum* virus capsid polypeptide was estimated as 130,000 using as standards *inter alia* β-galactosidase (assumed molecular weight 130,000) and phosphorylase a (assumed molecular weight 100,000).[53] Following sequence determinations, the molecular weights of β-galactosidase and phosphorylase a are now known to be 116,000 and 94,000. The value for *P. chrysogenum* virus capsid polypeptide is therefore too high and a redetermination is needed. In comparing molecular weight values for different viruses particular attention should be given to the values used for the standards. No thorough studies of the possibility that mycovirus proteins may bind SDS anomalously[134] or carry glycosyl, phosphate, or other modifications have been made. All values should therefore be treated with some caution.

D. Virion RNA Polymerases and Mode of RNA Replication

RNA polymerase activity has been found to be associated with virions of all dsRNA mycoviruses which have been examined for this property.[44] The most frequently described activity is transcription in which full length ssRNA copies of each dsRNA component are synthesized and released from the virus particles. The ssRNA molecules produced by transcription of *Saccharomyces cerevisiae* virus 1 L dsRNA has been shown by hybridization and in vitro translation experiments to be complementary to only one strand of the double-stranded template and to code for the 88,000 mol wt capsid protein.[135] The single-stranded RNA transcripts produced in vitro from the three dsRNA components of *Gaeumannomyces graminis* virus 38-4-A directed the synthesis of three different proteins in a reticulocyte lysate in vitro system, one of which was the virus capsid polypeptide,[45] i.e., in both these cases, it is the messenger RNA (mRNA) strand which is synthesized and released, as has been found also for transcription in vitro from core particles of reovirus[136] and from nucleocapsids of bacteriophage φ 6.[137] No evidence for capping and methylating activities have been found in mycovirions and it is noteworthy that the 5' ends of *S. cerevisiae* virus 1 L dsRNA transcripts, as well as both 5' ends of the L dsRNA template are ppp Gp.[135,138] This contrasts with reovirus in which the positive strand of virion dsRNA is capped and methylated[139] and ssRNA transcripts, synthesized in vitro by virus cores, are also capped and methylated by core enzymes.[140,141]

Transcription in S1a particles of *Aspergillus foetidus* virus has been shown to occur by a semiconservative displacement mechanism.[142,143] When an in vitro transcription reaction was carried out in the presence of radioactively labeled nucleoside triphosphate precursors, virion dsRNA became labeled in one strand and the first transcripts to be released, derived by displacement of one strand of dsRNA, were unlabeled. In second and subsequent rounds of transcription, the released transcripts were also labeled. The high level of labeling of genomic dsRNA during transcription in virions of several viruses from *Gaeumannomyces graminis*[92] and a *Phialophora* sp.[144] suggests that transcription in these viruses also occurs by a semiconservative displacement mechanism. Semiconservative transcription has also been shown to take place in bacteriophage φ 6.[137]

In contrast during transcription in *Saccharomyces cerevisiae* virus 1 L virions only 2% of the total radioactivity in newly synthesized RNA appeared in L dsRNA. This was interpreted to indicate that a maximum of 2% of the virus particles were engaged in semiconservative transcription.[135] It is possible that transcription was wholly conservative, as in reovirus,[136] and that the small amount of radioactivity incorporated into dsRNA was a result of the ss→dsRNA polymerase activity found in subviral particles isolated from log-phase yeast cells.[59] Some of these subviral particles could have persisted into the stationary-phase cells used for isolating particles for transcription studies. Alternatively, it is possible that only 2% of the yeast virus particles were active in transcription and that all of these were engaged in semiconservative transcription.

The distinction between semiconservative and conservative transcription and replication is a very fundamental one and would be extremely useful in the taxonomy of the isometric dsRNA mycoviruses, for which little data other than in vitro properties of the virus particles, are available. Since there are only two possibilities, i.e., semiconservative or conservative, and many mycovirus groups, this property may well be more useful in indicating fundamental differences between viruses, which may appear to be similar in other respects, than in suggesting relationships between viruses. The major problem is that while semiconservative transcription has been proved to occur in one virus and very likely in others, conservative transcription is more difficult to prove. The problem is compounded for mycoviruses by the occurrence of replicative intermediates in virus preparations, as discussed in Section III.A.2, so that synthesis of dsRNA as well as transcription may occur. Careful fractionation of virus preparations is necessary to overcome this problem.[121] It is clear that further investigations are required before mode of transcription of dsRNA can be of value in mycovirus taxonomy, and such information is only likely to become available for a small number of selected viruses.

The isolation from stationary phase yeast cells of virus particles with transcriptase activity and from log-phase yeast cells of subviral particles which contain dsRNA of less than unit length with single-stranded tails together with a ss→dsRNA polymerase activity able to complete dsRNA synthesis using the ssRNA tails as template, has led to the suggestion[146] that, like reovirus[147] and bacteriophage φ 6,[148] *Saccharomyces cerevisiae* virus 1 may replicate its dsRNA in two stages, i.e., ssRNA transcripts are synthesized and released from virions; they are then encapsidated and converted to dsRNA at a later time. Thus, the synthesis of the two strands of dsRNA occurs asynchronously. Another type of dsRNA replication has been found in *Penicillium stoloniferum* virus S. Since this apparent difference could be taxonomically important it warrants critical discussion. Virions of *P. stoloniferum* virus S catalyse in vitro the semiconservative replication of dsRNA and progeny dsRNA is not released so that particles containing two molecules of dsRNA within the capsid are produced.[121,145] However, whether this reaction occurs also in vivo is unknown. The first stage of the reaction is a semiconservative displacement transcription reaction; small number of particles containing a molecule of dsRNA and a full-length ssRNA molecule have been isolated from in vitro polymerase reactions and are also formed in vivo.[44] Initiation of a

Table 2
INFORMATION REQUIRED FOR THE TAXONOMY
OF ISOMETRIC dsRNA MYCOVIRUSES

Minimum Essential Criteria for placing viruses in families and
genera

Virion	Diameter measured by electron microscopy
	Sedimentation coefficients of particles containing dsRNA
dsRNA	Minimum number of segments required for virus replication[a]
	Molecular weights of segments
	Distribution of segments in different particles
Protein	Number of different polypeptide species
	Molecular weights of polypeptide species

Desirable additional criteria for placing viruses in families and
genera

Virion	Molecular weight
	Symmetry
	Number and arrangement of polypeptide molecules
dsRNA	Arrangement of genes
	Structure of 5′ and 3′ termini
	Conservative or semiconservative replication
Protein	Mode of synthesis of polypeptide species, eg., cleavage of a polyprotein, translation from genomic or subgenomic mRNAs, etc.

Additional criteria for delineating virus species

Virion	Serological relationships.
dsRNA	Extent of nucleotide sequence homology as measured by hybridization, fingerprinting, or direct sequence analysis
Protein	Extent of amino acid sequence homology as measured by peptide mapping or direct sequence analysis

[a]   Rigorous proof cannot be expected for every virus. For a detailed discussion of criteria which may be employed in evaluating this property see Section III.B.1; these criteria are summarized in Table 3 for individual families.

second round of transcription may require that the first transcript is released from the virion. Retention of the transcript, which may be an artifact of the in vitro system, would allow copying of the retained transcript to form dsRNA. The isolation from *P. stoloniferum* virus preparations of particles containing full length ssRNA transcripts (mRNAs), but no dsRNA,[48,44] suggests that this virus may replicate in vivo by a pathway similar to that of other dsRNA viruses. In this simple virus, it is likely that the same enzyme catalyses synthesis of ssRNA using a dsRNA template and synthesis of dsRNA using a ssRNA template. Thus, any ssRNA produced by transcription of dsRNA and not released from a particle could be copied to form another molecule of dsRNA. It is clear that investigations of *P. stoloniferum* virus S dsRNA replication in vivo are required to establish the significance, or otherwise, of the in vitro replication reaction.

Notwithstanding its possible significance in vivo it could be argued that the *Penicillium stoloniferum* virus S replication reaction in vitro is a useful feature to distinguish it taxonomically from other two component dsRNA viruses (see Tables 3, 6, 7, and 8). In this context it is interesting that *P. stoloniferum* virus F apparently catalyzes the same sort of replication reaction[120] (K. W. Buck, unpublished results). However, it would first need to be established that the difference was not due to differences in the stage of virus replication

Table 3
DISTINGUISHING FEATURES OF SUGGESTED FAMILIES AND GENERA OF ISOMETRIC dsRNA MYCOVIRUSES

FAMILY 1 (suggested name: Totoviridae).

Isometric particles, diameter 35 to 48 nm; one dsRNA component, molecular weight 3×10^6 to 6×10^6; the coding capacity of the dsRNA is two to three times as great as the molecular weight of the major capsid polypeptide species; sedimentation coefficients of particles containing dsRNA 160S to 283S; capsid composed of one major polypeptide species, molecular weight 70×10^3 to 120×10^3.

Genus 1 (suggested name: Saccerevirus).

Particle diameter 35 to 43 nm; dsRNA molecular weight 3.3×10^6 to 4.2×10^6; sedimentation coefficients of particles containing dsRNA 160S to 190S; capsid polypeptide molecular weight 70×10^3 to 90×10^3.

Genus 2 (suggested name: Helmayvirus).

Particle diameter 45 to 50 nm; dsRNA molecular weight 5.5×10^6 to 6.0×10^6; sedimentation coefficients of particles containing dsRNA 280S to 285S; capsid polypeptide molecular weight 115×10^3 to 125×10^3.

FAMILY 2 (suggested name: Bipartitoviridae).

Isometric particles, diameter 25 to 35 nm; two dsRNA components of similar molecular weight, each encapsidated in a separate particle; the coding capacity of one dsRNA component approximates the mol. wt. of the major capsid polypeptide species; molecular weight of dsRNA components in the range 0.9×10^6 to 1.6×10^6; sedimentation coefficients of particles containing dsRNA 100S to 145S; capsid composed of one major polypeptide species, molecular weight 42×10^3 to 73×10^3.

Genus 1 (suggested name: Penstolvirus)

Molecular weight of dsRNA components 0.9×10^6 to 1.1×10^6; sedimentation coefficients of particles containing dsRNA 100S to 105S; molecular weight of major capsid polypeptide 42×10^3 to 47×10^3.

Genus 2 (suggested name: Alphagaegramvirus).

Molecular weight of dsRNA components 1.14×10^6 to 1.30×10^6; sedimentation coefficients of particles containing dsRNA 109S to 128S; molecular weight of major capsid polypeptide 54×10^3 to 60×10^3.

Genus 3 (suggested name: Betagaegramvirus).

Molecular weight of dsRNA components 1.39×10^6 to 1.60×10^6; sedimentation coefficients of particles containing dsRNA 133S to 145S; molecular weight of major capsid polypeptide species 68×10^3 to 73×10^3.

FAMILY UNDESIGNATED
Genus (Suggested name: Penchrysvirus).

Three or four dsRNA components of similar molecular weight ca. 2×10^6, each encapsidated in a separate particle; the coding capacity of one dsRNA segment approximates the molecular weight of the major capsid polypeptide species; sedimentation coefficient of particles containing dsRNA 145S to 150S; capsid composed of one major polypeptide species, molecular weight 110×10^3 to 120×10^3.

Table 4

PROPERTIES OF VIRUSES ASSOCIATED WITH THE SUGGESTED SACCEREVIRUS GENUS OF FAMILY 1

	Dia. (nm)	$S_{20,w}^a$	dsRNA species (mol. wt $\times$ 10^{-6})	Major capsid polypeptide species (mol. wt. $\times$ 10^{-3})	Serological relationship to type species	Other characteristics	Key references
Type species *Saccharomyces cerevisiae* virus 1 (ScV-1)	35-40	161	3.3 (L1 dsRNA)	88	+	Virion RNA polymerase → full length ssRNA transcripts (mRNA for capsid protein). Killer strains of *S. cerevisiae* have additional satellite dsRNA species	54
Other member *Ustilago maydis* P1 virus (UmV-P1)	41-43	172	4.2 (H1 dsRNA)	73	−	Killer strains of *U. maydis* have additional satellite dsRNA species. The helper viruses in P4 and P6 killer strains containing H1 dsRNA are related.	52
Probable members *Mycogone pernicosia* virus (MpV)	42	ND	4.3	69	ND		155
Saccharomyces cerevisiae virus 2 (ScV-2)	35-40	ND	ca. 3 (L2 dsRNA)	ND	+	L1 and L2 dsRNAs are distinct but related. ScV-2 is a helper virus in k2 killer strains of *S. cerevisiae*	72,78,79,80
Possible members *Helminthosporium victoriae* (190S virus) (HvV-A)	35-40	190	3.0[b]	88,83[c]	ND		104
Thielaviopsis basicola d viruses (TbV)	40						106
Gaeumannomyces graminis virus F6-A (GgV-F6-A)	40	159	4.3[c],3.2	87,83,78[e]	ND	Serologically related to GgV 3bla-A	55
Aspergillus foetidus virus S (AfV-S)	40	176[f] 149[f]	4.1[e],2.6, 0.26	83,78[e]	−		50, 127
Aspergillus niger virus S (AnV-S)	40	ND	4.1[e],2.7,2.6,0.26	ND	ND	Serologically related to AfV-S	101, 127

Note: ND = Not determined.

a   Value for particles containing dsRNA in Svedbergs.

b   Almost certainly an underestimate, since the value was determined by linear extrapolation from reovirus dsRNA standards (see Section III.B.2).

c   The status of the second major capsid polypeptide species of HvV-A needs to be determined (degradation product?). Several minor capsid polypeptide species have been described for ScV-1.

d   A complex of five immunologically distinct viruses with five dsRNA components in the molecular weight range 4.5×10^6 to 2.7×10^6. It is likely, although unproven, that each virus contains one dsRNA molecule.

e   It is quite likely that only the largest dsRNA species is required for virus replication. There may also be only one major capsid polypeptide species, the smaller ones being derived from the largest by degradation (this has been proved for AfV-S).

f   Values for the particles containing the 4.1×10^6 and 2.6×10^6 mol wt species of dsRNA, respectively.

or physiological state of the fungus from which virus particles were isolated, or to a change in the virus after isolation, before a fundamental differences from other two component dsRNA mycoviruses (proposed Family 2, Table 3) could be made. Replication in vivo of all these viruses is probably semiconservative and asynchronous and differences between them may be artifacts of the systems in which they have been studied.

E. Minimum Essential Criteria Needed for Taxonomy

The major problems in the taxonomy of isometric dsRNA mycoviruses, discussed in Section III.A to D, may be divided into three categories: (1) paucity of data; (2) possible unreliability of some data; (3) doubts regarding the significance of some data. All three types of problems need further investigation to solve them. A list of the minimum information which should be available for every dsRNA mycovirus to enable it to be placed in a family or genus is given in Table 2. Emphasis has been placed on properties discussed in the present article and which are subject to the least doubt. The minimum evidence should be supplemented wherever possible by supplementary data listed in Table 2. Some other properties, with evaluation of their relative taxonomic importance by consensus opinion of the ICTV Fungal Virus Subcommittee, have been listed by Hollings.[12] Additional data required to delineate virus species are given in Table 2.

It could be argued, and indeed has, that viruses classified with such a small amount of information will need to be reclassified when further information, which could reveal unsuspected differences, is obtained. However, there are two main reasons for rejecting this argument; (1) comparisons with other families and genera of dsRNA viruses show that classification has been based on a similar number of properties, (2) taxonomy need not be regarded as a rigid classification system which, once laid down, must remain unchanged for all time. Rather it should be regarded as a system of grouping together viruses, which in the light of available information appear to share similar properties, and this system should be useful to both practicing virologists and teachers and students of virology at the present time. Ironically with the present rate of progress in fungal virus taxonomy a scheme accepted now is not likely to be modified for some considerable time. The alternative, unacceptable in the author's opinion, is to maintain indefinitely a jungle of unclassified mycoviruses.

F. Current Status of Taxonomy

Table 3 lists suggested families, genera, and groups of isometric dsRNA mycoviruses, which are under consideration by the ICTV Fungal Virus Subcommittee, together with a summary of their distinguishing features. Detailed properties of members, probable members, or possible members, are listed in Tables 4 to 9, together with comments on properties relevant to their taxonomy.

G. The Problem of Nomenclature

In view of the absence of distinguishing biological properties, mycoviruses have been named after the generic and species names of the host, together with numbers or letters, or combinations of both, to distinguish between viruses from the same host species. This system is well established and unlikely to change. Suggested generic names (Table 3) are also based on the name of the type virus species, with prefixes of alpha, beta, etc., when the type species of the two genera come from the same fungal species. The new rule 13 of the ICTV[20] that viruses should be named by the generic name followed by the species epithet becomes rather cumbersome, e.g., Penchrysvirus *Penicillium chrysogenum*, Saccerevirus *Ustilago maydis* or Penstolvirus *Penicillium stoloniferum* S. It is therefore suggested that abbreviations, which are well established in the literature for mycoviruses, e.g., PcV for *Penicillium chrysogenum* virus, ScV for *Saccharomyces cerevisiae* virus, PsV-S for *P. stoloniferum* virus S, etc., could be used for species names. In support of this, the naming of restriction

Table 5

PROPERTIES OF THE VIRUS IN THE SUGGESTED HELMAYVIRUS GENUS (MONOTYPIC) OF FAMILY 1

	Dia. (nm)	$S^a_{20,w}$	dsRNA species (mol wt $\times$ 10^{-6})	Major capsid polypeptide (mol wt $\times$ 10^{-3})	Key references
Type species	*Helminthosporium maydis* virus (HmV)				
	48	283	5.7	121	51, 52

[a] Value, in Svedbergs, for particles containing dsRNA.

Table 6

PROPERTIES OF VIRUSES ASSOCIATED WITH THE SUGGESTED PENSTOLVIRUS GENUS OF FAMILY 2

	Dia. (nm)	$S^b_{20,w}$	dsRNA species (mol wt × 10⁻⁶)	Major capsid polypeptide species (mol wt × 10⁻³)	Serological relationship to type species	Other characteristics	Key references
Type species *Penicillium stoloniferum* virus S (PsV-S)	25-30[a] 34[a]	101	1.11, 0.94	42	+	Virion RNA polymerase catalyses semiconservative replication of dsRNA in vitro. Minor virion polypeptide, mol wt 56,000, may be the RNA polymerase	5, 47, 49, 120, 145
Probable member *Diplocarpon rosae* virus	34	110[c]	ND	ND	+		156
Possible member *Penicillium stoloniferum* virus (PsV-F)	25-30[a] 34[a]	104	0.99, 0.89, 0.46	47	−	Major product of virion RNA polymerase is dsRNA. Minor virion polypeptide, mol wt 59,000, may be the RNA polymerase	5, 47, 49, 120

Note: ND = Not determined.

[a] The two values are from References 5 and 47.
[b] Values for particles containing dsRNA in Svedbergs.
[c] Fastest component of a multicomponent system.

Table 7
PROPERTIES OF VIRUSES ASSOCIATED WITH THE SUGGESTED ALPHAGAEGRAMVIRUS GENUS OF FAMILY 2

	Dia. (nm)	$S^a_{20,w}$	dsRNA species (mol wt $\times 10^{-6}$)	Major capsid polypeptide species (mol wt $\times 10^{-3}$)	Serological relationship to type species	Other characteristics	Key references
Type species *Gaeumannomyces graminis* virus 019/6-A	35	126	1.27, 1.19	60	+	Virion RNA polymerase→ full length ssRNA transcripts of each RNA (mRNAs), probably semiconservative; smaller dsRNA species encodes capsid polypeptide	45, 55, 92
Other members *G. graminis* viruses						01-1-4 and F6-C have been shown to have virion RNA polymerases similar to that of the type species	55, 92, 105
01-1-4-A	35	109	1.22, 1.14	55	+		
OgA-B	35	125	1.30, 1.22	55	+		
F6-C	35	128	1.27, 1.19	54	−	F6-C is serologically related to OgA-B	
45/101-D	35	113	1.14, 1.05	55	+		
Probable members *G. graminis* viruses	35					These viruses are included only as probable members because they have more than two dsRNA components. The third dsRNA in 38-4-A has the properties of a satellite. Nucleic acid hybridization has shown that 3bla-C and 45/9-A are related.	45, 55, 58
38-4-A	35	115	1.27, 1.19, 1.09	55	+		
3bla-C	35	115	1.27, 1.19, 1.11	55	+		
45/9-A		117	1.30, 1.22, 1.14, 1.11	55	ND		
Possible member *Phialophora* sp. virus 2-2-A	34-36	116	1.29, 1.22, 1.03	60	−	The virus has a virion RNA polymerase similar to that of the type species.	144, 157

Note: ND = Not determined.

[a] Value, in Svedbergs, for particles containing dsRNA.

Table 8

PROPERTIES OF VIRUSES ASSOCIATED WITH THE SUGGESTED BETAGAEGRAMVIRUS GENUS OF FAMILY 2

	Dia. (nm)	$S^a_{20,w}$	dsRNA species (mol wt × 10⁻⁶)	Major capsid polypeptide species (mol wt × 10⁻³)	Serological relationship to type species	Other characteristics	Key references
Type species *Gaeumannomyces graminis* virus T1-A	35	133	1.49,1.47	73	+		55
Other members *G. graminis* viruses OgA-A	35	135	1.49,1.39	68	+		55
3bla-B2[d]	35	140[b]	1.45,1.43	73[c]	−	3bla-B2 is serologically related to 3bla-Bl, OgA-A and F6-B	55, 58, 92
3bla-B1[d]	35	140[b]	1.60,1.54	73[c]	−		
Agaricus bisporus virus 4 (AbV-4 or MV-4)	35	140-145	1.5,1.4	69[e]	ND		26
Probable members *G. graminis* viruses F6-B	35	133	1.60,1.56,1.45	73	+	Included only as probable member because it has more than two dsRNAs; status of additional dsRNA unknown.	

Note: ND = Not determined.

[a] Value, in Svedbergs, for particles containing dsRNA.

[b] Determined with a mixture of 3b 1a-B1 and 3bla-B2.

[c] The value was incorrectly stated to be 68,000 in Reference 55.

[d] 3bla-B1 and 3bla-B2 were described in References 55 and 92 as a single virus, 3bla-B. This has recently been shown to be a mixture of closely related strains.[3]

[e] The capsid mol wt was reported to be 64,000 in Reference 26. However reference to this paper shows that the major capsid polypeptide species of AbV-4 (MV4) migrated just behind bovine serum albumin. In comparison to the values determined for the *G. graminis* polypeptides a value of 69,000 could be assigned to MV-4 polypeptide.

Table 9
PROPERTIES OF VIRUSES ASSOCIATED WITH THE SUGGESTED PENCHRYSVIRUS GENUS (FAMILY UNASSIGNED)

	Dia. (nm)	$S^a_{20}w$	dsRNA species (mol wt × 10^{-6})	Major capsid polypeptide species (mol wt × 10^{-3})	Serological relationship to type species	Other characteristics	Key references
Type species *Penicillium chrysogenum* virus (PcV)	35c 40e	145- 150	2.21,2.08,1.98b (2.18,199,1.89)	110-120c	+		53, 63
Other members *Penicillium brevi-compactum* virus (PbV)	40	147	2.18,1.99,1.89	ND	+		63, 124
Penicillium cyaneo-fulvum virus (Pc-fV)	35	145	2.21,2.08,1.98,1.93	110-120c	+		53
Possible member *Helminthosporium victoriae* 145S virus (HvV-B)d	35-40	145	2.4,2.2,2.1,2.0	106,97,92	ND	Multiple centrifugal components suggest that each RNA species is encapsidated in separate particles	104

Note: ND = Not determined.

a Value, in Svedbergs, for particles containing dsRNA.

b The three dsRN components of PcV, PbV, and Pc-fV (three largest) all have identical molecular weights. The two sets of values for PcV dsRNAs are those given in References 2 and 1, respectively.

c The value of 130,000 for the molecular weight of the major capsid polypeptide species of PcV and Pc-fV reported in Reference 53 is likely to be an overestimate since the assumed molecular weights of two of the standards, β-galactosidase (130,000) and phosphorylase A (100,000) used have since been redetermined by sequencing. The correct values are 116,000 and 94,000, respectively. The true molecular weight for PcV and Pc-fV capsid polypeptide species is likely to be between 110,000 and 120,000 but a redetermination is needed.

d HmV-B is included as a possible member on the basis of its particle size and S value, the molecular weights of its dsRNA components (probably encapsidated separately as with other members) and the molecular weight of its largest capsid polypeptide species. The status of the two smaller capsid polypeptide species is unknown, but they may well be degradation products (degradation products of PcV capsid polypeptide can be found using some protocols for virus isolation, K. W. Buck and R. F. Girvan, unpublished results).

e Values reported in References 53 and 63, respectively.

endonucleases in a similar way, e.g., Hae III, Sal I, etc., have proved to be very satisfactory. This would lead to more acceptable names, e.g., Penchrysvirus PcV, Saccerevirus ScV, Penstolvirus PsV-S, etc.

The names suggested for the two virus families are based on the number of dsRNA components of the type species: Totoviridae (from Latin, totus = undivided i.e., one component): Bipartitoviridae (from Latin, *bipartitus* = divided into two, i.e., two components). These names are not in use for any other virus family, but they do not serve to distinguish the families from all others, e.g., there are many viruses with undivided genomes. However, the same criticism can be levelled at accepted names e.g., *Togaviridae, Picornaviridae, Parvoviridae*, etc., and it would be difficult to find a simple name which would solve this problem. Names for mycovirus families will be urgently needed, because, for example, the name isometric dsRNA mycoviruses with undivided genomes is rather cumbersome, especially when it needs to be repeated several times in a paragraph (as the author has found in writing this article!).

It should be emphasized that the suggested families, genera, and groups of mycoviruses or their names have not yet been considered by the ICTV and they have no official status.

H. Cryptic dsRNA Viruses of Plants

Recently a number of small, isometric virus-like particles have been isolated from a number of apparently healthy plants. They have been named cryptic viruses and share the following properties: (1) presence in low concentrations in symptomless plants; (2) not mechanically, aphid or graft transmissible; (3) seed transmissible with high efficiency; (4) possess isometric particles about 29 nm in diameter. Examples are beet cryptic virus,[149] *Poinsettia* cryptic virus,[150] *Vicia* cryptic virus,[151] ryegrass spherical (or cryptic) virus,[152] and carnation cryptic virus.[153] Recently, carnation cryptic virus has been found to have four dsRNA segments of molecular weight 1.04×10^6, 0.88×10^6, and 0.84×10^6 and preliminary evidence indicated that ryegrass cryptic virus also has a genome of dsRNA.[154] These viruses bear some resemblance to the suggested Penstolvirus genus of dsRNA mycoviruses. Further data, such as proof of separate encapsidation, molecular weight of capsid polypeptide and significance of the four dsRNA components would be required to confirm the resemblance. However, the intracellular mode of transmission of these viruses suggests that they may suffer the same problem with regard to numbers of dsRNA components as do the dsRNA mycoviruses. The possibility that the cryptic dsRNA plant viruses may need to be grouped with the isometric dsRNA mycoviruses emphasizes the need to establish families and genera for these viruses with a nomenclature that avoids the term isometric dsRNA mycovirus.

IV. SUMMARY AND FUTURE OUTLOOK

The major problem in mycovirus taxonomy at the present time is undoubtedly shortage of information. This is particularly so for all the VLP's listed in Table 1 for many of which it is not even known if they contain nucleic acid. However, their taxonomy should not present any special difficulties once further information is available and some of them may well fall into existing groups of plant or animal viruses.

A number of isometric dsRNA mycoviruses have been sufficiently well characterized to allow useful progress with their taxonomy to be made. Nevertheless for most of these viruses more data are required to increase confidence in the taxonomic groupings which have been suggested. In particular more information is needed on noncapsid viral proteins, and on the organization and expression of the viral genome. The distinction between conservative and semiconservative replication is so fundamental that further work on the mechanism of dsRNA replication is also desirable. The available evidence suggests that viruses in suggested families

1 and 2 may replicate their genomes conservatively and semiconservatively, respectively. Confirmation of this possibility would be very valuable.

The intracellular mode of transmission of mycoviruses leads to the accumulation of mixtures of viruses, virus variants, defective RNAs, and satellite RNAs within a fungal mycelium. A serious problem which continues to impede the analysis of such systems is the absence of a reliable method for infection of fungi with purified virus particles. As a result indirect methods have to be employed to determine the minimum number of dsRNA segments necessary for infection. For some otherwise well characterized dsRNA mycoviruses this basic information is not available. With the advent in recent years of highly efficient methods for inducing uptake of nucleic acids and viruses into cultured animal cells[160] and plant protoplasts[161] it is likely that a fungal protoplast system could be developed which would overcome this problem.

Current work with mycoviruses is concerned with detailed investigations of rather few systems, e.g., the *Saccharomyces cerevisiae* and *Ustilago maydis* killer systems, and viruses of some fungi pathogenic for plants. Such studies, which include nucleotide sequence analysis of viral dsRNA, may enable the problem of the species concept to be tackled for a few viruses.

Future progress in mycoviruses taxonomy on a broader front will depend on more investigators being attracted to this field. This is only likely to happen if new biological properties of economic or medical importance, associated with mycoviruses, are discovered. The apparent similarity of the recently described plant cryptic dsRNA viruses to some dsRNA mycoviruses raises the possibility that some of these viruses may be able to replicate in both plants and fungi. If fungi pathogenic for plants could acquire viruses from their host plant with an effect on pathogenicity, this would obviously be of interest to plant pathologists. Hence a new area of mycovirus research could be created.

ACKNOWLEDGMENTS

I thank Drs. R. F. Bozarth, J. A., Bruenn, Y. Koltin, and H. A. Wood for useful discussions regarding mycovirus taxonomy and Drs. V. Lisa and R. G. Milne for interesting discussions concerning plant cryptic viruses.

REFERENCES

1. **Campbell, R. N.,** Fungal vectors of plant viruses, in *Fungal Viruses,* Molitoris, H. P., Hollings, M., and Wood, H. A., Eds., Springer-Verlag, Berlin, 1979, 8.
2. **Hollings, M.,** Viruses associated with a die-back disease of cultivated mushroom, *Nature (London),* 196, 962, 1962.
3. **Lampson, G. P., Tytell, A. A., Field, A. K., Nemes, M. M., and Hilleman, M. R.,** Inducers of interferon and host resistance. I. Double-stranded RNA from extracts of *Penicillium funiculosum, Proc. Nat. Acad. Sci. USA,* 58, 782, 1967.
4. **Ellis, L. F. and Kleinschmidt, W. J.,** Virus-like particles of a fraction of statolon, a mould product, *Nature (London),* 215, 649, 1967.
5. **Banks, G. T., Buck, K. W., Chain, E. B., Himmelweit, F., Marks, J. E., Tyler, J. M., Hollings, M., Last, F. T., and Stone, O. M.,** Viruses in fungi and interferon stimulation, *Nature (London),* 218, 542, 1968.
6. **Hollings, M.,** Mycoviruses-viruses that infect fungi, *Adv. Virus Res.,* 22, 3, 1978.
7. **Lemke, P. A., Ed.,** *Viruses and Plasmids in Fungi,* Marcel Dekker, New York, 1979.
8. **Saksena, K. N. and Lemke, P. A.,** Viruses in fungi, in *Comprehensive Virology,* Fraenkel-Conrat, H. and Wagner, R. R., Eds., Plenum Press, New York, 1978.

9. **Buck, K. W.,** Viruses and killer factors of fungi, in *The Eukaryotic Microbial Cell*, Gooday, G. W., Lloyd, D., and Trinci, A. P. J., Eds., Society for General Microbiology Symposium, Cambridge University Press, 1980.

10. **Ghabrial, S. A.,** Effects of fungal viruses on their hosts, *Annu. Rev. Phytopathol.*, 18, 441, 1980.

11. **Matthews, R. E. F.,** *Third Report of the International Committee on Taxonomy of Viruses*, S. Karger, Basel, 1979.

12. **Hollings, M.,** Taxonomy of fungal viruses, in *Fungal Viruses*, Molitoris, H. P., Hollings, M., and Wood, H. A., Eds., Springer-Verlag, Berlin, 1979.

13. **Bostian, K. A., Hopper, J. E., Rogers, D. T., and Tipper, D. J.,** Translational analysis of the killer-associated virus-like particle dsRNA genome of *Saccharomyces cerevisiae:* M dsRNA encodes toxin, *Cell*, 19, 403, 1980.

14. **Koltin, Y. and Day, P. R.,** Inheritance of killer phenotypes and double-stranded RNA in *Ustilago maydis*, *Proc. Natl. Acad. Sci. USA*, 73, 594, 1976.

15. **Day, P. R., Dodds, J. A., Elliston, J. E., Jaynes, R. A., and Anagnostakis, S. L.,** Double-stranded RNA in *Endothia parasitica, Phytopathology*, 67, 1393, 1977.

16. **Castanho, B., Butler, E. E., and Shephered, R. J.,** The association of double-stranded RNA with *Rhizoctonia* decline, *Phytopathology*, 68, 1515, 1978.

17. **Ghabrial, S. A., Sanderlin, R. S., and Calvert, L. A.,** Morphology and virus-like particle content of *Helminthosporium victoriae* colonies regenerated from protoplasts of normal and diseased isolates, *Phytopathology*, 69, 312, 1979.

18. **Passmore, E. L. and Frost, R. R.,** The detection and occurrence of virus-like particles in extracts of mushroom sporophores, *Phytopathol. Z.*, 95, 346, 1979.

19. **Le Coq, H., Boissonnet-Menes, M., and Delhotal, P.,** Infectivity and transmission of fungal viruses, in *Fungal Viruses*, Molitoris, H. P., Hollings, M., and Wood, H. A., Eds., Springer-Verlag, Berlin, 1979, 34.

20. **Matthews, R. E. F.,** *Fourth Report of the International Committee on Taxonomy of Viruses*, in press, 1982.

21. **Ushiyama, R.,** Fungal viruses in edible fungi, in *Fungal Viruses*, Molitoris, H. P., Hollings, M., and Wood, H. A., Eds., Springer-Verlag, Berlin, 1979, 25.

22. **Dieleman-Van Zaayen, A., Igesz, O., and Finch, J. T.,** Intracellular appearance and some morphological features of virus-like particles in an ascomycete fungus, *Virology*, 42, 534, 1970.

23. **Lapierre, H., Faivre-Amiot, A., Kusiak, C., and Molin, G.,** Particle de type viral associées au *Mycogone perniciosa* Magnus, agent d'une môles du champignon de couche, *C. R. Acad. Sci. (Paris) Series D*, 274, 1867, 1972.

24. **Van Zaayen, A.,** Mushroom viruses, in *Viruses and Plasmids in Fungi*, Lemke, P. A., Ed., Marcel Dekker, New York, 1979, 25.

25. **Yamashita, S., Doi, Y., and Yora, K.,** Electron microscopic study of several fungal viruses, in *Proc. First Intersect. Cong. Internat. Assoc. Microbiol. Soc.*, Hasegawa, T., Ed., Science Council of Japan, Tokyo, 1975.

26. **Barton, R. J. and Hollings, M.,** Purification and some properties of two viruses infecting the cultivated mushroom *Agaricus bisporus*, *J. Gen. Virol.*, 42, 231, 1979.

27. **Dodds, A. J.,** Association of type 1 viral-like dsRNA with club-shaped particles in hypovirulent strains of *Endothia parasitica, Virology*, 107, 1, 1980.

28. **Lesemann, D. E. and Koenig, R.,** Association of club-shaped virus-like particles with a severe disease of *Agaricus bisporus*, *Phytopathol. Z.*, 89, 161, 1977.

29. **Kuntzel, H., Barath, Z., Ali, I., Kind, J., and Althaus, N. H.,** Virus-like particles in an extranuclear mutant of *Neurospora crassa*, *Proc. Natl. Acad. Sci. USA*, 70, 1574, 1973.

30. **Kazama, F. Y. and Shornstein, K. L.,** Ultrastructure of a fungus herpes-type virus, *Virology*, 52, 478, 1973.

31. **Tikchonenko, T. I.,** Viruses of fungi capable of replication in bacteria (PB viruses), in *Comprehensive Virology*, Fraenkel-Conrat, H. and Wagner, R. R., Eds., Plenum Press, New York, 1978.

32. **Tuveson, R. W. and Peterson, J. F.,** Virus-like particles in certain slow-growing strains of *Neurospora crassa, Virology*, 47, 527, 1972.

33. **Tuveson, R. W., Sargent, M. L., and Bozarth, R. F.,** Purification of a small virus-like particle from strains of *Neurospora crassa, Abstr. Annu. Meet. Am. Soc. Microbiol.*, 216, 1975.

34. **Schnepf, E., Soeder, C. J., and Hegewald, E.,** Polyhedral virus-like particles lysing the aquatic phycomycete *Aphelidium* sp., a parasite of the green alga *Scenedesmus alatus, Virology*, 42, 482, 1970.

35. **Koltin, Y., Berick, R., Stamberg, J., and Ben-Shaul, Y.,** Virus-like particles and cytoplasmic inheritance of plaques in a higher fungus, *Nature New Biol. (London)*, 108, 1973.

36. **Shahriari, H., Kirkham, J. B., and Casselton, L. A.,** Virus-like particles in the fungus *Coprinus lagopus*, *Heredity*, 31, 428, 1973.

37. **Bozarth, R. F.**, The physicochemical properties of mycoviruses, in *Viruses and Plasmids in Fungi*, Lemke, P. A., Ed., Marcel Dekker, New York, 1979, 43.
38. **Wickner, R. B.**, The killer double-stranded RNA plasmids of yeast, *Plasmid*, 2, 303, 1979.
39. **Buck, K. W.**, Biochemical and biological implications of double-stranded RNA mycovirus, in *Biologically Active Substances - Exploration and Exploitation*, Hems, D. A., Ed., Wiley, Chichester, 1977, 121.
40. **Gunge, N. and Sakaguchi, K.**, Intergeneric transfer of deoxyribonucleic acid killer plasmids, pGK11 and pGK12 from *Kluyveromyces lactis* into *Saccharomyces cerevisiae* by cell fusion, *J. Bact.*, 147, 155, 1981.
41. **Hardy, K. G.**, Colicinogeny and related phenomena, *Bacteriol. Rev.*, 39, 464, 1975.
42. **Buck, K. W.**, Replication of double-stranded RNA mycoviruses, in *Viruses and Plasmids in Fungi*, Lemke, P. A., Ed., Marcel Dekker, New York, 1979, 93.
43. **Hopper, J. E., Bostian, K. A., Rowe, L. B., and Tipper, D. J.**, Translation of the L-species dsRNA genome of the killer-associated virus-like particle of *Saccharomyces cerevisiae*, *J. Biol. Chem.*, 252, 9010, 1977.
44. **Buck, K. W.**, Virion-associated RNA polymerases of double-stranded RNA mycoviruses, in *Fungal Viruses*, Molitoris, H. P., Hollings, M., and Wood, H. A., Eds., Springer-Verlag, Berlin, 1979, 62.
45. **Romanos, M. A., Buck, K. W., and Rawlinson, C. J.**, A satellite double-stranded RNA in a virus from *Gaeumannomyces graminis*, *J. Gen. Virol.*, 57, 375, 1981.
46. **Lemke, P. A.**, Coevolution of fungi and their viruses, in *Fungal Viruses*, Molitoris, H. P., Hollings, M., and Wood, H. A., Eds., Springer-Verlag, Berlin, 1979, 2.
47. **Bozarth, R. F., Wood, H. A., and Mandelbrot, A.**, The *Penicillium stoloniferum* virus complex: two similar double-stranded RNA virus-like particles in a single cell, *Virology*, 45, 516, 1971.
48. **Buck, K. W. and Kempson-Jones, G. F.**, Biophysical properties of *Penicillium stoloniferum* virus S, *J. Gen. Virol.*, 18, 223, 1973.
49. **Buck, K. W. and Kempson-Jones, G. F.**, Capsid polypeptides of two viruses isolated from *Penicillium stoloniferum*, *J. Gen. Virol.*, 22, 441, 1974.
50. **Buck, K. W. and Ratti, G.**, Biophysical and biochemical properties of two viruses isolated from *Aspergillus foetidus*, *J. Gen. Virol.*, 27, 211, 1975.
51. **Bozarth, R. F.**, Biophysical and biochemical characterization of virus-like particles containing a high molecular weight dsRNA from *Helminthosporium maydis*, *Virology*, 80, 149, 1977.
52. **Bozarth, R. F., Koltin, Y., Weissman, M. B., Parker, R. L., Dalton, R. E., and Steinlauf, R.**, The molecular weight and packaging of dsRNAs in the mycovirus from *Ustilago maydis* killer strains, *Virology*, 113, 492, 1981.
53. **Buck, K. W. and Girvan, R. F.**, Comparison of the biophysical and biochemical properties of *Penicillium cyaneo-fulvum* and *Penicillium chrysogenum* virus, *J. Gen. Virol.*, 34, 145, 1977.
54. **Bruenn, J. A.**, Virus-like particles of yeast, *Annu. Rev. Microbiol.*, 34, 49, 1980.
55. **Buck, K. W., Almond, M. R., McFadden, J. J. P., Romanos, M. A., and Rawlinson, C. J.**, Properties of thirteen viruses and virus variants obtained from eight isolates of the wheat take-all fungus, *Gaeumannomyces graminis* var. *tritici*, *J. Gen Virol.*, 57, 157, 1981.
56. **Hastie, N. D., Brennan, V., and Bruenn, J. A.**, No homology between double-stranded RNA and nuclear DNA of yeast, *J. Virol.*, 28, 1002, 1978.
57. **Vodkin, M.**, Homology between double-stranded RNA and nuclear DNA of yeast, *J. Virol.*, 21, 516, 1977.
58. **McFadden, J. J. P., Buck, K. W., and Rawlinson, C. J.**, *J. Gen. Virol.*, in press, 1983.
59. **Bevan, E. A. and Herring, A. J.**, The killer character in yeast; preliminary studies of virus-like particle replication in *Genetics, Biogenesis and Bioenergetics of Mitochondria*, Bandelow, W., Schwegen, R. J., Thomas, D. Y., and Kaudewitz, F., Eds., Walter de Gruyter, Berlin, 1976, 153.
60. **Wickner, R. B. and Leibowitz, M. J.**, Dominant chromosomal mutation by-passing chromosomal genes needed for killer RNA plasmid replication in yeast, *Genetics*, 87, 453, 1977.
61. **Warburg, O. and Christian, W.**, Isolierung und kristallisation des garungsferment enolase, *Biochem. Z.*, 310, 384, 1943.
62. **Nash, C. H., Douthart, R. J., Ellis, L. F., Van Frank, R. M., Burnett, J. P., and Lemke, P. A.**, On the mycophage of *Penicillium chrysogenum*, *Canad. J. Microbiol.*, 19, 97, 1973.
63. **Wood, H. A. and Bozarth, R. F.**, Properties of virus-like particles of *Penicillium chrysogenum*: one double-stranded RNA molecule per particle, *Virology*, 47, 604, 1972.
64. **Reichman, M. E.**, Determination of ribonucleic acid content of spherical viruses from sedimentation coefficients of full and empty particles, *Virology*, 24, 166, 1965.
65. **Lister, R. M.**, Serological screening for fungal viruses, in *Fungal Viruses*, Molitoris, H. P., Hollings, M., and Wood, H. A., Eds., Springer-Verlag, Berlin, 1979, 150.
66. **Moyer, J. W. and Smith, S. H.**, Partial purification and antiserum production to the 19 × 50 nm mushroom virus particle, *Phytopathology*, 66, 1260, 1976.
67. **Moyer, J. W. and Smith, S. H.**, Purification and serological detection of mushroom virus-like particles, *Phytopathology*, 67, 1207, 1977.

68. **Ikegami, M. and Francki, R. I. B.,** Presence of antibodies to double-stranded RNA in sera of rabbits immunized with rice dwarf and maize rough dwarf viruses, *Virology*, 56, 404, 1973.

69. **Andler, J. P. and Del Vecchio, V. G.,** Specialized assays for detection of fungal viruses and double-stranded RNA, in *Viruses and Plasmids in Fungi*, Lemke, P. A., Ed., Marcel Dekker, New York, 1979, 326.

70. **Ratti, G. and Buck, K. W.,** Virus particles in *Aspergillus foetidus:* a multi-component system, *J. Gen. Virol.*, 14, 165, 1972.

71. **Zaitlin, M.,** The RNAs of monopartite plant viruses, in *Nucleic Acids in Plants*, Hall, T. C. and Davies, J. W., Eds., CRC Press, Boca Raton, Florida, 1979, 32.

72. **Bevan, E. A. and Mitchell, D. J.,** The killer system in yeast, in *Viruses and Plasmids in Fungi*, Lemke, P. A., Ed., Marcel Dekker, New York, 1979, 161.

73. **Bruenn, J. and Kane, W.,** Relatedness of the double-stranded RNAs present in yeast virus-like particles, *J. Virol.*, 26, 762, 1978.

74. **Mossop, D. W. and Francki, R. I. B.,** Survival of a satellite RNA *in vivo* and its dependence on cucumber mosaic virus for its replication, *Virology*, 86, 562, 1978.

75. **Wickner, R. B.,** Twenty-six chromosomal genes needed to maintain the killer double-stranded RNA plasmid of *Saccharomyces cerevisiae*, *Genetics*, 88, 419, 1978.

76. **Welsh, J. D. and Leibowitz, M. J.,** Transcription of killer virion double-stranded RNA *in vitro*, *Nucl. Acids Res.*, 8, 2365, 1980.

77. **Welsh, J. D., Leibowitz, M. J., and Wickner, R. B.,** Virion DNA-independent RNA polymerase from *Saccharomyces cerevisiae*, *Nucl. Acids Res.*, 8, 2349, 1980.

78. **Young, T. W. and Yagiu, M.,** A comparison of the killer character in different yeasts and its classification, *Antonie van Leeuwenhoek J. Microbiol. Serol.*, 44, 59, 1978.

79. **Wickner, R. B.,** Plasmids controlling exclusion of the k2 killer double-stranded RNA plasmid of yeast, *Cell*, 21, 217, 1980.

80. **Brennan, V. E., Field, L., Cizdziel, P., and Bruenn, J. A.,** Sequences at the 3′ ends of yeast viral dsRNAs: proposed transcriptase and replicase initiation sites, *Nucl. Acids Res.*, 9, 4007, 1981.

81. **Day, P. R.,** Fungal virus populations in corn smut from Connecticut, *Mycologia*, 61, 379, 1981.

82. **Koltin, Y. and Day, P. R.,** Specificity of *Ustilago maydis* killer protein, *Appl. Microbiol.*, 31, 694, 1975.

83. **Wood, H. A. and Bozarth, R. F.,** Heterokaryon transfer of virus-like particles associated with a cytoplasmically inherited determinant in *Ustilago maydis*, *Phytopathology*, 63, 1019, 1973.

84. **Lentz, E.,** Physical, Molecular and Serological Characterisation of the Virus-Like Particles of the Corn Smut Fungus *Ustilago maydis*, M. A. thesis, Indiana State University, Terre Haute, 1977.

85. **Bozarth, R. F. and Lentz, E. T.,** Physicochemical properties and serology of *Ustilago maydis* viruses P1 and P4, *Abstr. Fourth Internat. Cong. Virol.*, *Centre for Agricultural Publishing and Documentation*, Wageningen, 1978, 172.

86. **Koltin, Y. and Kandel, J. S.,** Killer phenomenon in *Ustilago maydis:* The organisation of the viral genome, *Genetics*, 88, 267, 1978.

87. **Koltin, Y., Levine, R., and Peery, T.,** Assignment of functions to segments of the dsRNA genome of the *Ustilago* virus, *Mol. Gen. Genet.*, 178, 173, 1980.

88. **Koltin, Y., Mayer, I., and Steinlauf, R.,** Killer phenomenon in *Ustilago maydis:* Mapping viral functions, *Mol. Gen. Genet.*, 166, 181, 1978.

89. **Koltin, Y.,** Virus-like particles in *Ustilago maydis:* mutants with partial genomes, *Genetics*, 86, 527, 1977.

90. **Peery, T., Koltin, Y., and Tamarkin, A.,** Mapping the immunity function of the *Ustilago maydis* P1 virus, *Plasmid*, 7, 52, 1982.

91. **Day, P. R. and Dodds, J. A.,** Viruses of plant pathogenic fungi, in *Viruses and Plasmids in Fungi*, Lemke, P. A., Ed., Marcel Dekker, New York, 1979, 202.

92. **Buck, K. W., Romanos, M. A., McFadden, J. J. P., and Rawlinson, C. J.,** *In vitro* transcription of double-stranded RNA by virion-associated RNA polymerases of viruses from *Gaeumannomyces graminis*, *J. Gen. Virol.*, 57, 157, 1981.

93. **Fried, H. M. and Fink, G. R.,** Electron microscopic heteroduplex analysis of "killer" double-stranded RNA species from yeast, *Proc. Natl. Acad. Sci. USA*, 75, 4224, 1978.

94. **Bruenn, J. A. and Brennan, V. E.,** Yeast viral double-stranded RNAs have heterogeneous 3′ termini, *Cell*, 19, 923, 1980.

95. **Kane, W., Pietras, D., and Bruenn, J.,** The evolution of defective-interfering dsRNAs of the yeast killer virus, *J. Virol.*, 32, 692, 1979.

96. **Somers, J. M.,** Isolation of suppressive sensitive mutants from killer and neutral strains of *Saccharomyces cerevisiae*, *Genetics*, 74, 571, 1973.

97. **Holland, J. J., Grabau, E. A., Jones, C. L., and Semler, B. L.,** Evolution of multiple genome mutations during long-term persistent infection by vesicular stomatitis virus, *Cell*, 16, 495, 1979.

98. **Romanos, M. A.,** The Structure and Function of Some Viruses of *Gaeumannoyces graminis* var. *tritici*, Ph.D. thesis, University of London, 1981.

99. **Cox, R. A., Kanagalingam, K., and Sutherland, E. S.,** Double-helical character of ribonucleic acid from virus-like particles found in *Penicillium chrysogenum, Biochem. J.,* 120, 549, 1970.

100. **Cox, R. A., Kanagalingham, K., and Sutherland, E.,** Thermal denaturation in acidic solutions of double-helical ribonucleic acid from virus-like particles found in *Penicillium chrysogenum.* A spectrophotometric study, *Biochem. J.,* 125, 655, 1971.

101. **Buck, K. W., Girvan, R. F., and Ratti, G.,** Two serologically distinct double-stranded ribonucleic acid viruses isolated from *Aspergillus niger, Biochem. Soc. Trans.,* 1, 1138, 1973.

102. **Adler, J. P. and Mackenzie, D. W.,** Intrahyphal localisations of *Penicillium stoloniferum* viruses by fluorescent antibody, *Abstr. Annu. Meet. Am. Soc. Microbiol.,* 68, 1972.

103. **Demarini, D. M., Kurtman, C. P., Fennell, D. I., Worden, K. A., and Detroy, R. W.,** Transmission of PsV-F and PsV-S mycoviruses during conidiogenesis of *Penicillium stoloniferum, J. Gen. Microbiol.,* 100, 59, 1977.

104. **Sanderlin, R. S. and Ghabrial, S. A.,** Physicochemical properties of two distinct types of virus-like particles from *Helminthosporium victoriae, Virology,* 87, 142, 1978.

105. **Rawlinson, C. J. and Buck, K. W.,** Viruses in *Gaeumannomyces* and *Phialophora* spp., in *Biology and Control of Take-All,* Asher, M. J. C. and Shipton, P. J., Eds., Academic Press, London, 1972, 68.

106. **Bozarth, R. F. and Goenaga, A.,** A complex of virus-like particles from *Thielaviopsis basicola, J. Virol.,* 24, 846, 1977.

107. **Lhoas, P.,** Infection of protoplasts from *Penicillium stoloniferum* with double-stranded RNA viruses, *J. Gen. Virol.,* 13, 365, 1971.

108. **Pallett, I. H.,** Interactions between fungi and their viruses, in *Microbial and Plant Protoplasts,* Peberdy, J. F., Rose, A. H., Rogers, H. J., and Cocking, E. C., Eds., Academic Press, London, 1976, 107.

109. **Still, P. E., Detroy, R. W., and Hesseltine, C. W.,** *Penicillium stoloniferum* virus: altered replication in ultraviolet derived mutants, *J. Gen. Virol.,* 27, 275, 1975.

110. **Kamen, R. I.,** Structure and function of the Qβ RNA replicase, in *RNA Phages,* Zinder, N. D., Ed., Cold Spring Habor Laboratory, New York, 1975, 203.

111. **Baltimore, D., Ambros, V., Flanegan, J., Petterson, R., and Rae, J.,** Poliovirus 5' protein and viral replication, *Abstr. Fourth Internat. Cong. Virol.,* Centre for Agricultural Publishing and Documentation, Wageningen, 1978, 62.

112. **Banerjee, A. K., Abraham, G., and Colonno, R. J.,** Vesicular stomatitis virus: mode of transcription, *J. Gen. Virol.,* 34, 1, 1977.

113. **Cross, R. K. and Fields, B. N.,** Genetics of reoviruses, in *Comprehensive Virology,* Fraenkel-Conrat, H. and Wagner, R. R., Eds., Plenum Press, New York, 1977.

114. **Rimon, A. and Haselkorn, R.,** Temperature-sensitive mutants of bacteriophage φ 6 defective in both transcription and replication, *Virology,* 89, 218, 1978.

115. **Ikegami, M. and Fraenkel-Conrat, H.,** RNA-dependent RNA polymerase of tobacco plants, *Proc. Natl. Acad. Sci. USA,* 75, 2122, 1978.

116. **Dawson, W. O. and White, J. L.,** Characterisation of a temperature-sensitive mutant of tobacco mosaic virus deficient in the synthesis of all RNA species, *Virology,* 90, 209, 1978.

117. **Dawson, W. O. and White, J. L.,** A temperature-sensitive mutant of tobacco mosaic virus deficient in synthesis of single-stranded RNA, *Virology,* 90, 104, 1979.

118. **Reijnders, L.,** The origin of multicomponent small ribonucleoprotein viruses, *Adv. Virus Res.,* 23, 79, 1978.

119. **Kornberg, A.,** *DNA Replication,* W. H. Freeman, San Francisco, 1980.

120. **Chater, K. F. and Morgan, D. H.,** Ribonucleic acid synthesis by isolated viruses of *Penicillium stoloniferum, J. Gen. Virol.,* 24, 307, 1974.

121. **Buck, K. W.,** Replication of double-stranded RNA in particles of *Penicillium stoloniferum* virus S, *Nucl. Acids Res.,* 2, 1889, 1975.

122. **Banks, G. T., Buck, K. W., Chain, E. B., Darbyshire, J. E., and Himmelweit, F.,** Virus-like particles in penicillin-producing strains of *Penicillium chrysogenum, Nature (London),* 222, 89, 1969.

123. **Lemke, P. A., Nash, C. A., and Pieper, S. W.,** Lytic plaque formation and variation in virus titre among strains of *Penicillium chrysogenum, J. Gen. Microbiol.,* 76, 265, 1973.

124. **Wood, H. A., Bozarth, R. F., and Mislivec, P. B.,** Virus-like particles associated with an isolate of *Penicillium brevi-compactum, Virology,* 44, 592, 1971.

125. **Shatkin, A. J., Sipe, J. D., and Loh, P.,** Separation of ten reovirus genome segments by polyacrylamide gel electrophoresis, *J. Virol.,* 2, 986, 1968.

126. **Bozarth, R. F. and Harley, E. H.,** The electrophoretic mobility of double-stranded RNA in polyacrylamide gels as a function of molecular weight, *Biochim. Biophys. Acta,* 432, 329, 1976.

127. **Buck, K. W. and Ratti, G.,** Molecular weight of double-stranded RNA: a re-examination of *Aspergillus foetidus* virus S RNA components, *J. Gen. Virol.,* 37, 215, 1977.

128. **Kaper, J. M. and Diaz-Ruiz, J. R.,** Molecular weights of double-stranded RNAs of cucumber mosaic virus strain S and its associated RNA 5, *Virology,* 80, 214, 1977.

129. **Semancik, J. S., Vidaver, A. K., and Van Etten, J. L.,** Characterisation of a segmented double-helical RNA from bacteriophage φ 6, *J. Mol. Biol.,* 78, 617, 1973.

130. **Vanderwalle, M. J. and Siegel, A.,** A study of nucleotide sequence homology between strains of tobacco mosaic virus, *Virology,* 73, 413, 1976.

131. **Bozarth, R. F.,** Physicochemical properties of mycoviruses: an overview, in *Fungal Viruses,* Molitoris, H. P., Hollings, M., and Wood, H. A., Eds., Springer-Verlag, Berlin, 1979, 48.

132. **Khandjian, E. W., Turian, G., and Eisen, H.,** Characterisation of the RNA mycovirus infecting *Alloyces arbuscula, J. Gen. Virol.,* 35, 415, 1977.

133. **Hamilton, R. I., Edwardson, J. R., Francki, R. I. B., Hsu, H. T., Hull, R., Koenig, R., and Milne, R. G.,** Guidelines for the identification and characterisation of plant viruses, *J. Gen. Virol.,* 54, 223, 1981.

134. **Hedrick, J. L. and Smith, A. J.,** Size and charge isomer separation and estimation of molecular weights of proteins by disc gel electrophoresis, *Arch. Biochem. Biophys.,* 126, 155, 1968.

135. **Bruenn, J. A., Bobek, L., Brennan, V., and Held, W.,** Yeast viral RNA polymerase in a transcriptase, *Nucl. Acids Res.,* 8, 2985, 1980.

136. **Banerjee, A. K. and Shatkin, A. J.,** Transcription *in vitro* by reovirus-associated ribonucleic acid-dependent polymerase, *J. Virol.,* 6, 1980.

137. **Van Etten, J. L., Burbank, D. E., Cuppels, P. A., Lane, L. C., and Vidaver, A. K.,** Semiconservative synthesis of single-stranded RNA by φ 6 polymerase, *J. Virol.,* 33, 769, 1980.

138. **Bruenn, J. and Keitz, B.,** The 5′ ends of yeast killer factor RNAs are ppp Gp, *Nucl. Acids Res.,* 3, 2427, 1976.

139. **Furuichi, Y., Muthukrishnan, S., and Shatkin, A. J.,** 5′-terminal $m^7G(5')$ ppp(5′)GΣ p *in vivo:* identification in reovirus genome RNA, *Proc. Natl. Acad. Sci. USA,* 72, 742, 1975.

140. **Furuichi, Y., Morgan, M., Muthukrishnan, S., and Shatkin, A. J.,** Reovirus messenger RNA contains a methylated, blocked 5′-terminal structure: m^7 G(5′)ppp(5′)GΣpCp, *Proc. Natl. Acad. Sci. USA,* in press.

141. **Faust, M. and Millward, S.,** *In vitro* methylation of nascent reovirus mRNA by a virion-associated methyltransferase, *Nucl. Acids Res.,* 1, 1739, 1974.

142. **Ratti, G. and Buck, K. W.,** Semi-conservative transcription in particles of a double-stranded RNA mycovirus, *Nucl. Acids Res.,* 5, 3843, 1978.

143. **Ratti, G. and Buck, K. W.,** Transcription of double-stranded RNA in virions of *Aspergillus foetidus* virus S, *J. Gen. Virol.,* 42, 59, 1979.

144. **McGinty, R. M., Buck, K. W., and Rawlinson, C. J.,** Transcriptase activity associated with a type 2 double-stranded RNA mycovirus, *Biochem. Biophys. Res. Commun.,* 98, 501, 1981.

145. **Buck, K. W.,** Semi-conservative replication of double-stranded RNA by a virion-associated RNA polymerase, *Biochem. Biophys. Res. Comm.,* 84, 639, 1978.

146. **Herring, A. J. and Bevan, E. A.,** Yeast virus-like particles possess a capsid-associated single-stranded RNA polymerase, *Nature (London),* 268, 464, 1977.

147. **Silverstein, S. C., Christman, J. K., and Acs, G.,** The reovirus replicative cycle, *Annu. Rev. Biochem.,* 45, 375, 1976.

148. **Coplin, D. L., Van Etten, J. L., Koski, R. K., and Vidaver, A. K.,** Intermediates in the biosynthesis of double-stranded ribonucleic acids of bacteriophage φ 6, *Proc. Natl. Acad. Sci. USA,* 72, 849, 1975.

149. **Kassanis, B., White, R. F., and Woods, R. D.,** Beet cryptic virus, *Phytopathol. Z.,* 90, 350, 1980.

150. **Koenig, R. and Leseman, D. E.,** Two isometric viruses in poinsettias, *Plant Dis.,* 64, 782, 1980.

151. **Kenten, R. H., Cockbain, A. J., and Woods, R. D.,** Vicia cryptic virus (VCV). Report of Rothamsted Experimental Station for 1979, Part 1, 176, 1980.

152. **Milne, R. G.,** Electron microscopy of thin sections of Italian ryegrass infected with both ryegrass cryptic virus and oat sterile dwarf virus, *Microbiologia,* 3, 33, 1980.

153. **Lisa, V., Luisoni, E., and Milne, R. G.,** A possible virus cryptic in carnations, *Ann. Appl. Biol.,* 98, 431, 1981.

154. **Lisa, V., Boccardo, G., and Milne, R. G.,** Double-stranded ribonucleic acid from carnation cryptic virus, *Virology,* 115, 410, 1981.

155. **Barton, R. J.,** *Mycogone perniciosa* virus, *Rep. Glasshouse Crops Res. Inst.,* 133, 1978.

156. **Bozarth, R. F., Wood, H. A., and Goenaga, A.,** Virus-like particles from a culture of *Diplocarpon rosae, Phytopathology,* (Abstract) 62, 493, 1972.

157. **Buck, K. W., McGinty, R. M., and Rawlinson, C. J.,** Two serologically unrelated viruses isolated from a *Phialophora* sp., *J. Gen. Virol.,* 55, 235, 1981.

158. **Sweeney, T. K., Tate, A., and Fink, G. R.,** A study of the transmission and structure of double-stranded RNAs associated with the killer phenomenon in *Saccharomyces cerevisiae, Genetics,* 84, 27, 1976.

159. **Vodkin, M. H., Katterman, F., and Fink, G. R.,** Yeast killer mutants with altered double-stranded ribonucleic acids, *J. Bacteriol.,* 117, 681, 1974.

160. **Chu, G. and Sharp, P. A.,** SV40 DNA transfection of cells in suspension: analysis of the efficiency of transcription and translation of T antigen, *Gene,* 13, 197, 1981.

161. **Fukunaga, Y., Nagata, T., and Takebe, I.,** Liposome-mediated infection of plant protoplasts with tobacco mosaic virus, *Virology,* 113, 752, 1981.

Chapter 7

NEW VIRUSES OF EUKARYOTIC ALGAE AND PROTOZOA

J. Allan Dodds

TABLE OF CONTENTS

I. INTRODUCTION

Viruses may eventually be found in every type of organism, if sufficient effort is put into the search. How hard we look for viruses is determined more by practical reasons having to do with diseases and model systems, than by a belief that equal time should be given to each of the taxonomic groups of eukaryotic plants and animals. Algae and protozoa have definitely not been given equal time. There is a certain amount of bravery needed to even consider treating viruses from these eukaryotic microorganisms in a book with such meaty topics as those tackled in the other chapters. This is particularly the case when viral taxonomy is the topic. It is a subject that quickly becomes filled with nonscientific opinion when biological, biochemical, and biophysical characterizations of viral entities are incomplete. The characterization of viruses of algae and protozoa relies heavily on ultrastructural observations of virus-like particles in thin sections of numerous different host cells and draws lightly on biological, biochemical, and biophysical characterization of the particles detected by the electron microscopists. Substantial reviews have been provided by Brown,[1] Dodds,[2] Dutta,[3,4] Diamond and Mattern,[5] Lemke,[6] and Sherman and Brown,[7] and the reader wanting lists of hosts in which virus-like particles have been suspected, illustrations of representative particles, and further references should refer to these articles. It is not the intention of this chapter to re-write these excellent reviews.

In this current review I will draw some generalizations about the types of viruses and virus-like particles detected in protozoa and algae. The term "virus" will be used even for particles not proven to be viruses. The types of viruses with which the protozoan and algal viruses can be compared, but probably not grouped, will be pointed out. I will also review a short list of specific host-virus interactions, which have been studied in some detail in recent years. The information presented from these recent studies is as much biological as it is biochemical or biophysical. While it may be out of place in a volume on taxonomy of viruses, I have included it for three reasons. First, the data represent the state of investigation into the subject of this chapter. Second, the interactions are, for the most part, interesting. Third, lysogeny of DNA viruses in eukaryotic microbes is one possible explanation for some of the interesting biology.

Together, these topics may stimulate more virologists to enter this field and help generate the kind of data which is needed to characterize the viruses. The discovery of even more entirely new types of viruses in these previously overlooked organisms seems likely. This likelihood is reason enough for the inclusion of this chapter in a book on problems of viral taxonomy.

II. GENERAL REVIEW OF VIRUSES OF ALGAE AND PROTOZOA

The main line of evidence for viruses in eukaryotic algae is the accumulation over recent years of observations by electron microscopy of viruses in thin sections of numerous algae, both freshwater and marine.[8-30]

These observations have not normally been the result of a deliberate search for viruses but have been incidental to other studies. A few of the algae are available in culture, but most of the reports deal with field-collected materials no longer at hand. The following is a summary of the observations made in these studies.

Shape, structure, and size — Except for a rod shaped virus in *Chara corallina* (see later) the particles are all polygonal (5- or 6-sided) in outline and consist of a multilaminate membrane-like shell enclosing core material. One particle was further enclosed in a pear-shaped sac.[27] The reported diameters for the particles are from 22 to 390 nm. All particles are about the same size for a given alga except for two reports of more than one size in the same cell.[24,27] Certain size clusters are apparent in the Rhodophyceae (40 to 60 nm),[8-10,18] the Phaeophyceae (170 nm),[11,20,30] and the Chlorophyceae (200 to 240 nm).[16,17,23,26,29]

Location of the cells — Particles have been reported in the nucleus, in the cytoplasm or in both locations. Quasicrystalline arrangements of particles have been observed.

Inclusion bodies — Cells from cultures containing particles sometimes show features not associated with normal cells. These include amorphous[18,20] or concentric[10] inclusions in the nucleus or cytoplasm and unusual membranous activity in the cytoplasm.[20] There has been speculation that such inclusions may be involved in virus synthesis.

Effects on cells — Cells containing particles have been described as moribund with the expectation that such cells will ultimately lyse. Lysis has been reported for an infectious virus of *Micromonas pusilla*.[22] Some specific effects are proliferation of chloroplasts,[27] loss of the outer nuclear membrane,[25] general disruption of organelles,[26] a failure to produce a cell wall,[30] and abnormal development of surface scales.[27] Several of the reports which infer that the cytoplasm is the location of particles noted that infected cells lacked a nucleus.[11,12,17,27,29] One explanation for this, suggested by several micrographs, is that the particles were formed at the expense of the nucleus. In contrast to these adverse effects, other reports indicate that infected cells had a normal appearance except for the presence of particles.[25]

Location in tissues — The vegetative cells of multicellular algae appear not to contain virus. The exception is the virus in *Sirodotia*[18] which was observed in more than one stage in the algal life cycle. The more common observation is that particles are found only in reproductive cells or the germlings produced from them.[11,12,17,30] Such cells may be distinguishable from normal cells in the light microscope,[26] especially when appropriate DNA stains are used.[31]

Frequency of cells containing virus — Cells containing particles are generally infrequent in infected cultures, although 30% of *Platymonas* cells were visibly infected.[25] It is more usual to find that particles are present in only one or two cells out of hundreds examined.[12,15,24] Re-detection of viruses in other cultures or collections of the same alga has been negative in some searches[29] but positive in others.[9,12,18]

Types of algae with virus — The algae reviewed cover a wide range of vegetative types from many taxonomic groups.[2,6,7] Noticeably absent are the larger fucoid and laminarean algae.

What is still lacking in this summary of results is the evidence to establish these particles as true viruses, and an understanding of the part they play in the biology and ecology of the algae in which they are found. It is difficult to make this kind of progress because most of the specimens are not in culture, virus frequency is generally low, vegetative cells do not appear to be likely sources of virus for purification, and simple diagnostic tests which could be used to detect infections and to monitor virus purifications quantitatively have not been developed. Solution of this final practical problem has been a necessary prelude to progress in other fields of virology. Bacterial, animal, and plant virology have all relied heavily on plaque and lesion assays, or serological tests, for these purposes.

The literature on protozoal viruses is similar to, but not as extensive as that on algal viruses, consisting of ultrastructural observations of 35 to 100 nm diameter spherical particles, and some other unusual particles which are described in the next section. Excellent summaries have already been written.[3-6] Four hosts and their viruses deserve special mention; *Entamoeba histolytica, Naegleria gruberi, Leishmania hertigi*, and *Plasmodium* spp. They are discussed in the next section, which is a summary of the better studied viruses of eukaryotic microorganisms.

III. SPECIFIC EXAMPLES

A. *Chara corallina* Virus[32,33]

This virus is of interest for several reasons. It is one which can be justifiably compared to a well-characterized group of plant viruses (the tobamoviruses), it was the first algal virus to be described in detail, it is one of the few algal viruses for which infectivity has been demonstrated, and its algal host has one of the most complex vegetative and sexual morphologies of all algae.

The infectious particle is a rigid rod, 532 nm long, and 18 nm wide, with helical symmetry. It forms para-crystalline inclusions in the cytoplasm of infected cells. Any plant virologist would immediately compare it to tobacco mosaic virus (TMV) and related viruses (the tobamoviruses). The algal virus is almost twice as long as TMV (300 nm) and contains a correspondingly larger ss RNA genome (molecular weight = 3.6×10^6, molecular weight of TMV RNA = 2.1×10^6). Its capsid protein (molecular weight = 16,500) is smaller than TMV capsid protein (molecular weight = 17,500), a consistent observation in SDS polyacrylamide gels over a wide range of gel concentrations. The authors were unable to conclude that the pitch or the symmetry of the helix was identical to that of TMV particles. They did show that the amino acid composition of the capsid protein and base ratios of the RNA were more similar to tobamoviruses than to other groups of rod-shaped plant viruses. Their strongest evidence for relatedness with tobamoviruses was the observation that *Chara corallina* virus reacted with antisera to two strains of TMV, but not with antisera to 10 other strains. Reciprocal positive reactions using antisera against *Chara corallina* virus were not detected.

Further examples of algal viruses with TMV-like properties have not been forthcoming, nor were they found by the authors in several other species of the Charales, or in *Chara corallina* from distant locations.

B. *Cylindrocapsa geminella* Virus[16,17,31,34]

The host, *Cylindrocapsa geminella*, is a multicellular filamentous green alga that reproduces itself from single-celled zoospores which settle and germinate to give single-celled germlings which divide and elongate into new filaments. Virus particles have been observed in the single-celled germlings, but not in the zoospores or the cells of young or old filaments. The incidence of germlings with virus is low (5 to 10% at 20 hr after germination) and germlings producing virus are thought to lyse and die, because no infected germlings can be detected 48 hr after germination. Filaments which continue to grow can, in their turn, produce a low percentage of virus-producing germlings from zoospores. Two environmental factors, light and heat, have been shown to affect virus production. In high light intensities it is nearly impossible to detect virus-producing germlings. At low light intensities the incidence increases. The highest, most consistent incidence is produced when zoospores are given a 40°C heat shock for 6 hr.

The particles are 5- or 6-sided, 200 to 239 nm in diameter, and have a 14 to 16 nm thick multilaminate shell, and a dense fibrillar core. Particles without cores, cores without shells, and 2, 3, or 4-sided incomplete shells are also seen in infected cells. The particles are formed in a virogenic stroma consisting of bundles of fibrillar material and clusters of ribosomes. Cells with these particles and structures lack a nucleus. The virogenic stroma is located in the part of the cell where the nucleus is normally present.

Partially purified particles are also 200 to 230 nm in diameter, contain double-stranded DNA (molecular weight = $175\text{-}190 \times 10^6$ from gel electrophoresis) and up to 10 proteins (molecular weight = 10,000 - 160,000). These and other properties are summarized in Table 1. Infectivity of the purified virus has not been demonstrated.

C. *Uronema (Ulothrix) gigas* Virus[12,13,35]

The host, *Uronema gigas*, is another filamentous green alga that reproduces itself from single celled zoospores. Virus particles accumulate in liquid growth medium at a rate which parallels algal growth. Once algal growth ceases in old cultures, no new virus is released.

Table 1
PROPERTIES OF THREE POLYHEDRAL ALGAL VIRUSES

Virus property	Cylindrocapsa geminella virus[17,31]	Uronema gigas virus[12,35]	Hydra viridis, Chlorella virus[23,36]
Capsid diameter (nm)	200—230	390	185
Shell thickness (nm)	15	15	35
Tail length (nm)	ND[a]	1,000	ND
Sedimentation coefficient	?	6,340	2,600
Density	1.24 (sucrose)	1.30 (sucrose)	1.295 (CsCl)
dsDNA, mol wt ($\times$ 10^6)	175—190	8—72	136
dsDNA, density	?	1.719 (60% GC)	1.711 (52% GC)
Proteins, number	10	10	19
Proteins, mol wt	10,000— 160,000	26,000— 64,000	10,300— 82,000

[a] ND = not detected.

Virus is thought to be present in cells of germlings, but not in cells of mature filaments. Germlings selected from single zoospores eventually grow into cultures which yield either barely detectable quantities of virus or even higher quantities than the original culture. High virus production is correlated with poor, chlorotic growth of the host (disease). A heat shock treatment (38°C) applied to zoospores of high yielding subcultures increases virus yield sixfold from 14-day-old cultures.

The particles associated with *U. gigas* are 5- or 6-sided, 390 nm in diameter, and have a 15-nm thick multilaminate shell, and a core of two distinct materials. The inner layer of the shell can be digested with pronase. They are produced in a virogenic stroma in cells which lack a nucleus.

Partially purified particles are 390 to 575 nm in diameter. Swelling of unfixed particles, and expected loss of an angular outline to a spherical outline, is a function of increasing pH from 5.0 to 8.0. A most unusual, but distinctive feature is the presence of a 1-μm long tail with a distinctive fine structure, including a central swelling, attached to no more than 10% of partially purified particles. Tailed particles are seen, but rarely, in infected cells. Particles contain linear double-stranded DNA (molecular weight = 8-72 $\times$ 10^6 from length measurements of molecules which may have been fragments of full length molecules) and up to 10 proteins (molecular weight = 26,000-64,000, major protein = 45,000). Properties are summarized in Table 1. Infectivity of the purified virus has not been demonstrated.

D. *Hydra viridis/Chlorella* Virus

This virus is of particular interest because of the fascinating biology it represents. A *Chlorella*-like unicellular green alga which forms a symbiotic relationship with digestive cells of *Hydra viridis* is difficult to culture once isolated from its symbiotic host. It appears that the reason this is so is the rapid lysis of the algal cells caused by a virus. It is possible that the *Hydra* prevents the production of the virus when the algal cells are associated with its digestive cells, though more work is needed to conclusively prove this point. The *Chlorella* can apparently survive only when in association with the *Hydra*, and the algal virus may play an important role in the success of the symbiotic relationship.

The particles are 5- or 6-sided, measure 185 nm in diameter, and contain dense core material surrounded by a 35-nm wide multilaminate shell. All *Chlorella* cells contain large numbers of such particles 12 hr after separation from *H. viridis*. By 20 hr, all algal cells have lysed. The particles have not been detected in *H. viridis* cells, or in *Chlorella* cells while in association with *H. viridis* or immediately after separation from *H. viridis*. The virus particles accumulate in a virogenic stroma, which must form at the expense of the nucleus, which is not detected in virus infected cells, yet normally occupies a large portion of the cell. The particle contains double-stranded DNA (molecular weight = 136 × 10⁶ from the sum of restriction endonuclease fragments) and up to 19 proteins (molecular weight = 10,300 to 82,000). A single protein (molecular weight = 46,000) comprises a large percentage of the total protein. Properties are summarized in Table 1. Infectivity of the particle is suggested by its ability to attach to the outer surface of the cell wall of *Chlorella*.

E. *Entamoeba histolytica* Viruses[37-45]

The possibility that viral activity was the cause of spontaneous lysis in initial stages of axenization of *E. histolytica* led to the discovery of viruses in this protozoal host. The literature on this topic has been exhaustively reviewed,[5] and what follows is a summary of that review together with information from two more recent publications, one documenting a new and novel beaded virus,[43] and the other attempting to establish the effects of the viruses on virulence of the host.[44] The three types of viruses observed in *E. histolytica* have been abbreviated PV (polyhedral or icosohedral virus), FV (filamentous virus), and BV (beaded virus). They have all been detected by inoculation of indicator (tester) strains of *E. histolytica* with physical lysates of donor strains. Virus activity is indicated by lysis of indicator strains 24 to 48 hr after inoculation. Virus properties have been largely determined from observations made on indicator strains up to and after lysis. In fact it has not been possible to demonstrate the physical presence of virus particles in the cells of some donor strains used to prepare infectious physical lysates. Other strains constantly produce substantial and easily demonstrated quantities of virus.

The polyhedral (icosohedral) virus is the best studied virus of a protozoa. Particles with identical dimensions (75 to 80 nm diameter) and appearance (5- or 6-sided icosohedra) have been obtained from several donor sources. Specific antisera against the virus from each of three different donor strains are all able to neutralize the infectivity of the range of virus isolates that have been tested. Virus strains can be recognized by host range reactions on a series of indicator strains. Subcultures of donor cultures which release virus to the medium range from high to low yielders of virus (compare with the similar results for *Uronema gigas* subcultures). The yield of virus from some low yielding subcultures can be increased up to 100-fold by treatment with bromodeoxyuridine (compare with the similar results for *Naegleria gruberi*).

The particle is believed to have a DNA genome based on in vivo inhibitor, thymidine incorporation, and hybridization studies on virus infected and virus free cell lines. Elegant observation on the densely stained material making up the core within sectioned particles has resulted in a model for how the genome may be arranged within the particle. The particles are associated with nuclei prior to lysis and accumulate in the perinuclear space.

Purification of this particle has proved difficult, partly because of an association of particles with cell membranes after lysis. Such complexes are highly infectious, whereas free particles have low infectivity. The particle has not been as well characterized as the four algal viruses described earlier.

The filamentous virus is associated with the lysis of individual cells of some indicator strains. The nuclei of these cells contain massive whorls of 7-nm wide filamentous inclusions. The length of the filaments making up the inclusions is undetermined. The clusters of filaments are dispersed in the cytoplasm following lysis of the nucleus. Such cells never contain PV particles, even if other indicator cells treated with the same donor lysate do. Inhibitor studies of infected cells and acridine orange staining of nuclei suggest the involvement of double-stranded DNA in this host-virus interaction. More work is needed to prove that the filaments are virus particles. They have not been isolated and characterized, nor has their presence been demonstrated in infectious inocula.

The beaded virus (BV) is, from a taxonomic point of view, probably the most interesting virus described in the chapter. The virus accumulates in, and eventually lyses, a recipient amoeba strain 24 hr after inoculation with a physical lysate from a donor strain. Nuclei of infected cells are packed with clusters of beaded rods in parallel arrays. Such nuclei fluoresce yellow-green throughout when stained with acridine orange, whereas nuclei of uninfected cells do not. This suggests but does not prove that the beaded virus contains double-stranded nucleic acid. Analyses of partially purified preparations by equilibrium centrifugation in CsCl gradients and electron microscopy show that the virus particles are 235 nm long, 18 to 20 nm wide and consist of 14 beads each resembling a small spherical virus particle. The density of the particle is 1.34 g/cm.^{3} Two observations suggest the basic virion is the linear association of 14 beads; the next most frequent particle encountered in partially purified preparations is a dimer of 28 beads, and the parallel arrays in infected nuclei are 235 nm wide. The present data points to an entirely new class of viruses. Infectivity and genome assays are needed to confirm this possibility.

Attempts have been made to test the effects of virus isolates on nonvirulent strains of *E. histolytica*.[42] The donor strains had a range of pathogenicities to newborn hamster livers. The recipient cells used are the ones which recover after most of the virus inoculated cells have been lysed by virus from the donor cells. Such cultures are resistant to superinfection by the infecting virus isolate. They release infecting virus into their environment. The recipient amoebae have little or no initial virulence prior to inoculation. The virulence of such strains is increased, decreased, or unaltered by the virus treatment, but no consistent correlations could be made with the level of virulence of the donor strain. The changes could have as much to do with the selection pressure imposed by survival after near complete lysis, as with the effects of viruses on the recipient strains.

F. *Naegleria gruberi* Virus[45-50]

This is yet another example where very little can be said about the virus particle beyond the fact that it is 100 nm in diameter, but it has an abundance of interesting biology. The virus is seen in both the nucleus and cytoplasm of infected cells. Virus particles pass into the cytoplasm through tubes on the nuclear envelope. They then become enclosed in a membranous body. The bodies condense into spheres which can be released into the media and attach to other *Naegleria* cells. The presence of the virus is associated with some degenerative changes, including a failure to encyst, reduction of flagella formation, and cell swelling causing rupture of the plasma membrane, and frequently lysis.

The most interesting aspect to the biology of this virus-host interaction concerns the environmental factors which have been shown to influence virus synthesis. Axenic cultures fail to produce virus. Virus induction is achieved by supplying any of a wide variety of bacterial species as a food source. It is also achieved by treating axenically grown cells with the thymidine analog 5-bromo-2' deoxyuridine, a most interesting observation and a treatment which has been used to activate latent viruses in mammalian cell lines.

G. *Leishmania hertigi* Virus[51,52] and *Plasmodium* Viruses[53-58]

Small spherical virus-like particles 55 to 60 nm in diameter have been detected in the promastigote, but not the amastigote form, of several strains of *L. hertigi*. This suggests a latent phase for the virus in the amastigote form. In contrast to most other spherical particles of eukaryotic microorganisms, these have a 25 to 30 nm electron-translucent core. They accumulate in the cytoplasm of cells with normal nuclei but abnormal mitochondria, and are frequently associated with 20 to 24 nm external diameter tubular structures. A dense vesicular body with particles attached to its periphery has also been described. Transmission of this agent to sensitive cell lines or to other species of *Leishmania* has not been achieved, and the presence of the virus is not known to have any effect on the biology of the host.

Chemical or physical characterization of the particle or its genome has not been reported. This is one of the better studied protozoal viruses, yet little can be said about the nature of the virus or its relationship to other viruses. The same is true for reports of viruses in *Plasmodium* spp., and for this reason, they will not be discussed except to make a point about viruses of the hosts of protozoa. There are reports of well-characterized viruses of host cells within cells of parasitic protozoa. One of these concerns a cytoplasmic polyhedrosis virus of *Anopheles stephensi*.[56,57] This observation reinforces the need to prove replication of any suspected protozoal virus. It is not entirely impossible that some of the examples included in lists of protozoal viruses are actually internal accumulations of "foreign bodies" from the abiotic or biotic environments of the protozoa. The epidemiological consequences of protozoa harboring viruses they acquire from their environment is outside the scope of this chapter, but the topic has been reviewed.[59]

IV. LATENCY AND ECOLOGY

It is almost certain that viruses of eukaryotic microorganisms are more common than we know today. The list of examples is already long, and is sure to grow. All evidence to date points to a role for these viruses in the biology of their hosts, a claim which is more difficult to make for fungi. Questions which remain to be answered come to mind. Do the algae and protozoa act as both hosts and reservoirs for viruses of other organisms? Alternate hosts which share an environment could be fish, amphibia, insects, and, for parasitic protozoa, vertebrates including animals and humans. Are there other examples besides the virus-*Chlorella-Hydra* case where parasitism or symbiosis is favored or guaranteed by viruses? Can the viruses control microorganism populations, and is virus latency involved in this?

Most authors have been struck by the unusual biology they have had to work with, and have wondered whether virus latency and activation (lysogeny and lysis?) contribute to this observation. Examples which are summarized in Table 2 include:

1. The presence of virus in one, but not other stages of the life cycle of the host
2. The selection of high or low yielding subcultures from a starting culture, but a failure to cure cultures by single cell subculturing[5,12]
3. The lysis of tester strains when inoculated with virus from nonlysing donor strains[5]
4. The lysis or production of virus from low producer strains after treatment with thymidine analogs[5,50] or heat shocks[12,34] or after release from the influence of a symbiont[23,36]
5. Induction of virus accumulation by either changing from nonaxenic to axenic conditions (*Entamoeba*[5]) or by doing the opposite (*Naegleria*[48])

A reason for latency in algal viruses could be to ensure that algae survive virus infection and virus is released by lysis into an aqueous environment only when single celled zoospores are being released into that environment. The environmental trigger which induces zoosporogenesis and virus productivity may be the same. This creates an ideal situation for infection of zoospores, if these are the susceptible stage in the life cycle. This has yet to be proven.

The question of viral latency in protozoa is particularly tantalizing because protozoa are frequently human or animal parasites.[60] What part viruses might play in the survival or pathogenicity of protozoa remains to be worked out. The studies with *E. histolytica*[44] point out how difficult it is to answer these questions.

It is hoped that this review might stimulate more work on latency of viruses of eukaryotic microorganisms. The fact that most of them are probably double-stranded DNA viruses, together with the biological observations already described, suggest that this may be a profitable research topic. Contemporary interest in integrative replicating DNAs may have

Table 2
TREATMENTS WHICH CHANGE VIRUS
SYNTHESIS IN ALGAE AND PROTOZOA

Host	Treatment	Effect
Uronema gigas	Clonal selection	Low or high yielders selected. No cure.
	Heat shock	Sixfold yield increase.
Cylindrocapsa geminella	Heat shock	Conversion of nonproducer to producer.
Chlorella sp.	Release from symbiont	Conversion of nonproducer to lytic producer.
Entamoeba histolytica	Clonal selection	Low or high yielders selected. No cure.
	5-Bromodeoxyuridine	10 to 100-fold yield increase from low yielder.
Naegleria gruberi	5-Bromodeoxyuridine	Conversion of nonproducer to producer.
	Axenic to nonaxenic	Conversion of nonproducer to producer.

overlooked these host-virus interactions. The algal viruses are particularly attractive as models because of the shortage of different "vehicles" for genetic engineering of plants.

V. TAXONOMY

The pitfalls of taxonomic comparisons are exemplified by a report of a "rhabdovirus" in *E. histolytica*.[61] Bullet-shaped particles 250 nm long and 90 nm wide were observed scattered throughout the cytoplasm or condensed around a cytoplasmic body. The outer membranes of some particles were contiguous with cellular membranes. Such particles have been shown to differentiate into a more complex structure called T-bars and are probably a "normal" component of *E. histolytica*.[62] Physical characterization and pathogenicity studies are necessary before a particle of such common occurrence can be thought of as a virus-like entity.

The viruses that have been studied are clearly all new viruses and to date they each stand alone without any certain relationships to each other or to well-characterized viruses or virus groups from other hosts. Little will change until much more is known about them, especially the physical and chemical properties of the particles and their genomes. The only two groups which could accommodate some of these viruses are the tobamoviruses and the ICDV's (icosahedral cytoplasmic deoxyriboviruses) or iridoviruses.

The similarity of *C. corallina* virus to tobamoviruses has been discussed earlier. Common features of several of the viruses described in this chapter are as follows:

1. Particles are polyhedral, probably icosahedral. Two and three-sided incomplete particles are also seen in infected cells.
2. They are large (50 to 400 nm in diameter).
3. They have a 15 to 30-nm wide multilaminate shell, which can be partially digested by proteinases.
4. They contain dense core material that can be partially digested by deoxyribonucleases. Some particles may be empty or incomplete.
5. They have a low buoyant density.

6.   They contain numerous proteins (10-20), one of which (molecular weight about 50,000) is in greater abundance than the others.
7.   They possess a large (molecular weight 100-200 × 10^6), probably linear, double-stranded DNA genome.

These are all properties which are shared by viruses like Tipula iridescent virus or frog virus 3, which are grouped with the iridoviruses, or ICDV's.

The major differences have to do with biology, rather than physical or chemical properties. The algal and protozoal viruses of this kind are associated with the nucleus, rather than the cytoplasm, and there is moderately good evidence for latency being involved in the virus-host interaction.

The *Uronema gigas* virus has two characters which set it apart.[12,35] It is larger than reported ICDVs, and indeed is one of, if not the largest virus known. It also has a 1 μm long tail which, because of its consistent and distinctive architecture (fine structure), must surely have some biological function. No other virus of a eukaryote has been described with such an appendage. It is important details such as this which make it difficult to propose the inclusion of these viruses in currently accepted virus groups.

The beaded virus of *E. histolytica* clearly stands alone even though it is only partially characterized.[43] Other viruses are too poorly characterized to merit discussion.

VI. CONCLUSION

It is hoped that this chapter will draw attention to the new viruses which are being discovered in eukaryotic algae and protozoa. I have tried to emphasize the need for better virus characterization. The reason progress has been slow in this area has to do with difficulties in obtaining the amounts of purified virus particles necessary. The recognition that latency is a factor in the virus-host interactions and that appropriate stresses or treatments can induce virus synthesis is a clear way to rectify the problem of supply. The use of tester host strains which can be lysed is another important approach. The bringing together of interested microbiologists, electron microscopists and virologists to propose fundable research is another obvious need. Those willing to probe have an interesting field to work on; new viruses, fascinating biology, the possibility of integration of DNA genomes and perhaps a new tool to investigate the phylogeny and classification of eukaryotic algae and protozoa. On this last point I have two related questions with which I will finish this chapter. Does the discovery of RNA viruses in the Charales point to a closer taxonomic link to higher plants, which have similar viruses, than to other algae, which do not? Why is it that large DNA viruses of the type so common in algae have not been found in higher plants? The answer to these questions may be that viruses adapt to the hosts they find in their ecosystems, and may not remain with hosts which evolve and adapt to new ecosystems. On this basis I predict that some viruses of eukaryotic algae and protozoa will be found to have nonmicrobial alternate hosts.

REFERENCES

1. **Brown, R. M., Jr.,** Algal viruses, *Adv. Virus Res.,* 17, 243, 1972.
2. **Dodds, J. A.,** Viruses of marine algae, *Experientia,* 35, 440, 1979.
3. **Dutta, G. P.,** Recent advances in cytochemistry and ultrastructure of cytoplasmic inclusions of Sarcodina (Protozoa): Part 1 Anaerobic amoeba, *J. Sci. Ind. Res.,* 36, 226, 1977.

4. **Dutta, G. P.**, Recent advances in cytochemistry and ultrastructure of cytoplasmic inclusions in Sarcodina (Protozoa): Part 2 Pathogenic and nonpathogenic aerobic free-living amoebae, *J. Sci. Ind. Res.*, 38, 199, 1979.

5. **Diamond, L. S. and Mattern, C. F. T.**, Protozoal viruses, *Adv. Virus. Res.*, 20, 87, 1976.

6. **Lemke, P. A.**, Viruses of eucaryotic microorganisms, *Ann. Rev. Microbiol.*, 30, 105, 1976.

7. **Sherman, L. A. and Brown, R. M.**, Cyanophages and viruses of eukaryotic algae, in *Comprehensive Virology*, Vol 12, Fraenkel-Conrat, H. and Wagner, R. B., Eds., Plenum Press, New York, 1978, 145.

8. **Chapman, R. L.**, Further ultrastructural studies of nuclear inclusions in *Porphyridium*, *J. Phycol.*, 8, (Supplement), 14, 1972.

9. **Chapman, R. L.**, The presence of virus-like particles and centrosomes in the M. B. Allen strain of *Porphyridium purpureum*, *J. Phycol.*, 9, (Supplement), 16, 1973.

10. **Chapman, R. L. and Lang, N. I.**, Virus-like particles and nuclear inclusions in the red alga *Porphyridium purpureum* (Bary) Drew et Ross, *J. Phycol.*, 9, 117, 1973.

11. **Clitheroe, S. B. and Evans, L. V.**, Virus-like particles in the brown alga *Ectocarpus*, *J. Ultrastruct. Res.*, 49, 211, 1974.

12. **Dodds, J. A. and Cole, A.**, Microscopy and biology of *Uronema gigas*, a filamentous eucaryotic green alga, and its associated tailed virus-like particle, *Virology*, 100, 156, 1980.

13. **Dodds, J. A., Stein, J. R., and Haber, S.**, Characterization of a virus-like particle in a filamentous green algae, *Proc. Canadian Phytopath. Soc.*, 42, 26, 1975.

14. **Franca, S.**, On the presence of virus-like particles in the dinoflagellate *Gyrodinium* resplendens (Hulbert), *Protistologica*, 12, 425, 1976.

15. **Hoffman, L.**, Virus-like particles in *Hydrurus* (Chrysophyceae), *J. Phycol.*, 14, 110, 1978.

16. **Hoffman, L. R. and Stanker, L. H.**, Virus-like particles in the green alga *Cylindrocapsa*, *J. Phycol.*, 11, (Supplement), 7, 1975.

17. **Hoffman, L. R. and Stanker, L. H.**, Virus-like particles in the green alga *Cylindrocapsa*, *Can. J. Bot.*, 54, 2827, 1976.

18. **Lee, R. E.**, Systemic viral material in the cells of the freshwater red alga *Sirodotia tenuissima* (Holden) Skuja, *J. Cell Sci.*, 8, 623, 1971.

19. **Manton, I. and Leadbeater, B. S. C.**, Fine structural observations on six species of *Chrysochromulina* from wild Danish marine nanoplankton as a whole, *Dan. Vidensk. Sel. Biol. Skr.*, 20, 1, 1974.

20. **Markey, D. R.**, A possible virus infection in the brown alga *Pylaiella littoralis*, *Protoplasma*, 80, 223, 1974.

21. **Mattox, K. R., Stewart, F., and Floyd, J.**, Probable virus infections in four genera of green alga, *Can. J. Microbiol.*, 18, 1620, 1974.

22. **Mayer, J. A. and Taylor, F. J. R.**, A virus which lyses the marine nanoflagellate *Micromonas pusilla*, *Nature* (London), 281, 299, 1979.

23. **Meints, R. H., Van Etten, J. L., Kuczmarski, D., Lee, K., and Ang, B.**, Viral infection of the symbiotic *Chlorella*-like alga present in *Hydra viridis*, *Virology*, 113, 698, 1981.

24. **Moestrup, D., and Thompson, H. R.**, An ultrastructural study of the flagellate *Pyramimonas orientalis* with particular emphasis on Golgi apparatus activity and the flagellar apparatus, *Protoplasma*, 81, 247, 1974.

25. **Pearson, B. R. and Norris, R. E.**, Intranuclear virus-like particles in the marine alga Platymonas sp. (Chlorophyta-Prasinophyceae), *Phycologia*, 13, 5, 1974.

26. **Pickett-Heaps, J. D.**, A possible virus infection in the green alga *Oedogonium*, *J. Phycol.*, 8, 44, 1972.

27. **Pienaar, R. N.**, Virus-like particles in three species of phytoplankton from the San Juan Island, Washington, *Phycologia*, 15, 185, 1976.

28. **Soyer, M. O.**, Virus-like particles and trichocystoid filaments in dinoflagellates, *Protistologica*, 14, 53, 1978.

29. **Swale, E. M. F. and Belcher, J. J.**, A light and electron microscopic study of the colourless flagellate *Aulacomonas* Skuja, *Arch. Mikrobiol.*, 92, 91, 1973.

30. **Toth, R. and Wilce, R. T.**, Virus-like particles in the marine alga *Chorda tomentosa*, Lyngbye (Phaeophyceae), *J. Phycol.*, 8, 126, 1972.

31. **Stanker, L. H. and Hoffman, L. R.**, A simple histological assay to detect virus-like particles in the green alga *Cylindrocapsa* (Chlorophyta), *Can. J. Botany*, 57, 838, 1979.

32. **Gibbs, A. J., Skotnicki, A. H., Gardiner, J. E., Walker, E. S., and Hollings, M.**, A tobamovirus of a green alga, *Virology*, 64, 571, 1975.

33. **Skotnicki, A., Gibbs, A. J., and Wrigley, N. G.**, Further studies on *Chara corallina* virus, *Virology*, 75, 457, 1976.

34. **Stanker, L. H., Hoffman, L. R., and MacLeod, R.**, Isolation and partial chemical characterization of a virus-like particle from a eukaryotic alga, *Virology*, 114, 357, 1981.

35. **Cole, A., Dodds, J. A., and Hamilton, R. I.,** Purification and some properties of a double-stranded DNA containing virus-like particles from *Uronema gigas*, a filamentous eucaryotic green alga, *Virology*, 100, 166, 1980.

36. **Van Etten, J. L., Meints, R. H., Burbank, D. E., Kuczmarski, D., Cuppels, D. A., and Lane, L. C.,** Isolation and characterization of a virus from the intracellular green alga symbiotic with *Hydra viridis*, *Virology*, 113, 704, 1981.

37. **Diamond, L. S., Mattern, C. F. T., and Bartgis, I. L.,** Viruses of *Entamoeba histolytica*. I. Identification of transmissible virus-like agents, *J. Virol.*, 9, 326, 1972.

38. **Mattern, C. F. T., Diamond, L. S., and Daniel, W. A.,** Viruses of *Entamoeba histolytica*. II. Morphogenesis of the polyhedral particle (ABRM$_2$ HK-9) HB-301 and the filamentous agent (ABRM)$_2$ HK-9, *J. Virol.*, 9, 342, 1972.

39. **Hruska, J. F., Mattern, C. F. T., Diamond, L. S., and Keister, D. B.,** Viruses of *Entamoeba histolytica*. III. Properties of the polyhedral virus of the HB-301 strain, *J. Virol.*, 11, 129, 1973.

40. **Hruska, J. F., Mattern, C. F. T., and Diamond, L. S.,** Viruses of *Entamoeba histolytica*. IV. Studies on the nucleic acids of the filamentous and polyhedral viruses, *J. Virol.*, 13, 205, 1974.

41. **Mattern, C. F. T., Hruska, J. F., and Diamond, L. S.,** Viruses of *Entamoeba histolytica*. V. Ultrastructure of the polyhedral virus, V. 301, *J. Virol.*, 13, 247, 1974.

42. **Diamond, L. S., Mattern, C. F. T., Bartgis, I. L., Daniel, W. A., and Keister, D. B.,** Viruses of *Entamoeba histolytica*. VI. A study of host range, in Proc. Int. Conf. on Amebiasis, Mexico City, 1977.

43. **Mattern, C. F. T., Keister, D. B., Daniel, W. A., Diamond, L. S., and Kontonis, A. L.,** Viruses of *Entamoeba histolytica*. VII. A novel beaded virus, *J. Virol.*, 23, 685, 1977.

44. **Mattern, C. F. T., Keister, D. B., and Diamond, L. S.,** Experimental amebiasis. IV. Amebal viruses and the virulence of *Entamoeba histolytica*, *Am. J. Trop. Med. Hygiene*, 28, 653, 1979.

45. **Schuster, F. L. and Dunnebacke, T. H.,** Formation of bodies associated with virus-like particles in the amoebo-flagellate *Naegleria gruberi*, *J. Ultrastructure Res.*, 36, 659, 1971.

46. **Schuster, F. L. and Dunnebache, T. H.,** Growth at 37°C of the EG^s strain of the amoebo-flagellate *Naegleria gruberi* containing virus-like particles, *J. Invert. Pathol.*, 23, 182, 1974.

47. **Schuster, F. L. and Dunnebacke, T. H.,** Virus-like particles and an unassociated infectious agent in amoebae of the genus *Naegleria*, *Ann. Soc. Belge Med. Trop.*, 54, 359, 1974.

48. **Schuster, F. L. and Dunnebacke, T. H.,** Virus-like particles in the amoeba flagellate *Naegleria gruberi* EGS, *Am. Soc. Microbiol.*, 579, 1977.

49. **Schuster, F. L. and Dunnebacke, T. H.,** Development and release of virus-like particles in *Naegleria gruberi* EGS, *Cytobiologie*, 14, 131, 1976.

50. **Schuster, F. L. and Clemente, J. S.,** 5-bromodeoyuridine induced formation of virus-like particles in *Naegleria gruberi*, *J. Cell Sci.*, 26, 359, 1977.

51. **Molyneux, D. H.,** Virus-like particles in Leishmania parasites, *Nature* (London), 249, 588, 1974.

52. **Croft, S. L. and Molyneux, D. H.,** Studies on the ultrastructure of virus-like particles and infectivity of Leishmania hertigi, *Ann. Trop. Med. Parasitol.*, 73, 213, 1979.

53. **Garnham, P. C. C., Bird, R. G., and Bakr, J. R.,** Electron microscope studies of motile stages of malaria parasites. III. The ookinetes of *Haemamoeba* and *Plasmodium*, *Trans. R. Soc. Trop. Med. Hyg.*, 56, 116, 1962.

54. **Dasgupta, B.,** A possible virus disease of the malaria parasite, *Trans. R. Soc. Trop. Med. Hyg.*, 62, 730, 1968.

55. **Terzakis, J. A.,** A protozoan virus, *Mil. Med.*, 134, 916, 1969.

56. **Davies, E. E., Howells, R. E., and Venters, D.,** Microbial infections associated with plasmodial development in *Anopheles stephensi*, *Ann. Trop. Med. Parasitol.*, 65, 403, 1971.

57. **Bird, R. G., Draper, C. C., and Ellis, D. S.,** A cytoplasmic polyhedrosis virus in midgut cells of *Anopheles stephensi* and in sporogonic stages of *Plasmodium berghei yoelii*, *Bull. W. H. O.*, 46, 337, 1972.

58. **Terzakis, J. A.,** Virus-like particles in the Malaria parasite, *J. Parasitol.*, 62, 366, 1976.

59. **Perez Prieto, S. I. and Garcia Gancedo, A.,** Viral infections of protozoa and their possible ecological and epidemiological implications. Current state of the problem, *Rev. Sanid. Hig. Publica*, 53, 1517, 1979.

60. **Bird, R. G.,** Protozoa and Viruses, *Aust. Soc. Parasitol.*, 2, 39, 1980.

61. **Bird, R. G. and McCaul, T. F.,** The rhabdoviruses of *Entamoeba histolytica* and *E. invadens*, *Ann Trop. Med. Parasitol.*, 70, 81, 1976.

62. **Mattern, C. F. T., Diamond, S., Daniel, W. A., and Keister, D. B.,** Unusual variation of "normal" cytoplasmic bars of *Entamoeba histolytica*, in *Proc. Int. Conf. Amebiasis*, Mexico, 1977, 346.

Chapter 8

VIRUS DATA — PROBLEMS IN COLLECTION, STORAGE, AND RETRIEVAL

J. G. Atherton, I. R. Holmes, and E. H. Jobbins

TABLE OF CONTENTS

I. INTRODUCTION

At the first meeting of the Executive Committee of the ICNV at Moscow in 1966 a committee was set up to examine the utility of the virus cryptogram. As outlined in Chapter 1 this committee has changed in its function and name in the intervening years. It is now the ICTV "Standing subcommittee for virus data" and is associated with the World Data Centre for Microorganisms in Brisbane.

Because of the diversity and amount of virus data now available and with the increase in information produced daily from laboratories around the world, it is no longer practicable to store virus data manually. Computer storage is the only alternative and it offers many advantages. Information is never "lost" since all data stored in a computer are retained and can be retrieved with an appropriate query to the computer at any time. Information can be accessed with greater rapidity and ease and this leads to an increased use of the data, compared with the impracticable manual system, which in turn can pinpoint gaps in knowledge.

The virus data collected for computer storage need to be in a special format. The ICTV and its various Subcommittees have produced "A Code for the Description of Virus Characters". This is written in a data sheet questionnaire format. This allows the computer to prompt the operator by displaying a cue for the next item of data.

The advantages of a universal code are that it provides for:

- Systematic and uniform storage of data
- Standard descriptions of characters
- Easy transformation of data into languages other than English
- Flexibility of data output
- Indication of gaps in virus knowledge

One of the main uses of computer stored virus data will be for taxonomic purposes, either in virus identification or in elucidation of virus relationships.

Computers may be programmed to:

- Generate lists of viruses with like characters
- Generate hierarchical identification keys
- Investigate the relationship between viruses via linkage analysis, percentage similarity and dendrogram generation

Computer stored data has been widely used for taxonomic studies in hierarchial identification keys,[1] bacterial taxonomy,[2] ant taxonomy,[3-4] and grass taxonomy.[5] The use of the computer for taxonomic investigation allows for a more detailed analysis than would ever be possible by manual means.

In developing a computer system for the storage of virus data and its subsequent analysis, two approaches may be followed. A complete system may be developed with programs written to store and retrieve the data and further programs written to analyze the data for taxonomic purposes. Alternatively, a commercial data storage and retrieval system may be used and matched with an existing suite of programs for taxonomic analysis. The latter approach only requires that a linking program be developed to allow the transfer of data between the storage system and the processing system. An example of this type of program is the DELTA system which has been used for ant and grass taxonomy investigations.[6] The major advantage of such an approach is that use can be made of already developed and refined taxonomic analysis systems, such as the Australian CSIRO's TAXON suite of programs. A fully operational computer system for the storage, retrieval, and analysis of viral data will be of greater taxonomic use for those virus genera with large numbers of species and strains. In such genera, the computer may help in delineating species. Computer stored data is unlikely to be needed in delineating a new virus family.

II. GENERAL PRINCIPLES AND ADVANTAGES OF COMPUTER STORAGE

The most obvious advantage of computer storage of virus data is the ease with which the data may be accessed compared with a manual system of organization. The easier it is to access data, the more that data will be used. A more subtle advantage is that the computer can help to point out gaps in the knowledge about viruses. There is relatively little data available for a large proportion of viruses, but even so, no one person could hope to be *au fait* with all present virus knowledge, or to be able to manually record and index it and retrieve specific data on demand. The situation can only deteriorate as time progresses unless a suitable computer system is developed for virus data storage and retrieval. The aim of the development of a Code for the description of virus characters is to provide a base on which to build a suitable computer system. In addition the Code should be able to be used by any efficient database management computer system.

An efficient data processing system should embrace the following general principles:

1. It should be as simple as possible to use, i.e., interactive and self-guiding.
2. It should be easily understood by operators with minimal training.
3. It should provide multiple means of access to the stored data to allow greater flexibility. It is insufficient to be able to access the data by only one character such as virus name. Ideally, data should be accessible through any character.
4. It should be able to work efficiently with large and small amounts of data.
5. It should be relatively simple to amend the stored data and the overall structure of the data.
6. It should be flexible enough to accept data from questionnaires of varying format and to output data in various formats.
7. The program software and computer should both be maintained and regularly updated by their manufacturers and not lapse into obsolescence.

III. A CODE FOR THE DESCRIPTION OF VIRUS CHARACTERS

The Basic Data Sheet (BDS) and Supplementary Data Sheet (SDS) have been written as questionnaires because such a format adapts readily and efficiently to a computerized data processing system.

There are three general types of questionnaire which can be developed. An example of the first is the Rogosa[2] code for bacteria which consists of an exhaustive series of questions which admit only yes/no answers. The two major problems with such a format are the overall length of the questionnaire and the lack of provision for numerical answers such as molecular weights, etc. The second format involves a much smaller set of questions to which free-form plain language answers may be given. While such a format is very simple, it can create problems if the system is to be computerized due to the nonstandard nature of the answers, e.g., a computer will regard ribonucleic acid and RNA as entirely different. The third approach was the one followed in the development of the BDS and SDS. The principle of a relatively small set of questions, as with the plain language format, was adhered to but instead of allowing any answer to be given, an exhaustive list of all possible specific answers was supplied. This has the advantage of reducing the number of questions as compared to the binary format and at the same time rigorously defines the data which will be stored in the computer. It also decreases the amount of computer storage required. Provision has been made for numerical values and also for questions for which a set of specific answers cannot be supplied, but which require plain language answers, e.g., detailed host range. Such a format combines the best features of the binary and plain language formats.

Because many terms are used in the Code, an obvious necessity is to make available a set of definitions of these terms. A document ''Definitions in Virology'' is in preparation, and will be forwarded to the Coordination Subcommittee of the ICTV for approval before publication.

IV. ORGANIZATION OF THE CODE

A. Summary
The Code consists of 11 sections, as follows:

- Section 1: General details (source of information) — This is used to record information such as name and address of the person submitting the data, date of submission and other relevant details.
- Section 2: Specific strain data - taxonomic descriptions — Reserved for insertion of the descriptions of the family, genus, and type species as compiled by the relevant Subcommittees of the ICTV.
- Section 3: Basic data — Records important basic information on the virus in brief form, e.g., salient features of the physicochemical properties of the particle, replication mechanism and host-virus relationship. This information is probably available at the present time for most well-known viruses.
- Section 4: Supplementary data — Records further information on ''general characters'' in full detail. It is doubtful if all of this information is known at the present time for any one virus. Ultimately, however, most of this data will become known for many viruses and systematic storage is therefore warranted. The basic and supplementary data will ultimately be linked.
- Section 5: Reference list — Records Author(s), year, title and journal, for review and special articles pertaining to Sections 3 and 4.
- Sections 6-11: Host subcommittee data — Virologists interested in different host groups attach importance to different data within their respective group, in addition to the basic and supplementary data. Hence, further sections specifically relating to vertebrate, invertebrate, plant, bacterial, fungal and algal viruses, and perhaps other specialized information groups, are necessary.

B. The Basic Data Sheet
This section has been written after extensive consultation within the Code and Data Subcommittee of the ICTV. All members of the ICTV were given an opportunity to contribute to the BDS. It includes provision for information on characters common to viruses irrespective of host affinity. Information contained in this section should provide a satisfactory description of a virus to the generic and often to the specific level. Characters applicable only to viruses of specific host groups (e.g., plant viruses) have been reserved for inclusion in the later section relating to viruses of that host affinity (Sections 6-11). The basic data section contains data in 12 categories:

3.1   Nucleic acid
3.2   Protein
3.3   Lipid
3.4   Carbohydrate
3.5   Physicochemical Properties

3.6   Morphology
3.7   Antigenic Properties
3.8   Biological Properties
3.9   Replication
3.10 Virus Induced Changes in Host Cell
3.11 Host Range
3.12 Transmission

The full Basic Data Sheet is given in the Appendix.

C. The Supplementary Data Sheet

This section contains provision for more detailed information, which will be available only for some viruses. It is written using the same categories as the basic data section.

V. THE COMPUTER SYSTEM IN USE IN BRISBANE

The computer system which has been used to store basic data is System 1022*, a state-of-the-art database management system developed by Software House, Cambridge, Massachusetts.

At the University of Queensland, 1022 is running on a DEC PDP-1090 (KL10), generally regarded as one of the best interactive computers available. It is now possible to access the KL10 through the CSIRO computer network (CSIRONET), which is Australia-wide, and international through the Australian Overseas Telecommunication Commission's MIDAS link.

System 1022 has the following important general features:

● Easy to use interactively with visual display (VDU) or hardcopy terminals.
● Creates and updates large and small data collections.
● A very fast retrieval capability gives immediate access to data in the largest collections of records
● A comprehensive internal language allows simple programming for output of data in any desired format.

System 1022 stores data in a particularly efficient manner and makes use of Key Tables for rapid access to specific subsets of the data. Key Tables may be likened to the index of a book. They allow system 1022 to rapidly find, for example, all viruses whose nucleic acid is RNA by referring to RNA in the "index" without having to leaf through the "book" searching for all occurrences of RNA.

The entry of data into a 1022 dataset is simple as the computer will prompt for the required information. The construction of the Key Tables is done automatically. If desired, it is possible to enter data in large amounts rather than one item at a time. Data update, deletion or amendment are all simple and rapid irrespective of the number of items involved. Data retrieval is also simple and rapid. Any item, whether it is virus name or protein molecular weight, can be used to obtain subsets of data. The FIND and SEARCH commands, in conjunction with various modifiers, are used to select subsets of the data.

1022 allows for the creation of a number of related datasets which can be open and accessed simultaneously. In the case of viruses the related datasets could be Basic Data, Supplementary Data, References, and Specific Host Group Data.

1022 has an internal data programming language (DPL) which is an easy to use English-

* System 1022 (TM) is a proprietary product of Software House Inc., Cambridge, Massachusetts.

based language. DPL programs may be written to output data in any desired format and data stored numerically can be translated into plain language. In addition 1022 has interfaces to cater for interrogation of the database by programs written in FORTRAN, MACRO, or COBOL languages. By using any of these languages it is possible to develop a totally interactive 1022 data processing system. Such a system will be usable by people relatively inexperienced in the operation of computers.

VI. PROBLEMS IN THE COLLECTION, STORAGE AND RETRIEVAL OF VIRUS DATA

The work done to date on the Code and its integration with system 1022 have revealed a number of problems with the collection, storage and retrieval of virus data.

A. Collection

Cooperation by virologists — The major difficulty in the collection of data is in obtaining it from individual researchers around the world. It is hoped that virologists will realize the importance and value of a uniform collection of all known viral data and that they will cooperate fully in completing the Code data sheets.

Differences in data for different viruses — The decision as to which data should be placed in which section of the Code presents difficulties. Those sections most affected are 3 — Basic Data; 4 — Supplementary Data; and 6-11 — Host Subcommittee Data. This problem arises because of the way in which virology is sectionalized on the basis of host groups. It is obviously impossible to produce a data sheet which will satisfy virologists of all persuasions but which does not contain questions some virologists would regard as redundant or unnecessary. The concept of separate data sheets for each host group is patently undesirable and the simplest solution is for virologists to ignore those questions which do not apply to the virus(es) for which they are submitting data.

The amount of detail needed — In how much detail should data be collected in each section? As stated in the introduction, sections 3 and 4, combined, allow for the collection in great detail of data common to all viruses. Data relevant to only particular virus groups is collected in far less detail. This data is then collected in detail in the relevant host data section.

Reliability and relevance of data — The relevance of the data collected is a problem which arises from the nonstandard nature of the tests used by virologists. As stated by Ackermann et al.,[7] much effort could be devoted to the collection of data that had been obtained using obsolete methods or to the input of obsolete parameters, which, for lack of standard methods, cannot be compared with existing data. This is detrimental to comparative virology and there is obviously a need for the publication of a standard description of tests used for the collection of data. As mentioned by Rogosa et al.,[2] the assumption made at present is that, regardless of how determined, the test result is acceptable for coding as long as the method is reproducible and reliable within available knowledge and technology.

The number of possible answers in the code — The nature of the format decided upon in the construction of the Code (Section III) raises the question of how many alternative answers need to be supplied to cover all possibilities. An exhaustive study of the information available and international consultation and cooperation are necessary so that all known answers are supplied. However, provision must also be made for new answers to be added as discoveries are made in the future.

Standardization in the form of input data — Inconsistencies in the form of the data being collected must be avoided when using a computer system for its storage. It is essential that input to the computer be consistent in form, i.e., RNA must always be referred to as RNA and not RNA sometimes and ribonucleic acid at other times. The computer will recognize each form as different items. This problem can be overcome by the careful construction of the Code initially so that no inconsistencies occur.

Ambiguities in questions and answers — The questions asked and answers given must be clear, concise, and free of ambiguity. If not, uncertainty in the answers given and possibly incorrect answers could result. A clear set of definitions of terms used in each section of the Code will help to overcome this difficulty. Once again this problem should be overcome by the careful construction of the Code initially and checking with trial runs to pinpoint any problems.

Languages other than English — The potential problem of languages other than English must be considered. To overcome this, data information sheets could be written in a number of different languages. Because much of the data thus obtained is stored in a coded numerical form rather than as text, its output in a language other than English is relatively simple. The internal programming language of system 1022 (Section V) allows the coded data to be output as text and the text could be in any language.

B. Storage

No difficulties should arise in the storage of information obtained using the Code data sheets provided that all of the potential problems of data collection, listed above, are noted and solved. This will be the continuing responsibility of the Subcommittee for Virus Data. The essential feature of the storage system used is that it must be flexible enough to accept data from questionnaires of varying format and be able to output data in the format presented for imput. Experience with system 1022 has shown it to be well suited to the task.

C. Retrieval

Difficulties in the retrieval of data relate mainly to the computer system used and would arise if it were

1. Inflexible, i.e., if data could be output in only one format.
2. Not easy to access, i.e., if access to data could be obtained only through one character such as the virus name.
3. Slow, i.e., if the computer must search through all of the information in the database each time a subset of the data is required.

These problems are all overcome by the use of system 1022 and details of retrieval using this system can be found earlier in Section V.

VII. VIRUS DATA CENTERS RECOGNIZED BY THE ICTV

The World Data Centre for Microorganisms, supported by UNESCO, at the Department of Microbiology, University of Queensland, St. Lucia, Queensland, Australia, under its Director, Professor V.B.D. Skerman, has been recognized by the ICTV as the center for the storage and retrieval of data relating to all viruses. This center will also act as a back-up center for the collections holding only the more specialized data, such as specific host-group data.

It is the intention of the ICTV to foster the establishment and maintenance of other centers for virus data relating to particular groups of viruses. As experience with computer data storage and processing progresses, centers with particular interest and expertise in specialized data will be recognized by the ICTV. One such center has been recognized so far. The ICTV resolved that the WHO Collaborating Centre for the Collection and Evaluation of Data on Comparative Virology, Munich, shall be recognized as the center for data on viruses of vertebrates. This center, under the direction of Dr. Peter Bachmann, has developed a virus data questionnaire adopted by the ECICTV. This questionnaire is in the plain language format and is the core of a virus database. The database is managed by a number of programs — written in Pascal — which allow input, update, and retrieval of data. Data may be retrieved in whole or in part and can be output as hard copy or microfiche.

VIII. PRESENT STATE AND FUTURE PLANS

At present, data sheets have been written for two main sections of the Code, Section 3: Basic Data Sheet, and Section 4: Supplementary Data Sheet. The Basic Data Sheet is reproduced in the Appendix.

We have access to data on vertebrate viruses from Dr. P. Bachmann, on invertebrate viruses from Dr. D. L. Knudson, Dr. R. R. Granados, and Mr. J. F. Longworth, on plant viruses from Dr. A. Gibbs, and on bacterial viruses from Dr. H.-W. Ackermann. Other viral data are yet to be obtained.

Data should be channelled through the Chairman of the ICTV Host Subcommittee from whom it will be formally transmitted to the data center. Any virologist wishing to obtain a copy of the Code may receive a microfiche copy by writing to the principal author. It is proposed that when a significant amount of viral data has been stored, virologists may request data. This may be sent on magnetic tape or microfiche.

ACKNOWLEDGMENTS

The authors wish to acknowledge gratefully the financial support of the Clive and Vera Ramaciotti Foundation, Sydney, Australia for this project.

The authors also wish to acknowledge the assistance of Dr. H.-W. Ackermann, Dr. P. Bachmann, Chairman of the Standing Committee on Code and Data, ICTV, Dr. A. Eisenstark, Dr. R. I. B. Francki, Dr. A. J. Gibbs, Dr. N. Karabatos, Dr. D. L. Knudson, Dr. F. A. Murphy, Dr. T. W. Tinsley, and Professor C. Vago who have contributed to the work of the Code and Data Subcommittee.

APPENDIX: ICTV BASIC DATA SHEET

CONTENTS

1. Standard answers
2. Standard questions
3. References to literature
4. Computer abbreviations

VERSION VC OF BASIC DATA SHEET

Compiled 2.04.1981

THE BASIC DATA SHEET HAS BEEN DESIGNED TO ACCEPT
INFORMATION FOR VIRUSES FROM ANY HOST GROUP. SOME
QUESTIONS WILL NOT BE APPLICABLE FOR SOME VIRUSES AND AS
SUCH SHOULD BE IGNORED.

J. G. Atherton and I. R. Holmes

Dept. of Microbiology
University of Queensland
St. Lucia, Queensland 4067
AUSTRALIA.

SECTION 1: GENERAL DETAILS

01. Date of response (DRESP):
Day (01-31)   Month (01-12)   Year(00-99)

02. Edition of questionnaire used (from title page) (BDSED):

03. Name of person submitting information (PERSN):
Surname (Family name):
.... (30)
First name and other names and/or initials
(Leave one space between each name):
.... (30)
Title or rank (eg. Professor, Doctor, etc.) (TITLE):
.... (15)

04. Collection number in World Directory of Culture
Collections (if known) (COLNM):

(If none write NONE)

05. Postal address (Leave one space between each word or number. If applicable, use
exact World Directory address) (PADDR):
....
....
....
....
.... (100)

06. Category of virus being described (HOST):
    VV = Virus-vertebrate
    VIA = Virus-invertebrate-arthropod
    VIN = Virus-invertebrate-nonarthropod
    VP = Virus-plant
    VB = Virus-bacterium
    VA = Virus-algal
    VF = Virus-fungal

07. Virus name (NAME):
    .... (50)
08. Identity of virus has been independently confirmed (IDCON):
    [If Yes give full reference and 3-6 keywords]
    1. No
    2. Yes
    .... (20)
09. Virus other names (VONAM):
    .... (40)
    [If none write NONE]
10. Virus originally described by (DESBY):
    [Full reference and 3-6 keywords]
    .... (20)
11. Has virus family/group description been published
    by the ICTV (VFDES):
    [If Yes give full reference and 3-6 keywords]
    1. No
    2. Yes
    .... (20)
12. ICTV approved virus name (APVNM):
    .... (50)
    [If none write NONE. If proposed enclose name in "?"]
13. Reference to approved name (RFAVN):
    .... (20)
    [If none write NONE]
14. Has virus been classified in current ICTV report (CLASS):
    1. No
    2. Yes
15. Is it a viroid (naked nucleic acid) VROID):
    1. No
    2. Yes
16. Type species of genus (TYPE):
    .... (50)
    [If none write NONE. If proposed enclose in "?"]
17. Family or group name (FAMIL):
    .... (20)
    [If none write NONE. If proposed enclose in "?"]
18. Subfamily name- if any (SBFAM):
    .... (20)
    [If none write NONE. If proposed enclose in "?"]
19. Genus or subgroup name (GENAM):
    .... (20)
    [If none write NONE. If proposed enclose in "?"]

20. Subgenus name- if any (SBGEN):
.... (20)
[If none write NONE. If proposed enclose in ''?'']
21. Has a type culture ever been established for the type species (TCTS):
    1. No
    2. Yes
22. Where was the virus first isolated (FISOL):
    1. Africa
    2. Australasia
    3. China/Korea
    4. Eastern Europe
    5. Japan
    6. Middle East
    7. North America
    8. Pacific Islands
    9. S.E. Asia
    10. South America
    11. U.K./Ireland
    12. USSR
    !Give further details (region, town) if known:
.... (20)
23. What is the current geographical distribution of the virus- any combination of the above or W = world-wide (GDIST):
.... (20)
24. Give full references and 3-6 keywords each (REFGD):
....
....
....
....
.... (100)

SECTION 2: SPECIFIC STRAIN DATA - TAXONOMIC DESCRIPTIONS

THIS SECTION IS RESERVED FOR INFORMATION WHICH MAY BE
REQUIRED AT SOME FUTURE DATE

SECTION 3.1: NUCLEIC ACID

01. Nucleic acid type (NATYP):
    1. DNA
    2. RNA
02. Nucleic acid form (NAFOR):
    1. SS
    2. DS
    3. DS with SS regions
    4. DS with nicks
03. Genome is (GTYPE):
    1. Circular
    2. Linear

3. Circular, supercoiled
4. Twisted

04.   Number of genome segments (SEGMN):
....(2)

05.   Molecular weight of entire genome (TOTMW) $\times$ 10^44
....(5)

06.   Molecular weight(s) of genome segment(s) $\times$ 10^44 (NMW??):
    1. OLWTSEG1.. (5)
    2. OLWTSEG2.. (5)
    3. OLWTSEG3.. (5)
    4. OLWTSEG4.. (5)
    5. OLWTSEG5.. (5)
    6. OLWTSEG6.. (5)
    7. OLWTSEG7.. (5)
    8. OLWTSEG8.. (5)
    9. OLWTSEG9.. (5)
    10. OLWTSEG10 .... (5)
    11. OLWTSEG11 .... (5)
    12. OLWTSEG12 .... (5)
    13. OLWTSEG13 .... (5)
    14. OLWTSEG14 .... (5)
    15. OLWTSEG15 .... (5)

07.   Percent weight of virion, or each particle if different $\times$ 10^{-1}-1
    1. PWTV1.. (3)
    2. PWTV2.. (3)
    3. PWTV3.. (3)
    4. PWTV4.. (3)
    5. PWTV5.. (3)

08.   Sedimentation coefficient(s) S20,w
    1. S1 .... (3)
    2. S2 .... (3)
    3. S3 .... (3)
    4. S4 .... (3)
    5. S5 .... (3)
    6. S6 .... (3)
    7. S7 .... (3)
    8. S8 .... (3)
    9. S9 .... (3)
    10. S10 .... (3)
    11. S11 .... (3)
    12. S12 .... (3)
    13. S13 .... (3)
    14. S14 .... (3)
    15. S15 .... (3)

09.   Base composition % GC
    1. GC1 .... (2)
    2. GC2 .... (2)
    3. GC3 .... (2)
    4. GC4 .... (2)
    5. GC5 .... (2)
    6. GC6 .... (2)

10.   Percent base pairing of nucleic acid in virion (BPVIR):

    ....(2)

11.   Percent base pairing of nucleic acid in 1M SSC (BPFRE):

    ....(2)

12.   Nature of 5′ end of nucleic acid (NAT5E):

    1. Not known

    2. Cap

    3. Protein

    4. Other (specify) ....

13.   Nucleic acid is infective (NAIN):

    1. No

    2. Yes

14.   Genome contains unusual bases (GENUB):

    1. No

    2. Yes

15.   Genome contains unusual sugars (GENUS):

    1. No

    2. Yes

16.   Genome contains poly-A regions (GENPA):

    1. No

    2. Yes

17.   Genome able to circularize by SS end regions (GENSS):

    1. No

    2. Yes

18.   Give full references and 3-6 keywords each (REFNA):

    ....

    ....

    ....

    ....

    ....

    ....

    .... (100)

SECTION 3.2: PROTEINS

01.   Percent weight of virion (PWTVR):

    ....(2)

02.   Number of chemically distinct polypeptides
    in the virion (PPN)

    ....(2)

03.   Molecular weights of the protein(s) $\times 10^2 2$

    1. PMW1 .... (4)

     2. PMW2 . . . . (4)
     3. PMW3 . . . . (4)
     4. PMW4 . . . . (4)
     5. PMW5 . . . . (4)
     6. PMW6 . . . . (4)
     7. PMW7 . . . . (4)
     8. PMW8 . . . . (4)
     9. PMW9 . . . . (4)
    10. PMW10 . . . . (4)
    11. PMW11 . . . . (4)
    12. PMW12 . . . .(4)
    !It is possible to record the M.W.'s of up to 99 proteins.

04. Number of distinct structural proteins in the virion (STRPT):
     . . . .(2)

05. Number of different enzymes in the virion (ENZN):
     . . . .(2)

06. Number of other functional proteins (eg. haemagglutinin)
     in the virion (FUNCP):
     . . . .(2)

07. Give full references and 3-6 keywords each (REFPR):
     . . . .
     . . . .
     . . . .
     . . . .
     . . . .
     . . . .
     . . . . (100)

SECTION 3.3: LIPIDS

01. Does the virion contain lipid(s) (LIPCO):
     1. No
     2. Yes

02. Percent weight of virion (LWTVR):
     . . . .(2)

03. Number of chemically distinct lipid molecules per virion (LIPNUM): . . . .(4)

04. Molecular weight(s) of lipid molecules $\times 10^2 2$
     1. LMW1 . . . . (4)
     2. LMW2 . . . . (4)
     3. LMW3 . . . . (4)
     4. LMW4 . . . . (4)
     5. LMW5 . . . . (4)
     !It is possible to record the M.W.'s of up to 99 lipids.

05. Lipid component(s) located in the virion envelope (LIPEN):
     1. No
     2. Yes

06. Lipid component(s) located in the virion external coat (LIPCT):
     1. No
     2. Yes

07. Lipid component(s) located in some other region of the virion (LIPOT):
    1. No
    2. Yes (specify)..

08. Does the virion contain phospholipid(s) (CPXPL):
    1. No
    2. Yes

09. Does the virion contain glycolipid(s) (CPXGL):
    1. No
    2. Yes

10. Give full references and 3-6 keywords each (REFLI):
    ....
    ....
    ....
    ....
    .... (100)

SECTION 3.4: CARBOHYDRATES

01. Does the virion contain carbohydrate(s) (CACON):
    1. No
    2. Yes
    3. Not known

02. Percent weight of virion (CWTVR):
    .... (2)

03. Carbohydrate location:
    1. Not known (CNKN)
    2. Virion envelope (CENV)
    3. Virion envelope glycoproteins (CVGP)
    4. Virion envelope glycolipids (CVGL)
    5. Fiber glycoproteins (CBGP)
    6. All structural proteins glycosylated (CASPG)
    7. Some structural proteins glycosylated (CSSPG)
    8. Surface projections (CPROJ)
    9. Other (specify) COTHR) (....)
    !NOTE ANSWER NO OR YES-IF NOT KNOWN LEAVE BLANK!

04. Give full references and 3-6 keywords each (REFCA):
    ....
    ....
    ....
    ....
    .... (100)

SECTION 3.5: VIRION PHYSICOCHEMICAL PROPERTIES

01. Molecular weight(s) of virion(s) $\times$ 10^4:
    1. VRMW1 .... (5)
    2. VRMW2 .... (5)
    3. VRMW3 .... (5)
    4. VRMW4 .... (5)
    5. VRMW5 .... (5)

02. Sedimentation coefficient(s) of virion(s)-S20,w:
    1. VIRS1 .... (4)
    2. VIRS2 .... (4)
    3. VIRS3 .... (4)
    4. VIRS4 .... (4)
    5. VIRS5 .... (4)
03. Buoyant density(ies) in CsCl in g/cc $\times 10^{-2}$-2:
    1. BDCC1 .... (3)
    2. BDCC2 .... (3)
    3. BDCC3 .... (3)
    4. BDCC4 .... (3)
    5. BDCC5 .... (3)
04. Buoyant density(ies) in Cs2SO4 in g/cc $\times 10^{-2}$-2:
    1. BDCS1.. (3)
    2. BDCS2.. (3)
    3. BDCS3.. (3)
    4. BDCS4.. (3)
    5. BDCS5.. (3)
05. Buoyant density(ies) in sucrose in g/cc $\times 10^{-2}$-2:
    1. BDS1 .... (3)
    2. BDS2 .... (3)
    3. BDS3 .... (3)
    4. BDS4 .... (3)
    5. BDS5 .... (3)
06. Partial specific volume(s) of virion(s) in cc/g $\times 10^{-2}$-2:
    1. V1 .... (2)
    2. V2 .... (2)
    3. V3 .... (2)
    4. V4 .... (2)
    5. V5 .... (2)
07. Diffusion coefficient(s) of virion(s) in m²/sec $\times 10^{-13}$-13:
    1. DIF1 .... (2)
    2. DIF2 .... (2)
    3. DIF3 .... (2)
    4. DIF4 .... (2)
    5. DIF5 .... (2)
08. Isoelectric point(s) of virion(s) $\times 10^{-1}$-1:
    1. IEP1 .... (2)
    2. IEP2 .... (2)
    3. IEP3 .... (2)
    4. IEP4 .... (2)
    5. IEP5 .... (2)
09. Electrophoretic mobility(ies) of virion(s) at pH 7.0 in cm²/sec/volt $\times 10^{-6}$-6 [if determined at pH other than 7.0 answer the appropriate question in Supplementary Data Sheet]:
    1. EMOB1 .... (4)
    2. EMOB2 .... (4)
    3. EMOB3 .... (4)
    4. EMOB4 .... (4)
    5. EMOB5 .... (4)

10. Absorbance at 260nm of purified virion suspension (1mg/mℓ) $\times 10^{-2}$-2 (A260):

....(4)

11. A260/A280 ratio of purified virion $\times 10^{-2}$-2 (ABRTO):

....(3)

12. Virion infectivity affected by 20% chloroform (VSCHL):
    1. No
    2. Yes

13. Virion infectivity affected by 20% ether (VSETH):
    1. No
    2. Yes

14. Virion infectivity affected by 1M CaCl$_2$ (VSCC):
    1. No
    2. Yes

15. Virion infectivity affected by heat (56C) (VSHET):
    1. No
    2. Yes

16. Virion infectivity affected by pH-give range of stability (VSPH):
    1. No
    2. Yes

17. Virion infectivity affected by UV radiation (VSUV):
    1. No
    2. Yes

18. Virion infectivity affected by visible light (VSLIT):
    1. No
    2. Yes

19. Virion infectivity affected by SDS (VSSDS):
    1. No
    2. Yes

20. Virion infectivity affected by EDTA (VSEDT):
    1. No
    2. Yes

21. Dilution end point of virus in plant sap,$10^{-?}$-? (DEP):

....(4)

22. Thermal inactivation point of virus in plant sap (TIP):

....(4)

23. Give full references and 3-6 keywords each (REFPP):

....

....

....

....

....

....

.... (100)

SECTION 3.6: VIRION MORPHOLOGY

01. Virion shape is (VIRSH):
    1. Isometric
    2. Bacilliform [both ends of virion rounded]
    3. Bullet-shaped [one end of virion rounded, other square]
    4. Geminate

    5. Rod-rigid

    6. Rod-filamentous-straight

    7. Rod-filamentous-flexuous

    8. Elongate-prolate

    9. Spheroidal

  10. Ovoid

  11. Pleomorphic

  12. Other (specify)...

  13. Not known

02. Virion tail is (VT):

    1. Absent

    2. Contractile

    3. Noncontractile long

    4. Short

    5. Sheathed

    6. Sheathed and contractile

    7. Other (specify) ....

03. Virion possesses outer membrane(envelope) (VENV):

    1. No

    2. Lipid unit membrane

    3. Lipid bilaminar membrane

    4. Lipoprotein membrane

    5. Outer coat

    6. Other (specify) ....

04. Virion possesses core membrane (VCMEM):

    1. No

    2. Lipid unit membrane

    3. Lipid bilaminar membrane

    4. Lipoprotein membrane

    5. Other (specify) ....

05. Virion occluded in crystalline protein inclusion (VIROC):

    1. No

    2. Yes-one virion per inclusion

    3. Yes-more than one virion per inclusion

06. Nucleocapsid symmetry is (NCPSY):

    1. Cubic-icosahedral

    2. Cubic-other (specify)...

    3. Helical

    4. Other (specify) ....

07. Nucleocapsid surface projections are (NCPPR):

    1. Absent

    2. Spikes—knob-like

    3. Spikes—brush-like

    4. Filaments

    5. Other (specify) ....

08. Envelope (outer membrane) surface projections are (ENVPR):

    1. Absent

    2. Spikes

    3. Tubular

    4. Globular

    5. Spherical

6. Club-shaped
7. Filamentous
8. Other (specify) . . . .
09. Virion possesses additional tail structures:
   1. Core (TSTCR) . . . .
   2. Collar (TSTCL) . . . .
   3. Neck (TSTNK). . . . .
   4. Sheath (TSTSH) . . . .
   5. Base plate (TSTBP). . . . .
   6. Fibers (TSTFB) . . . .
   7. Spikes (TSTSP) . . . .
   8. Other(s) (specify) (. . . .)
10. Dimensions of spheroidal virion(s), diameter in nm:
   1. SVD1 . . . . (3)
   2. SVD2 . . . . (3)
   3. SVD3 . . . . (3)
   4. SVD4 . . . . (3)
   5. SVD5 . . . . (3)
11. Dimensions of elongated particles, length × breadth in nm (EPDIM):
   : . . . . (20)
12. Dimensions of geminate virions, diameter × length in nm (GVDM):
   : . . . . (2:3)
13. Dimensions of rods and filaments:
   1. Length × breadth in nm (RFLB):
   : . . . . (5:3)
   2. Diameter of axial channel in nm $\times 10^{-1}$-1 (RFDAC):
   . . . .(3)
   3. Pitch of basic helix in nm $\times 10^{-1}$-1 (PITCH):
   . . . .(2)
14. Dimensions of tailed particles:
   1. Length × breadth overall in nm (TPLB):
   : . . . . (10)
   2. Length × breadth of head in nm (TPLBH):
   : . . . . (10)
   3. Length × breadth of tail in nm (TPBLT):
   : . . . . (10)
15. Dimensions of enveloped or coated virions:
   1. Diameter-where regular-in nm (ECVD):
   . . . .(3)
   2. Length × breadth-where regular-in nm (ECVLB):
   . . . .(7)
16. Dimensions of nucleocapsid of enveloped or coated virion:
   1. Nucleocapsid diameter in nm (NCPD):
   . . . .(3)
   2. Length × breadth in nm (NCPLB):
   : . . . . (7)
   3. Pitch of basic helix in nm $\times 10^{-1} - 1$ (NCPBH):
   . . . .(2)
17. Envelope (outer membrane) composition:
   1. Percent wt. lipid (L):
   . . . .(2)

2. Percent wt. protein (P):

....(2)

3. Percent wt. carbohydrate (C):

....(2)

4. Percent wt. phospholipid (PHOS):

....(2)

5. Percent wt. other (specify) (O):

....(2)

!Note: If % unknown but absence/presence known answer N or Y!

18. Envelope (outer membrane) description:
    1. Thickness in nm (ENVTH):

....(2)

    2. Number of layers (LAYR):

....(2)

    3. Envelope loose-fitting (LSE):
    1. No
    2. Yes
    4. Envelope close-fitting (CLSE):
        1. No
    2. Yes
19. Number of capsomeres making up the capsid (CAPRN):

....(5)

20. Diameter of capsomeres [negative-stain EM] in nm (CAPDI):

....(2)

20. Dimensions of surface projections, length × breadth in nm (PRJDM):

:..(5)

22. Give full references and 3-6 keywords each (REFVM:

....
....
....
....
....
....
.... (100)

SECTION 3.7: ANTIGENIC PROPERTIES

01. Number of distinct antigenic determinants (ANTS):

....(2)

02. Type(s) of virion antigens:
    1. Antigens which elicit neutralizing antibodies (NEUAG):
        1. No
        2. Yes
    2. Antigens which elicit haemagglutinating antibodies (HAGAG):
        1. No
        2. Yes
    3. Antigens which elicit complement-fixing antibodies (CFAG):
        1. No
        2. Yes

    4. Antigens which elicit other (specify) antibodies (OTHAN):
      1. No
      2. Yes ....

03. Position of antigenic determinants:
    1. Surface projections (PRJAG):
      1. No
      2. Yes
    2. Envelope (outer membrane (ENVAG) :
      1. No
      2. Yes
    3. Capsid (CAPAG):
    1. No
    2. Yes
    4. Internal (INTAG):
      1. No
      2. Yes
    5. Other (specify) (OTHR):
    1. No
    2. Yes ....

04. Are monoclonal antibodies available for this virus (MONO):
    [If yes give appropriate reference in 05]
    1. No
    2. Yes

05. Give full references and 3-6 keywords each (REFAP)
    ....
    ....
    ....
    ....
    .... (100)

SECTION 3.8: VIRION BIOLOGICAL PROPERTIES

01. Purified particles or semipurified particles in dense suspensions cause:
    1. Lysis of cells (LYSIS):
      1. No
      2. Haemolysis
      3. Other(s) (specify) ....
    2. Agglutination of cells (AGG):
      1. No
      2. Haemagglutination
      3. Other(s) (specify) ....
    3. Fusion of cells (FUS):
      1. No
      2. Yes (specify) ....

02. Give full references and 3-6 keywords each (REFBP):
    ....
    ....
    ....
    ....
    .... (100)

SECTION 3.9: REPLICATION

01.   Site of adsorption to host cell (ADSIT):
      1. Not known
      2. Cell wall
      3. Pili
      4. Flagella
      5. Poles
      6. Other (specify) . . . .
02.   Mode of penetration of host cell (CEPEN):
      1. Unknown
      2. Injection of nucleic acid
      3. Engulfment of whole virion
      4. Direct penetration by whole virion
03.   Site of nucleic acid accumulation (NAACC):
      1. Unknown
      2. Nucleus
      3. Cytoplasm
      4. Both
      5. Other-prokaryotic cell
04.   Site of virion protein accumulation (PRACC):
      1. Unknown
      2. Nucleus
      3. Cytoplasm
      4. Both
      5. Other-prokaryotic cell
05.   Site of nucleocapsid assembly (NCPAS):
      1. Unknown
      2. Nucleus
      3. Cytoplasm
      4. Both
      5. Cell membrane
      6. Other-prokaryotic cell
06.   Site of intracellular virion maturation or accumulation (MATAC):
      1. Unknown
      2. Nucleus
      3. Cytoplasm
      4. Both
      5. Intracellular membrane
      6. Other-prokaryotic cell
07.   Mode of virion release (VRREL):
      1. Traversing of special channels
      2. Extrusion or budding through membranes
      3. Expulsion through gaps in ruptured cells
      4. Other (specify) . . . .
08.   Effect on host RNA synthesis (RNAEF):
      1. Not known
      2. Not affected
      3. Enhanced
      4. Partial shut-down
      5. Complete shut-down

09. Effect on host DNA synthesis (DNAEF):
    1. Not known
    2. Not affected
    3. Enhanced
    4. Partial shut-down
    5. Complete shut-down
10. Effect on host protein synthesis (PROEF):
    1. Not known
    2. Not affected
    3. Enhanced
    4. Partial shut-down
    5. Complete shut-down
11. Nucleic acid replicating enzymes present in virion:
    1. RNA-dependent RNA polymerase (RDR):
        1. No
        2. Yes
    2. DNA-dependent RNA polymerase (DDR):
        1. No
        2. Yes
    3. RNA-dependent DNA polymerase (RDD):
        1. No
        2. Yes
    4. DNA-dependent DNA polymerase (DDD):
        1. No
        2. Yes
    5. Other enzyme(s) (OTENZ):
        1. No
        2. Yes (specify) ....
12. mRNA transcribed from viral genome by (TRANS):
    1. Virion enzyme
    2. Virus specified non-virion enzyme
    3. Host enzyme
    4. Mixed enzyme
13. 5' Terminal of mRNA capped by (RNA5T):
    1. Not known
    2. Nucleotide
    3. Protein
    4. Absent
14. Virion envelope required for infectivity (ERINF):
    1. No
    2. Yes
15. Comparison of infectivity of enveloped and nonenveloped virions (CIENE):
    1. Envelope does not affect infectivity
    2. Envelope enhances infectivity
16. Virus replication has host cell nucleus dependent phase (VRCND):
    1. No
    2. Yes
17. Host-virus relationship is (HVREL):
    1. Virulent
    2. Temperate

18. Provirus state (PROST):
    1. Not known
    2. Provirus is integrated
    3. Provirus is plasmid
    4. Provirus is inducible
19. Replication affected by:
    1. Actinomycin D (RAFAD)
    2. FUDR (RAFFU)
    3. IUDR (RAFIU)
    4. 5-fluorouracil (RAF5F)
    5. Rifampicin (RAFRF)
    6. Guanidine (RAFGU)
    7. 1-methylisatin-beta-thiosemicarbazone (RAFIB)
    8. Cycloheximide (RAFCX)
    9. Chloramphenicol (RAFCL)
    10. 2-thiouracil (RAF2T)
    11. 8-azaguanine (RAF8A)
    12. Other(s) (specify) (RAFOT) (...)
20. Co-operative interactions:
    1. Genetic recombination (ITREC)
    2. Genetic complementation (ITCOM)
    3. Multiplicity reactivation (ITMUL)
    4. Nongenetic reactivation (ITNON)
    5. Phenotypic mixing (ITMIX)
    6. Genomic masking (ITGEM)
    7. Mutual exclusion (ITMUX)
    8. Defective, interfering particles (ITDEF)
    9. Transcapsidation (TTTRA)
    10. Helper function (ITHEL)
    11. Helper required in replication (ITHER)
    12. Helper required in vector transmission (ITHET)
    13. Satellitism (ITSAT)
    14. Other(s) (specify) (ITOTH) (...)
21. Give full references and 3-6 keywords each (REFRE):
    ....
    ....
    ....
    ....
    ....
    ....
    .... (100)

SECTION 3.10: VIRUS INDUCED CHANGES IN HOST CELL

01. Virus inclusion bodies are (VIRIB):
    1. Absent
    2. Present
    3. Not known
02. Virus inclusion bodies are restricted to particular tissues (VIBRE):
    1. No
    2. Yes

03. Inclusion bodies contain virus (ICOVR):
    1. No
    2. Yes
    3. Not known
04. Inclusion bodies contain (IBCON):
    1. 1 virion per body
    2. >1 virion per body, each separately enveloped
    3. >1 virion per body, several enveloped by one membrane
05. Dimensions, where regular, in nm:
    1. Diameter (INCDI):
    ....(3)
    2. Length × breadth (INCLB):
    ....(7)
06. Inclusions are enveloped (INCEN):
    1. No
    2. Yes
07. Changes induced by the virus are located in:
    1. Nucleus (INLNU)
    2. Nucleolus (INLN)
    3. Perinuclear space (INLPN)
    4. Cytoplasm (INLCY)
    5. Mitotic spindle fibers (INLMS)
    6. Chloroplasts (INLCH)
    7. Mitochondria (INLMO)
    8. Other (specify) (INLOT) (....................................)
08. Shape of inclusion bodies is:
    1. Amorphous (INSAM)
    2. Round (INSRO)
    3. Crescentic (INSCR)
    4. Pinwheel (INSPN)
    5. Paracrystalline (INSPA)
    6. Polyhedral (INSPO)
    7. Other (specify) (....)
09. Inclusion bodies contain:
    1. Subviral particles (ICOSV)
    2. Viral protein(s) (ICOVP)
    3. DNA (ICODA)
    4. RNA (ICORA)
    5. Cellular protein(s) (ICOCP)
    6. Other(s) (ICOTH) (....)
10. Viral coded antigens appear in host cell (VAAHC):
    1. In nucleus
    2. In cytoplasm
11. Cell coded antigens appear in virus infected cells (CCAVC):
    1. In nucleus
    2. In cytoplasm
    3. At cell surface
12. Give full references and 3-6 keywords each (REFIC):
    ....
    ....
    ....

....
....(100)

SECTION 3.11: HOST RANGE

01.   Host range - natural:
    1. Vertebrate (HRNVB)
    2. Invertebrate-Arthropod (HRNIA)
    3. Invertebrate-Nonarthropod (HRNNA)
    4. Angiosperm (HRNAG)
    5. Gymnosperm (HRNGY)
    6. Pteriodophyte (HRNPT)
    7. Bryophyte (HRNBR)
    8. Algae (except blue-green) (HRNAL)
    9. Basidiomycete (HRNBS)
    10. Ascomycete (HRNAS)
    11. Phycomycete (HRNPH)
    12. Fungi imperfecti (HRNIM)
    13. True bacteria (HRNBA)
    14. Rickettsia (HRNRC)
    15. Mycoplasma (HRNMY)
    16. Blue-green algae (HRNBG)
    17. Other (specify) (....)
02.   Host range - experimental:
    1. Vertebrate (HREVB)
    2. Invertebrate-Arthropod (HREIA)
    3. Invertebrate-Nonarthropod (HRENA)
    4. Angiosperm (HREAG)
    5. Gymnosperm (HREGY)
    6. Pteriodophyte (HREPT)
    7. Bryophyte (HREBR)
    8. Algae (except blue-green) (HREAL)
    9. Basidiomycete (HREBS)
    10. Ascomycete (HREAS)
    11. Phycomycete (HREPH)
    12. Fungi imperfecti (HREIM)
    13. True bacteria (HREBA)
    14. Rickettsia (HRERC)
    15. Mycoplasma (HREMY)
    16. Blue-green algae (HREBG)
    17. Other (specify) (....)
    18. Organ culture (HREOC)
    19. Cell culture (HRECC)
    20. Developing embryo in ovo (HREEM)
03.   Virus produces recognizable symptoms of infection in natural host(s) (RSINH):
    1. No
    2. Symptoms in only some hosts
    3. Symptoms in all hosts
04.   Morphological symptoms in natural host (SYMNH) 200 plain language characters
including spaces and punctuation:

....

....
....
....
....
....

05. Virus produces recognizable symptoms of infection in experimental host(s) RSIEH):
    1. No
    2. Symptoms in only some hosts
    3. Symptoms in all hosts

06. Morphological symptoms in experimental hosts (SYMEX) 500 plain language characters including spaces and punctuation:

    ....
    ....
    ....
    ....
    ....
    ....
    ....
    ....
    ....
    ....

07. Virus has wide natural host range (WNHR):
    1. No
    2. Yes

08. Virus has wide experimental host range (WEHR):
    1. No
    2. Yes

09. Virion infects following host groups:
    1. Vertebrate (VIVER)
    2. Arthropod (VIART)
    3. Nonarthropod-invertebrate (VINAI)
    4. Plant (VIPLA)
    5. Fungal (VIFUN)
    6. Bacterial (VIBAC)

10. Host range - list of genera and/or species (HOSTS) 500 plain language characters including spaces and punctuation:

    ....
    ....
    ....
    ....
    ....
    ....
    ....
    ....
    ....
    ....

11. Give full references and 3-6 keywords each (REFHR):

    ....
    ....
    ....
    ....

....
....
....(100)

SECTION 3.12: MODES OF TRANSMISSION

01. Transmission without vectors:
    1. No (TNVNO)
    2. Contact (TNVCO)
    3. Mechanical (TNVME)
    4. Eggs (TNVEG)
    5. Seed (TNVSE)
    6. Spores (TNVSP)
    7. Pollen (TNVPO)
    8. Graft (TNVGR)
    9. Other(s) (specify) (...)
02. Transmission by vectors:
    1. No (TRVNO)
    2. Insects (TRVIN)
    3. Arachnids (TRVAR)
    4. Nematodes (TRVNE)
    5. Fungi (TRVFU)
    6. Dodder (TRVDO)
    7. Other(s) (specify) (...)
03. Virus vector relationship (VVREL):
    1. Nonpersistent (hours)
    2. Semipersistent (days)
    3. Persistent (weeks)
04. Is virus circulative within vector (VCIRV):
    1. No
    2. Yes
05. Virus multiplies in vector (VMULV):
    1. No
    2. Yes
06. Vectors (list species) (VECTR) 500 plain language characters including spaces and punctuation:
    ....
    ....
    ....
    ....
    ....
    ....
    ....
    ....
    ....
    ....
07. Give full references and 3-6 keywords each (REFMT):
    ....
    ....
    ....
    ....

. . . .
. . . .
. . . .
. . . .
. . . .
. . . . (100)

Entry Date of Entry   /   /
Entered by . . . .

REFERENCES

1. **Dallwitz, M. J.,** A flexible program for generating identification keys, *Syst. Zool.,* 23, 50, 1974.
2. **Rogosa, M., Krichevsky, M. I., and Colwell, R. R.,** Method for coding data on microbial strains for computers (edition AB), *Int. J. Syst. Bacteriol.,* A, 21, 1, 1971.
3. **Taylor, R. W.,** A taxonomic guide to the ant genus *Orectognathus* (Hymenoptera: Formicidae), *Australian Division of Entomology, CSIRO,* Report No. 3, 1978.
4. **Taylor, R. W.,** New Australian ants of the genus *Orectognathus,* with summary descriptions of the twenty-nine known species, *Aust. J. Zool.,* 27, 773, 1979.
5. **Watson, L. and Milne, P.,** A flexible system for automatic generation of special-purpose dichotomous keys, and its application to Australian grass genera, *Aust. J. Bot.,* 20, 331, 1972.
6. **Dallwitz, M. J.,** User's guide to the DELTA system, *Australian Division of Entomology, CSIRO,* Report No. 13, 1980.
7. **Ackermann, H.-W., Audurier, A., Berthiaume, L., Jones, L. A., Mayo, J. A., and Vidaver, A. K.,** Guidelines for bacteriophage characterization, *Adv. Virus Res.,* 23, 1, 1978.

Chapter 9

FUTURE PROSPECTS FOR VIRAL TAXONOMY

R. E. F. Matthews

TABLE OF CONTENTS

I. INTRODUCTION

Before considering particular aspects concerning the future of viral taxonomy it may be instructive to compare briefly the state of viral taxonomy with the situation in bacteriology. The first moves towards international cooperation in developing a taxonomy for viruses were made in 1930 (Chapter 1). In the same year the International Society for Microbiology held its first meeting in Paris and constituted a Commission on Nomenclature and Taxonomy.[1] Bacteriologists moved much more rapidly than virologists and in 1947 a bacteriological Code of Nomenclature was published.

A major problem confronting bacteriologists has been the law of priority. This required extensive and frequently difficult searches of the literature merely to determine the earliest name used for a particular taxon. This problem has been overcome in part by establishing a new starting date (1 January 1980) for the nomenclature of bacteria. Approved lists of bacterial names have been compiled for descriptions up to that date. After 1 January 1980, only new names validated by publication in the *International Journal of Systematic Bacteriology* need be considered.

Although many problems in virus taxonomy remain for the future, virologists have for various reasons and particularly through some of the rules adopted in 1966 (see Chapter 1) escaped the worst difficulties in bacteriology. For example virologists escaped the problem of literature searching from the start through Rule 6 — "The law of priority shall not be observed" — and in addition by publishing an officially approved triennial summary of the state of viral taxonomy, the ICTV provides for working virologists a list and descriptions of all the approved taxa within a single volume. Continuation of this system should greatly facilitate the dissemination and use of viral taxonomy as it continues to be developed by the ICTV.

Another significant problem has been that any individual bacteriologist can publish a taxonomic proposal that becomes official as long as he abides by the international code of nomenclature, whether or not he has had any training in taxonomy. It is true that most virologists do not have any formal training in taxonomic concepts and procedures. However taxonomic proposals for viruses are considered first by a study group — a panel of experts for the viruses concerned — then by a host-oriented subcommittee, then by the Executive Committee and finally by the ICTV itself. This cooperative effort by many individuals smooths out the idiosyncracies and limitations of individuals or small groups, and is likely to produce a much more stable and consistent approach to taxonomy than one which is dependent to a significant degree on individuals.

II. VIRUS FAMILIES

A. How Many More Virus Families?

The data in Figure 5, Chapter 1, might suggest that most of the existing virus families have already been delineated. Indeed this is probably true for the viruses infecting vertebrates and higher plants. Over recent years, most newly described viruses from vertebrates or higher plants have been placed in one of the established families or plant virus groups without difficulty. Considering the number of host species and families in the vertebrates and higher plants, the number of virus families seems surprisingly small.

For the invertebrate viruses, the almost static situation over the past few years is more likely to reflect lack of knowledge. It is probable that many more families of viruses will be found to infect invertebrates.

Most viruses infecting prokaryotes have been described from enterobacteria. When other groups of prokaryotes and the archaebacteria have been studied in more detail, new types of virus are likely to be found.

Virus-like particles or viruses have been reported from over 100 species of fungi but no families of viruses infecting fungi have yet been established. We can expect several families to be delineated in the near future and more at a later date. A significant proportion of fungal isolates sampled at random contain virus-like particles. Hollings[2] estimated that at least 5000 species of fungi are probably infected with viruses. The lower plant groups — algae, byrophytes, psilophytes, lycopods, and ferns — have barely been examined at all for the presence of viruses. We can expect some new types to be found in these groups and also in a wide range of unicellular eukaryotes. Marine organisms of all host groups have also been examined only in a sporadic and superficial way.

Table 1 summarizes the number of virus families and plant groups with members infecting more than one of the major host groups. An interesting question for the future will be the extent to which further families will be found to have member viruses that infect different major host groups; or even like the *Rhabdoviridae* that have individual viruses replicating in members of two of the major host groups.

Past experience has shown that close morphological similarities between viruses from different kinds of host are very likely to reflect real relationships, as within the *Reoviridae* and *Rhabdoviridae* families. Present indications are that more such families will be found. For example various virus-like particles in fungi have morphologies quite similar to those of certain viruses infecting plants or vertebrates.[3-7] Various examples are given in Table 1 and Section III.H, Chapter 5.

However virus families infecting prokaryotes appear to be quite distinct from those infecting eukaryotes, and it seems unlikely that any virus families will emerge with members infecting both.

B. The Plant Virus Groups

The present taxonomy of viruses infecting plants remains anomolous, in that 24 of the 26 taxa are in the indefinite category of "groups".

Even such an outstanding plant virologist and critic of early taxonomic schemes as F. C. Bawden does not appear to have been, in principle, against the idea of virus families and genera. Criticizing Holmes,[8] he wrote: "The great failing of Holmes' scheme is the paucity of genera within families and the setting up of some families on insufficient and artificial grounds!"[9] However it can now be seen that many of these groups are equivalent to the family rank used by other virologists. For example, the *Geminivirus, Caulimovirus* and tomato spotted wilt virus groups each have a cluster of properties that set them well apart from other viruses. They could well be given family status just as the vertebrate virologists in 1975 elevated many of their genera to families. Even some of the small icosahedral plant

Table 1

NUMBERS OF VIRUS FAMILIES AND PLANT VIRUS GROUPS WITH MEMBERS REPLICATING IN MORE THAN ONE OF THE MAJOR HOST GROUPS

Major host group	Total families	Families also infecting				
		Vertebrate	Invertebrate	Plant	Fungal	Bacterial
Vertebrate	17	—	8	2	0	0
Invertebrate	11	8	—	2	0	0
Plant	26	2	2	—	0	0
Fungal	0	0	0	0	—	0
Bacterial	11	0	0	0	0	—

virus groups appear to be ready for family status. For example the *Tymovirus* and *Luteovirus* groups each have a set of properties that clearly demarcate them from the other small icosahedral viruses infecting plants. The seven rod-shaped plant virus groups have quite distinctive particle morphologies. Sufficient is also known about their genome strategies for some of these to be given family status.

Some of the small icosahedral plant viruses present greater difficulties in the present state of knowledge. For example representatives of the *Tombusvirus* and *Sobemovirus* groups have recently been shown to have some strikingly similar structural features in their icosahedral shells, but it is not yet clear whether or not they should be placed in a single family.

Four plant virus groups, the *Bromovirus, Cucumovirus, Ilarvirus*, and alfalfa mosaic virus groups have many properties in common including tripartite ssRNA genomes. It has recently been proposed that these four groups should constitute a single family with the name *Tricornaviridae*.[10] The *Bromovirus* and *Cucumovirus* groups are very closely related in many of their properties as are the *Ilarvirus* and alfalfa mosaic virus groups. There are significant differences between these two pairs, however, and some virologists will argue that they should remain separate. With the objective of a uniform taxonomy for all viruses in mind, the time is ripe for plant virologists to reconsider the status of many of their groups.

C. Subfamilies

Subfamilies have been used to split several complex families of viruses infecting vertebrates namely the *Poxviridae, Herpesviridae*, and *Retroviridae* (Table 3, Chapter 1). This level of division may be necessitated by the nature of the viruses within a family, but I believe that it is a device that should be used very sparingly. It adds significantly to the complexity of a classification scheme.

D. The Evolutionary Significance of Virus Families

Looking at the field of virology as a whole, it now appears that most of the established virus families, particularly those infecting vertebrates, have very distinctive sets of characters, based not only on particle morphology but also genome organization and mode of replication. Thus, at the family level, the taxonomy of viruses appears to be set for a relatively stable future.

As discussed in the next section, we have no knowledge concerning the evolutionary origins of virus families and of possible relationships between them. However there are probably few virologists who would now dispute the view that the viruses within most of the established families represent entities with a common evolutionary origin — even those with members infecting several host groups. For example, in the family *Reoviridae* there are members infecting plants and invertebrates, or vertebrates and invertebrates, but these viruses share so many basic properties that a common evolutionary origin for them seems almost certain.

In some families, such as those for the tailed bacterial viruses and the *Baculoviridae* (Chapter 6, Figure 1) it is possible to discern morphological series that may have evolutionary significance within the family.

It may soon be appropriate to expand ICTV Rule 21 along the following lines — "A family is a group of genera sharing a set of characters such that a common evolutionary origin for all the viruses within the family appears highly probable."

III. CATEGORIES ABOVE THE LEVEL OF FAMILY

As outlined in Chapter 1, the Provisional Committee for Nomenclature of Viruses (PCNV) set up in 1963, proposed a complete hierarchical system for the classification of viruses based on a set of arbitrarily chosen characters, following the suggestions of Lwoff et al.[11] This scheme was not adopted by the ICNV in 1966; and since the idea was laid to rest in Wildy's term of office as President, there has been no move within the ICTV to consider classification above the level of the Family (Chapter 1). It seems unlikely that this aspect of taxonomy will be given formal consideration for some time to come. Many families appear to be quite unrelated to each other. Consider the three virus families with dsDNA genomes and enveloped particles — *Poxviridae, Baculoviridae,* and *Herpesviridae.* If virus orders were erected, these families would certainly rate an order each.

On the other hand, some clusters of virus families and plant virus groups give indications of relatedness. Three examples might be (1) the families of tailed bacterial viruses, (2) some of the families containing enveloped viruses with ssRNA genomes, and (3) some of the groups containing rod-shaped plant viruses with ssRNA genomes. However, there is as yet no scientific basis for creating such higher taxa, nor is there a pressing practical need. A classification into families is sufficient for most needs.

However a problem remains for teachers, for those who write textbooks of virology, and for those who write the Reports of the ICTV. If all the virus families or many of them are to be considered, they have to be dealt with in *some* order however arbitrary that may be. They could be considered in alphabetical order, in order of increasing or decreasing size of genome or virus particle, or arranged in some arbitrary hierarchical manner as with the scheme of Lwoff et al.[11] where kind of nucleic acid — DNA or RNA — was the basis for the highest level of subdivision.

I have previously suggested an empirical procedure whereby the most informative way of arranging the virus families can be determined.[12] Briefly two of the most basic properties of a virus that can be expressed as single numbers are its size and the size of its genome. Genome size "sums up" many other properties of a virus. The size of the virus itself is related in a general way to the structural complexity of the particle. Larger viruses tend to have many components and complex structures while small viruses have a simple structure with few components.

If the size of the virus (measured as dry mass or as volume) is plotted against MW of the genome on a log-log-plot, a scatter diagram is obtained. Clustering of other properties can then be explored using minimum convex polygons. These have been used in various fields of biology to delineate groups based on variation in two physical characters. A minimum convex polygon is a polygon of minimal area with its vertices on data points and with no internal angles greater than 180°. Examples are provided in Figures 1 to 5 in which the molecular weight of the genome is plotted against dry mass of the virus particle using recent data.

The polygons in Figure 1 show, as would be expected, that there is substantial overlap for virus families and groups infecting the major kinds of host. The polygon for viruses infecting vertebrates is very similar to that for those infecting invertebrates, probably reflecting a single origin for many of the families infecting animals.

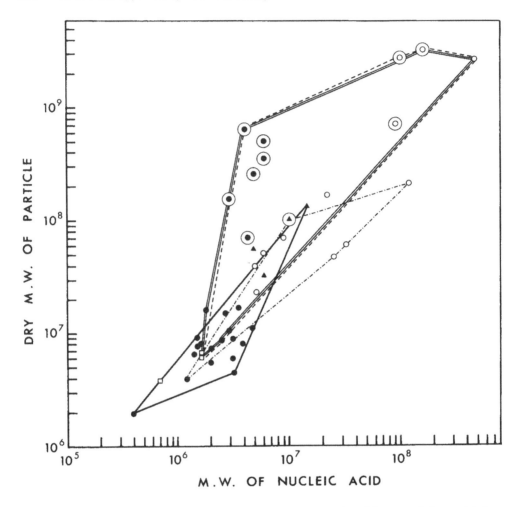

FIGURE 1.   Relationship between molecular weight of genome nucleic acid and dry mass of virus particle for a single representative from each of 43 families and groups of viruses: ○ = geometric viruses with dsDNA; ▲ = geometric viruses with dsRNA; □ = geometric viruses with ssDNA; ● = geometric viruses with ssRNA; ◯ = enveloped viruses with dsDNA; ▲ = enveloped viruses with dsRNA; ⊙ = enveloped viruses with ssRNA; ——— = polygon enclosing viruses infecting plants; - - - = polygon enclosing viruses infecting vertebrates; —●— = polygon enclosing viruses infecting bacteria; == = polygon enclosing viruses infecting invertebrates.

Many authors have used DNA versus RNA as the primary criterion for arranging virus families. Figure 2 shows that this also gives rise to polygons with substantial overlap, whereas there is much less overlap when strandedness of nucleic acid is used (Figure 3). No overlap occurs when the presence or absence of a lipoprotein envelope is used to delineate two groups (Figure 4). (I have defined an enveloped virus as one having a bounding lipoprotein bilayer membrane necessary for infectivity in normal conditions; and geometric viruses as those without such an envelope. Some members of the *Iridoviridae* have a host-derived outer membrane that is not necessary for infectivity. I have included this family as geometric).

Figure 5 shows the four nonoverlapping polygons obtained when the criteria of Figures 3 and 4 are combined.

When volume of particle rather than dry mass of particle is plotted against genome size, two nonoverlapping polygons can be formed around the enveloped and the geometric viruses very similar to those shown in Figure 4; and four nonoverlapping clusters are obtained by combining the enveloped versus geometric criterion with a division based on ds or ss nucleic acid as in Figure 5.

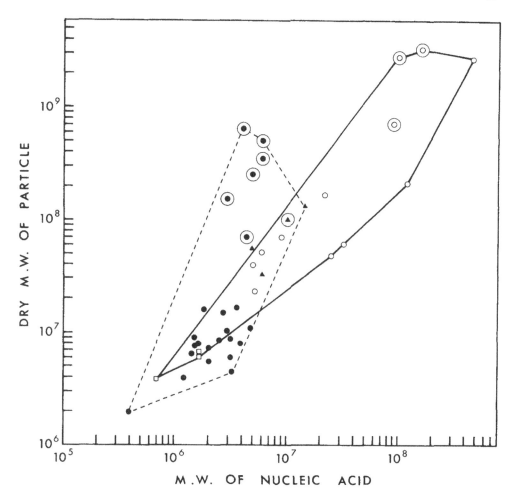

FIGURE 2. Polygons delineate DNA viruses (————) and RNA viruses (- - -). ○ = geometric viruses with dsDNA; ▲ = geometric viruses with dsRNA; □ = geometric viruses with ssDNA; ● = geometric viruses with ssRNA; ◎ = enveloped viruses with dsDNA; ⊛ = enveloped viruses with dsRNA; ⊙ = enveloped viruses with ssRNA.

The enveloped virus families have the following related properties: (1) about 40×10^6 of particle molecular weight per 10^6 molecular weight of genome nucleic acid; (2) a particle volume of about 2×10^5 nm^3 per 10^6 molecular weight of nucleic acid; (3) most have about 20 to 30% by weight of lipid, and (4) the few enveloped viruses that have been examined contain about 3 to 4 g of water per g dry matter. These properties are congruent with those of prokaryote cells.[12] By contrast the geometric viruses have particles with roughly one-tenth the anhydrous mass per unit of nucleic acid and one-twentieth the particle volume per unit of nucleic acid. They contain no lipid except the families *Iridoviridae* and *Corticoviridae* which have some lipid internal to an icosahedral shell. Last, the geometric viruses that have been examined contain 0.7 to 1.5 g of water per g dry weight. This is in the range for water bound by proteins.

Thus, I suggest that the most informative and predictive criterion for making an arbitrary primary division of the virus families and groups is the presence or absence of a bounding lipoprotein membrane (as used in Figures 1 to 4, Chapter 1) rather than the DNA versus RNA criterion usually employed.[13]

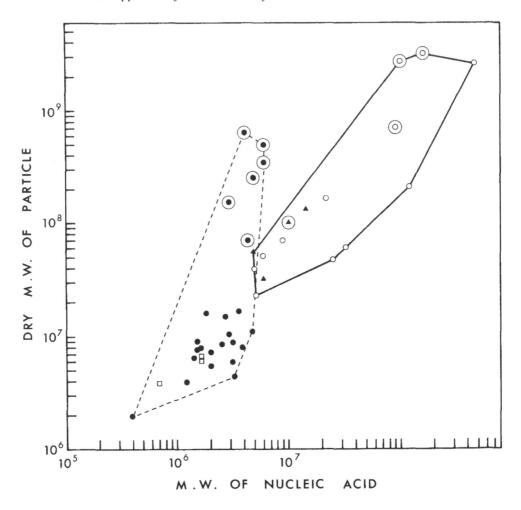

FIGURE 3.   Polygons delineate viruses with ds nucleic acid (————) or ss (- - -). ○ = geometric viruses with dsDNA; ▲ = geometric viruses with dsRNA; □ = geometric viruses with ssDNA; ● = geometric viruses with ssRNA; ⊚ = enveloped viruses with dsDNA; ⊛ = enveloped viruses with dsRNA; ⊙ = enveloped viruses with ssRNA.

IV. GENERA AND SUBGENERA

In 1975 many vertebrate virus taxa that had been given the rank of genus in 1970 were elevated to family rank. This made possible the delineation of genera within the new families as sufficient data about the viruses became available. Since 1975 there has been a steady development of genera and this can be expected to continue for some years ahead (Figure 5, Chapter 1).

All but one of the ten families of viruses infecting bacteria at present have only one genus, so that this category has not been of much value up to the present. This may be due partly to the properties of the viruses themselves, and partly to the lack of well-based data about many of the viruses. Taxonomy of viruses infecting invertebrates is in a similar position in this respect.

Among plant viruses there are two genera, *Phytoreovirus* and *Fijivirus,* both in the family *Reoviridae.* It may be that genera will also emerge fairly soon among the plant members of the *Rhabdoviridae.* Otherwise, there is no possibility of utilizing the genus category until some existing groups are converted to families or several existing groups are placed in a

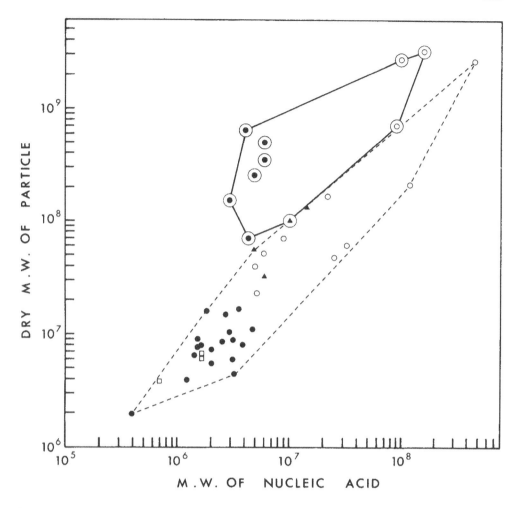

FIGURE 4. Polygons delineate enveloped viruses (———) and geometric viruses (- - -). ○ = geometric viruses with dsDNA; ▲ = geometric viruses with dsRNA; □ = geometric viruses with ssDNA; ● = geometric viruses with ssRNA; ◎ = enveloped viruses with dsDNA; ⊕ = enveloped viruses with dsRNA; ⊙ = enveloped viruses with ssRNA.

single family. The suggestion for a family *Tricornaviridae* with four genera is the first positive step in this direction.[10] The *Potyvirus* group includes a large number of viruses with similar rod-shaped particle morphology. Most are transmitted by aphids. However some viruses tentatively assigned to this group are transmitted by whiteflies, mites, or fungi. Further work may show that these viruses could usefully form several genera within a *Potyviridae* family.

The subgenus category has been employed in one difficult family of vertebrate viruses — the *Retroviridae*. The unnamed genus that includes the type C oncovirus group has four subgenera: mammalian, avian and reptilian type C oncoviruses, and the type B oncovirus group. Even more than the subfamily the subgenus complicates the taxonomy considerably, and I believe that it should be used in the future only after the closest scrutiny. It is perhaps a little incongruous that neither the subfamily nor the subgenus are defined in the current Rules of Virus Nomenclature.

V. DELINEATION OF VIRUS SPECIES

As already summarized in Chapter 1, the ICTV has developed a substantial taxonomic

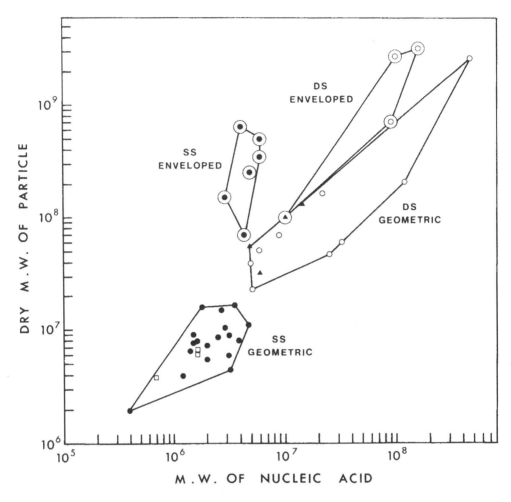

FIGURE 5. Four nonoverlapping polygons delineating clusters of virus families and groups based on presence or absence of an envelope and ds or ss nucleic acid. ○ = geometric viruses with dsDNA; ▲ = geometric viruses with dsRNA; □ = geometric viruses with ssDNA; ● = geometric viruses with ssRNA; ◎ = enveloped viruses with dsDNA; ◬ = enveloped viruses with dsRNA; ⊙ = enveloped viruses with ssRNA.

framework at the family level (assuming for the moment that most plant virus groups equate best with the family). Genera are being delineated steadily, at least for the viruses infecting vertebrates. For some medically or agriculturally important viruses, specialist groups develop and maintain methods for typing and naming variants (e.g., WHO and influenza viruses). These names do not, however, constitute a complete formally approved identification of a virus. The category equivalent to species, lying between genus and strain or variant is missing, except for the Adenoviruses discussed below. Type species have, of course, been allocated to genera, but in general these have not been clearly delineated and carry English vernacular names. Thus the taxonomy developed so far has a major gap. Until the classification and nomenclature is complete in this respect, so that the approved name together with a strain designation gives a positive identification of the virus involved, use of the approved taxonomy in journals and elsewhere is likely to remain optional and incomplete. Moreover, if we look beyond the English language to journals printed in many other languages the situation with respect to virus names remains confused to a substantial degree. Thus the problem of delineating and formally naming virus species is a major task facing the ICTV over the coming years.

From talking with colleagues I have the impression that many virologists lack any significant formal training in taxonomic ideas. For this reason I will first consider briefly some aspects of the species concept as it relates to eukaryotes and prokaryotes. This is followed by a discussion of the species concept as applied to viruses.

A. Species Concepts and their Application to Eukaryotes

The species is wide regarded as the most important taxonomic category. It is the basic unit in taxonomy, evolutionary biology, and ecology, and yet there is no universally applicable definition of what constitutes a species. There have been innumerable attempts to make such a definition. All of these can be placed in one of three groups, according to Mayr.[14]

1. The Typological Species Concept

On this view observed diversity is considered to reflect the existence of a limited number of unchanging "universals" or types. Variation merely represents imperfect manifestation of the unchanging reality of each species. In terms of practical taxonomy this leads to the definition of species based on morphological characters. The typological concept of species was used by Linneaus, who introduced the genus and species system of taxonomy we now use for plants, animals, and microorganisms. He believed that each species was created individually by a supreme being. He made an explicit distinction between species and varieties: "Today there are two kinds of difference between plants: one a true difference, the diversity produced by the all-wise hand of the Almighty, but the other, variation within an outward shell, the work of nature in a sportive mood... And so I distinguish the species of the Almighty Creator which are true from the abnormal varieties of the gardener".[15]

There are two good reasons why the typological concept of species as defined at the beginning of the previous paragraph is now widely rejected in the taxonomy of plants and animals. (1) Individuals that undoubtedly belong to the same breeding population may vary greatly in morphology due for example to sexual dimorphism, age, etc., and (2) many examples of sibling species exist. These are species which may be very similar indeed by morphological criteria but which are good biological species in that they do not interbreed.

2. The Nominalistic Species Concept

On this concept only individuals exist. Species are manmade abstractions. It is an old idea and still had adherents in this century,[16] but its influence today is negligible.

3. The Biological Species Concept

Both the above concepts of species developed in medieval times. About 200 years ago a new concept of species began to emerge based on the distinctive properties of living things rather than on those of inanimate objects. On the basis of this biologically oriented view the following definition has been derived for animal species.[14] *Species are groups of interbreeding natural populations that are reproductively isolated from other such groups.*

Although the idea of reproductive isolation is central to this definition, a taxonomist using this concept would normally take into account all available information — morphological, biochemical, genetic, and ecological in assessing the reality of any particular species.

The biological species concept is widely used by modern taxonomists but there are limitations to its application, and I will briefly mention three of these. The first is with respect to higher plants. Most of these are relatively immobile, and even when populations of the same species are separated by rather small distances gene flow between them may be minimal or absent.[17] In some angiosperm species interspecific and even intergeneric hybridization may be occurring in nature on any extensive scale. The existence of such hybrid swarms may make the concept of species very difficult to apply. The problem of hybridization blurring species boundaries is very much less common in higher animals than in plants. Raven[18] has discussed the difficulties in applying the concept of species to higher plants.

The second problem arises for paleontologists attempting to delineate fossil species. They cannot use the interbreeding and reproductive isolation criteria, but must rely primarily on morphology and on paleoecology.[19] In theory, fossils form a continuous sequence in the sedimentary record but in practice various discontinuities in the record allow species to be delineated.

A third problem for the biological concept of species arises with asexual (or uniparental) populations. This form of reproduction, brought about by a variety of mechanisms, occurs in widely different groups of organisms. With these groups, taxonomists are forced to place reliance on morphological criteria and such features as ecological role. In most such groups, there are usually well-defined morphological discontinuities that allow species to be defined.

To summarize the species situation for eukaryotes, if that is possible, we can conclude (1) that the "biospecies" concept of an interbreeding population gives an objective criterion for defining a species, but the criterion very often cannot be applied; (2) that by far the most common and prominent criteria for delineating species have been morphological, and that such criteria generally fit with the biological species concept; and (3) to quote from Davis and Heywood[20] discussing angiosperm taxonomy, "there is no universally correct definition and progress in understanding the species problem will only be reached if we concentrate on the problem of what we shall treat as a species for any particular purpose."

B. The Species Concept in Prokaryotes

The taxonomy of bacteria has been made difficult in the past as the result of two main factors: (1) a lack of knowledge concerning evolutionary relationships and (2) the consequent empirical definition of taxa using a selection of phenotypic characters (e.g., serological typing, chemical tests for enzyme activities, morphology, and pathogenicity). New improved methods for classifying bacteria have led to a continual reorganizing of taxa especially at the species level.

The problems of an empirical system were summarized by Stanier and van Niel[21] "....an empirical system is largely unmodifiable because the differential characters employed are arbitrarily chosen and usually cannot be altered to any great extent without disrupting the whole system. Its sole ostensible advantage is its greater immediate practical utility; but if the differential characters used are not mutually exclusive.... even this advantage disappears."

A system that is based even in part on phylogenetic relatedness is better than a completely arbitrary one, as Stanier and van Niel[21] point out: "Even granting that the true course of evolution can never be known and that any phylogenetic system has to be based to some extent on hypothesis, there is good reason to prefer an admittedly imperfect natural system to a purely empirical one. A phylogenetic system has at least a rational basis, and can be altered and improved as new facts come to light; its very weaknesses will suggest the type of experimental work necessary for improvement."

The ability to sequence DNAs, RNAs, and proteins has opened up new possibilities for the natural classification of microorganisms even down to the species category. Complete sequencing may not be necessary for a start to be made. Catalogues of oligonucleotide frequencies can provide useful information,[22] and a large amount of data on nucleic acid hybridization has become available. Within the well-studied *Enterobacteriaceae* it is beginning to be possible to discern trends in genome evolution and to compare species on the basis of genome properties such as gene order and the presence or absence of particular genes.[23]

C. Application of the Species Concept to Viruses

If we equate the concept of a virus species with a "distinct virus" or a "kind of virus" some viruses can readily be seen to be species. For example in the *Paramyxoviridae* the human mumps virus in the genus *Paramyxovirus* and the human measles virus in the genus *Morbillivirus*, are clearly distinct species. At the other extreme some virus families may present quite severe problems in the delineation of stable sets of species within established genera, for example the *Togaviridae, Rhabdoviridae,* and *Bunyaviridae* and also many of the viruses within such plant virus groups as the *Potyvirus* group.

1. The Problems Are Not Unique to Viruses

I hope that the brief discussion in the previous sections will have shown that there are problems in defining species for all groups of organisms. I have heard it said by some virologists that "the species concept is not applicable to viruses and should be used only for sexually reproducing organisms". As seen in the previous section there is no single "species concept" or satisfactory definition of the category species. Furthermore there are many groups of asexually or uniparentally reproducing animals, plants, and prokaryotes to which the concept has been applied usefully.

The ease with which the species concept can be applied varies widely with different families and genera of plants and animals, as it will do with viruses.

2. The Practical Necessity for Delineating Virus Species

In different areas of virology over many years numerous viral entities have been described and named. This process has been a practical necessity for virologists involved in the diagnosis of diseases of medical, veterinary, or agricultural importance, for teachers of virology and writers of textbooks, and also to a more limited extent for those involved in basic virological research on particular viruses. In fact, these viruses often equate with virus species even if the species has not been explicitly and unambiguously defined.

For example, many of the poxvirus names shown in the *Nomina conservanda* (see Chapter 1) belong to viruses which had been recognized as distinct entities for many years. The *Descriptions of Plant Viruses* published by the Commonwealth Agricultural Bureau and the Association of Applied Biologists (CMI/AAB) include about 240 viruses up to 1981. Each description is written by an expert on the particular virus. However, the editors for the series must decide whether a particular virus isolate described in the literature is merely a strain or a synonym for one that has already been described, or whether it warrants a separate description as a distinct entity. Whenever they invite a contributor to write a description for a "new virus" they are, in my view, unofficially delineating another virus species.

This excellent series of virus descriptions is widely used by plant virologists in most countries. Their existence emphasizes the point that virus species delineation is a practical necessity.

3. The Evolutionary Implications of Virus Species

As discussed earlier in this chapter, there is at present no sound phylogenetic basis for creating categories of viral taxa higher than the family, except perhaps for a few families, such as those containing the tailed bacterial viruses of enterobacteria. Many virologists now believe that the virus families are of very ancient evolutionary origin and that viruses within a given family have coevolved with their hosts and vectors. Thus many virologists would now agree that the viruses contained with a given family have had a common evolutionary origin. For example, the individual members of the *Poxviridae* have so many characters in common that it is most unlikely that they had a polyphyletic origin.

Burnet[24] suggested that herpes simplex virus and B virus *(Cercopithecid herpesvirus* 1) probably began separating in the Miocene period about 25×10^6 years ago or earlier, when the genera *Homo* and *Macaca* began to diverge from their common ancestor. Herpesviruses share many distinctive properties and it is reasonable to assume a common ancestor. Thus they have probably taken some 400×10^6 years to evolve.[25]

Full nucleotide sequences have been determined for three DNA viruses belonging to the *Papovaviridae* and infecting man, monkey, or mouse. Evolutionary trees based on (1) the fossil records of these host species, (2) the globin amino acid sequences of these species, and (3) the nucleotide sequences of three viral genes were in very good agreement.[26] This provides good evidence that the three viruses have coevolved with their hosts. On this assumption, the common ancestor for these viruses must have existed at least 80×10^6 years ago, the period at which the order Rodentia diverged from the order Primates.

In the *Tymovirus* group of small icosahedral plant viruses eleven viruses have been described in the CMI/AAB Descriptions. They are connected by a web of serological relationships (depending on the single coat protein). They have closely similar particle structures and a close and distinctive association with the chloroplasts in their intracellular replication. They show specificity with respect to the Angiosperm family they infect, their beetle vector species, and geographically, they each tend to be confined to one continent. It seems very likely that the various tymoviruses coevolved with their host families and beetle vectors since Cretaceous times.[27]

Thus, there is now good reason to believe that distinct viruses (virus species) within a virus family do have evolutionary significance. Many of the distinct viruses existing now presumably represent adaptive peaks of evolution within the virus family. Adaptation will be centered on such factors as a particular host species, or tissue or organ site within a species, or a particular host-vector combination. The ICTV guideline No. 7 for the delineation and naming of species is as follows: "Virus taxonomy at its present stage has no evolutionary or phylogenetic implications." I believe that this guideline should be altered with respect to the species category to allow possible evolutionary considerations to be given some weight in delineating virus species.

4. Criteria to be Used for Delineating Virus Species

Three generalizations can be made. *First, relatively stable characters of viruses such as particle morphology, genome organization and replication strategy will usually be of little use in delineating species.* Important specific characteristics will mainly depend on the more variable features noted below.

Second, the relative importance of various criteria will vary with different families of viruses. For example the significance to be placed on serological relationships may depend on whether or not the virus must contend directly with the immune response of the host it infects.

Third, in some families and genera the delineation of species will be relatively easy, while in others it may be very difficult. Difficulties will arise for at least two reasons — lack of sufficient data at the present time, and an evolutionary situation within the genus or family where the adaptive peaks are not clearly defined.

The current ICTV definition of a virus species is "*.... a concept that will normally be represented by a cluster of strains from a variety of sources or a population of strains from a particular source, which have in common a set or pattern of correlating stable properties that separates the cluster from other clusters of strains.*"

There are three types of property that might be used to delineate viral species — structural, serological, and biological. Space does not permit a detailed discussion of these criteria, but briefly they are as follows:

a. Structural Properties of the Genome Nucleic Acid

These might include size, base ratios, oligonucleotide frequencies, restriction enzyme maps, hybridization data, organization of repeated sequences, heteroduplex mapping, and base sequence analysis. For viruses with dsDNA, particularly those with large genomes, restriction endonuclease mapping of the DNA gives very useful information for virus classification.

A full base sequence analysis would include all the other properties listed, but it has two limitations. First, in spite of modern techniques, full base sequences have been established so far for relatively few clusters of related viruses. Nevertheless, indications are that comparison of base sequences over even a few hundred bases reflects closely the overall base sequence relationships within a related group of viruses. The second limitation is more fundamental. It was once thought that a virus classification scheme based on base sequences would be the ultimate aim.[28] It is now apparent that we cannot use these sequences to the full without other information. For example we could not locate the coat protein recognition site in tobacco mosaic virus RNA just by inspection of the base sequence of the RNA. The significance to be placed on viral nucleic acid base sequence data can be judged properly only in conjunction with a knowledge of the organization of the genome, and the functions of its parts and products.

b. Structural Properties of the Viral Protein or Proteins

For the small nonenveloped viruses the coat protein is of particular relevance to the delineation of species. Besides its intrinsic properties (size, tryptic peptide map, amino acid sequence, secondary and tertiary structure) many other measurable properties of a small virus depend largely or entirely on this protein. These include serological specificity, electrophoretic mobility, cation binding, and stability to various agents. Coat protein sequences are potentially a valuable criterion for delineation of virus species, but the information is available at present for only a few clusters of related viruses. One of the largest sets is that for tobacco mosaic virus coat protein. Gibbs[29] estimated relationships between 13 members of the *Tobamovirus* group using dissimilarity between their coat protein sequences. It is encouraging that although the coat protein gene represents only a small fraction of the tobacco mosaic virus genome, RNA hybridization studies suggest that it reflects differences in the rest of the genome.[30]

c. Serological Relationships

Serological relationships reflecting the extent of similarities between the coat protein or proteins of different geometric viruses or the envelope proteins of enveloped viruses are one of the most important and useful criteria for delineating virus species, and variants within species. This is particularly so for those viruses of medical and veterinary importance where sera from convalescent or recovered individuals provide directly a diagnostic tool. The techniques used vary widely and the significance to be placed on any set of data can be assessed adequately only by experts in a particular family. Furthermore the use of monoclonal antibodies is just beginning to revolutionize the application of serological methods to viruses.

d. Biological Relationships

Differences and similarities in the biological properties of viruses may involve functions of the viral genome not concerned with structural properties of the virus particle. Biological properties can be expected to show up differences between virus isolates that no physical or chemical method applied to the virus itself could detect. In addition, biological properties tend to be more unstable than properties of the virus particle. Many biological properties are more appropriate for distinguishing variants or strains within a species than for delineating species themselves. Biological properties that might be useful include host range, disease development in particular host species and genotypes, including cytopathological effects; methods of transmission including species and races of invertebrate vector. The plant viruses with genomes divided between two to three particles afford the possibility of testing mixtures of particles from different isolates thus giving a functional biological test of relationships. Crossing over between the genomes of viruses with dsDNA may give another genetically based criterion.

5. The Provisionally Approved Species of the Adenoviridae

Among taxonomists working with all groups of organisms there have been disagreements between "lumpers" — those who would create few taxa — (e.g., species within a particular genus) and "splitters" who would create more.

The issue was stated well by Bessey[16] — "If we multiply our species unduly we approach too near the individuals, and we are as much burdened as though we had not invented species. On the other hand, if we make too few species those we do make include so many variations from the type that confusion results again. We must steer a course between these two extremes.... We are in danger of destroying the usefulness of taxonomy in our zeal for describing every differing form as a separate species. We have lost sight of the primitive reason for the formation of species, namely, that we should have fewer things to hold in mind." I wish to raise this issue briefly with respect to the adenovirus species provisionally approved by the ICTV.

In 1981 the ICTV provisionally approved sets of virus species in the genera *Mastadenovirus* and *Aviadenovirus* (Table 1, Chapter 2). These are the first such species to be approved. The definition of these species is as follows: "*A species (formerly type) is defined on the basis of its immunological distinctiveness, as determined by quantitative neutralization with animal antisera. A species either has no cross-reaction with others or shows a homologous-to-heterologous titer ratio of >16 in both directions. If neutralization shows a certain degree of cross-reaction between two viruses in either or both directions (homologous-to-heterologous titer ratio of 8 or 16) distinctiveness of species is assumed if (1) the hemagglutinins are unrelated as shown by lack of cross-reaction on haemagglutination inhibition, or (2) substantial biophysical/biochemical differences of the DNAs exist.*"

In addition, the 34 human adenovirus species were classified according to a set of properties into five subgenera (formerly subgroups A-E). The properties were DNA homology, % G + C in DNA; tumors in newborn hamsters; cell transformation; transforming region homology; T-antigen group, and HA group.[31,32] The extent of DNA homology measured by a liquid hybridization method was the main criterion and the others generally fitted well.

It appears to me that the subgenera A-E have a set of "correlating stable properties" such as is called for in the ICTV definition of species, while the human species 1 to 34 are in fact delineated mainly on the basis of a single serological test. On the "lumper" philosophy the five subgenera might have made a set of stable species while delineation of the serotypes could have remained the responsibility of a specialist group dealing with the *Adenoviridae*.

D. Summary and Conclusions

Problems in the application of the idea of species to viruses are by no means unique to viruses. In the absence of sexual reproduction the most appropriate species concept for viruses is one depending mainly on morphological properties but, of course, including any other useful criteria that might be available in particular virus families.

As can be seen in the virological literature there is a practical necessity for delineating distinct viruses (species) wherever it is possible to do so.

It is now reasonable to suppose that the different viruses within a single virus family share a common evolutionary ancestry. Therefore in delineating virus species using structural, serological, and biological criteria, evolutionary implications should be an additional and significant factor in the process. Two groups of properties will probably emerge as generally the most important in delineating species: first, serological cross-reactions, and later — as information accumulates — amino acid sequences and three dimensional structure of proteins; second, nucleic acid hybridization data and related information — and at a later stage, genome organization and base sequence analyses.

As is well known, both serological cross-reactions and nucleic acid hybridization have their limitations. As far as serotypes are concerned (1) cross-reactions not infrequently occur between isolates that otherwise appear unrelated; (2) antisera can be chosen that have either broad or narrow specificity; (3) neutralization of infectivity tests may depend on the three dimensional structure of a relatively small part of one gene product; (4) over the next few years the use of monoclonal antibodies may revolutionize our approach to the immunological properties of viruses.

The answer obtained in nucleic acid hybridization tests depends upon the kind of test and the stringency of the conditions used. In addition the use of both serological cross-reactions and nucleic acid hybridization require arbitrary decisions regarding the "cut-off points" to be used. Nevertheless, as the information becomes available these two criteria together with any others appropriate to a particular family or genus, should allow populations of strains with a "common set or pattern of correlating stable properties" to be delineated realistically and usefully as virus species.

VI. NAMING OF VIRUS SPECIES

The delineation of virus species is surely a more important scientific exercise than giving them names once they are established; yet I believe that questions relating to the naming of viruses have raised more heat among virologists than the more basic problem of deciding what are species in any particular family. ICTV Rules 2-9, 12-16, and guidelines 2, 4 and 5 relate to the naming of species.

A. Good Qualities in a Species Epithet

I believe that the following are desirable (or necessary) properties for a species name:

- A single word
- As short as possible
- Easily pronounced
- Easily remembered
- Distinctive; not duplicated by any other virus species epithet within the same family
- The name should not have an offensive or ridiculous meaning in any major language

B. The Disadvantages of Numbers and Letters

There is no reason why a few well established and widely used number-letter names should not be used as "conserved" species names, e.g., *Polyomavirus* SV$_{40}$; and this is allowed under ICTV Rule 15.

However I believe that it would be a grave error to set up a new series of species names in a systematic way based on numbers (and/or letters). Indeed such series are forbidden by ICTV Rule 16. Such systems are easy to set up in the first instance but would lead to chaos later, as was demonstrated in the early period of bacteriological taxonomy. Some disadvantages of numbers, or numbers following letters are as follows:

- Numbers are not easily remembered.
- They are prone to error in thinking, speaking, writing, and typing (e.g., 21 easily becomes 12).
- They could lead to confusion and error when several unrelated virus genera infecting the same host have the same serial numbers.
- There would be even more confusion if two genera in the same family were given the same number series; and then species had to be rearranged between the genera, or in newly created genera. Letters from a host name followed by a series of numbers, as with the *Adenoviridae* species (Table 1, Chapter 2) will lead to the same kind of confusion as numbers. For example, what happens to the series if, to take a hypothetical

example, further work shows that *Mastadenovirus sus3* is much more closely related to *Mastadenovirus* h31 than to any other *sus* virus. This example should be considered in relation to Section G below.

In the past, newly described viruses infecting bacteria have been given names consisting of numbers, letters (particularly Greek letters) superscripts, hyphens, etc. ICTV guideline No. 4 suggests that such devices should be avoided in future virus species names. The last entries in the virus names index of the 4th Report of the ICTV[33] set out in Table 2 illustrate some of the undesirable features of letter and number series. They are mainly for bacterial viruses, and present an incoherent jumble. Number and letter series are incomplete, and unrelated viruses may be included in the same series. Correction of this situation will involve giving new species epithets to many viruses (see Section G).

C. English Vernacular Names as International Approved Species Epithets

There are a few English vernacular names that could be used directly as internationally approved species names with little or no modification, for example *Enterovirus polio* 1. However there are a number of English speaking virologists, particularly in the U.S., who believe that English vernacular names should be adopted as internationally approved species names more or less across the board. The main reasons given for this view are (1) English is the established international language of modern science, (2) the English vernacular names are entrenched in spoken and written usage, (3) new virus species names invented by the ICTV would not be used.

This question is, I believe, one of the more difficult problems facing the ICTV over the next few years. For viruses infecting vertebrates, Murphy has brought the consequences of using English vernacular names into clear focus in Table 2 of Chapter 2. Apart from contravening ICTV Rule 14 many of the names would be preposterous as internationally approved species names, for example *Alphavirus Venezuelan equine encephalomyelitis* or *Paramyxovirus finch paramyxovirus*.

Furthermore, there are other major languages that have entrenched names in vernacular use, both in speech and in virological journals in that language. Some examples are given in Table 3.

With respect to this problem it is perhaps unfortunate that the majority of virologists contributing to the development of virus taxonomy through serving on the specialist study groups of the ICTV come from English speaking countries, while virologists who will make use of that taxonomy are found in all countries.

There are two groups of user virologists who have very little representation or voice in the deliberations of the ICTV. The first of these consists of people in universities whose primary function is to teach virology to medical, veterinary, agricultural, or science students. The second group comprises the large number of virologists distributed through most countries of the world who work in medical, veterinary, or plant disease diagnostic laboratories. It is these two large groups — the teachers and diagnostic virologists — who have the greatest need for an internationally based and approved species nomenclature, one that, together with a strain designation, would give positive identification with maximum information retrieval.

I am not wishing to suggest that English names or words should not be adopted as species names. I believe that a single English or anglicized word with the properties listed in Section A above would be quite appropriate, as would an appropriate word from any other language. It would be important that such names were used in their approved form in papers, and not transliterated into the language in which the paper was written.

D. Latinization of Species Epithets

As outlined in Chapter 1, the question of a compulsory Latin binomal system for viruses

Table 2
VIRUS NAMES FROM THE END OF THE VIRUS INDEX TO THE FOURTH REPORT OF ICTV[33]

ZG/1 levivirus
ZG/2 'inovirus'
ZG/3A stylovirus
Zika flavivirus
Zlk/1 levivirus
Zirqa nairovirus
ZJ/1 levivirus
ZJ/2 'inovirus'
ZL/3 levivirus
ZS/3 levivirus
Zwogerziekte lentivirus
Zygocactus 'potexvirus'

α stylovirus
α1 myovirus
α3, α10 microviruses
α15 levivirus
β levivirus
β4 stylovirus
δA 'inovirus'
δ1 microvirus
ζ3 microvirus
η8 microvirus
λ stylovirus
μ2 levivirus
o6 microvirus
φA, φR microviruses
φAg8010 stylovirus

φC stylovirus
φCb5, φCb8r, φCb12r, φCB23r leviviruses
φCP2, φCP18 leviviruses
φCr14, φCr28 leviviruses
φD328 stylovirus
φI, φII podoviruses
φKZ myovirus
φNS11 tectivirus
φW-14 myovirus
φX174 microvirus
φ6 cystovirus
φ17, φ29, φ2042 podoviruses
φ80, φ105 styloviruses
χ stylovirus
Ω8 podovirus

2, 114, 182 podoviruses
3 myovirus
3A stylovirus
3ML stylovirus
3T+ myovirus
7s levivirus
7-7-7 stylovirus
7-11 podovirus
9/0 myovirus
06N 58P 'corticovirus'
10tur plectrovirus
11F,66F myoviruses
12S myovirus
13 myovirus
16-6-12 stylovirus
16-19 myovirus
24, 77, 107, 119, 187, 317 styloviruses
50 myovirus
63U11 'bunyavirus'
75V-2374, 75V-2621, 78V-2441 'bunyaviruses'
108/106 myovirus
121 myovirus
1307 plasmavirus
5845, 8893 myoviruses
7480b podovirus
8764 stylovirus
9266 myovirus
1φ1, 1φ3, 1φ7, 1φ9 microviruses

was one of the major causes of disagreement among the virologists involved in beginning a system of internationally approved viral taxonomy. Rule 4 now says only that ''an effort will be made towards a Latinized nomenclature'' and Rule 5 states ''existing Latinized names shall be retained wherever feasible.'' Latinized names have several advantages:

- Universality. There is no nationality problem since Latin is not used to a significant extent in the modern world. There is little tendency to transliterate Latin names, so that the possibility of transliteration errors is greatly reduced.
- Regularity with respect to spelling, because the rules for Latinization are quite precise.
- Uniformity with other biological systems of nomenclature.
- Pronounciation. Many Latinized names are easily pronounced.

Table 3
ENTRENCHED VERNACULAR NAMES IN SOME WESTERN EUROPEAN LANGUAGES FOR A SELECTION OF WELL-KNOWN VIRUS DISEASES

English	French	Spanish	German	Italian
Measles (rubeola)	Rougeole	Sarampion	Masern	Morbillo
Rubella (German measles)	Rubéole	Rubéola	Röteln	Rosolia
Mumps	Oreillons	Paperas	Mumps	Orecchioni (gattoni)
Foot-and-mouth disease	Fièvre aphteuse	Fiebre aftosa	Moul-und-klauen Seuche	Afta epizootica
Grapevine fan leaf	Court-noué de la vigne	—	Reisigkrankheit der Weinrebe	Arriciamento
Tomato bushy stunt	Rabougrissement buissonneux de la tomate	Achaparrado peludo del tomate	—	—

Certain virus diseases of medical importance have been recognized as disease entities for centuries and thus have well-entrenched names in various languages. Some examples are given in Table 3. Finding acceptable species names for the viruses causing such diseases may be particularly difficult. Latinization could be a solution. For example the virus causing mumps might be called *Paramyxovirus parotitis* (or *parotitidis*). This specific name would have an appropriate connotation for physicians.

E. Place Names
The English vernacular names of some viruses consist of names of places in various countries. Where these consist of a single word they might readily be adopted as the internationally approved name, e.g., *Alphavirus sindbis*. Other place names might be readily abbreviated or condensed, for example semliki forest virus in the *Alphavirus* genus might become *Alphavirus semliki*. I think that there would be wide support for the adoption of place names where these are appropriate, but unfortunately they are applicable to only a minority of viruses.

F. Usage of Approved Names
I believe that there are two fairly widely held misconceptions about the future usage of approved species names. If a new virus name is being coined it is often considered that this name should reflect, or remind the user of some property of the virus, otherwise the name will not be used. The point to be borne in mind here is that the virus taxonomy being developed is not just for today or this decade, but for generations ahead. Names with some particular meaning when first coined often lose that meaning with time and gain new connotations; or a name with little meaning at first, gains significance after a period.

The second misconception is that officially approved ICTV names would never replace current vernacular usage, and that therefore official names should be the current vernacular (English) or there should be no official species names. I believe that those who hold this view misconceive the proper function of approved names.

The officially approved taxonomic names for species of animals, plants and microorganisms do not replace established vernacular names either in speech or writing, in English or in any other language. There is no reason to suppose that the situation will be any different for officially approved virus species names. For example, whatever species epithet is given to measles virus by the ICTV, in vernacular use it will almost certainly still be called measles in English, virus del sarampion in Spanish, and virus de la rougeole in French.

The major purpose for internationally approved species (virus) names is that together with the genus name (and a strain identification, if necessary) they would give an unequivocal identification of the virus in question. In scientific papers, this name would need to be used only once, say in the Materials and Methods section, and perhaps the title or keyword list. Elsewhere the vernacular name would be used in a paper, as is current practice for many organisms in most journals. The use of a complete international name in the title of a paper and/or the keyword section would greatly facilitate computer-based data retrieval. The question of usage in papers is really an issue involving communication and consultation between the Executive Committee of the ICTV and the Editorial Boards of leading international journals publishing papers in the field. Several years ago I carried out a survey among the editors of the three main journals and five others who publish substantial numbers of papers in the field of virology. The answer then was quite clear. If the ICTV, acting in a democratic and responsible manner, introduces official species (virus) names, most journals would amend their instructions to authors to require the use of the official international name at least once in a paper — and that is all that is really necessary.

G. Continued Association of a Given Epithet with a Particular Species

To avoid confusion in the future, it is most important that a species naming system should allow for continuity from the first definition of a species. Once a virus species is named, that epithet should stay with that virus whatever other taxonomic rearrangements may be made later (e.g., a rearrangement of species between or within genera), unless shown to be a synonym. As noted earlier, it is here that, given time, serial number systems for species names are very likely to prove defective.

Where two or more named virus species are to be condensed into one, then the name first used should be adopted.

The idea of type cultures being held should be fostered where this is possible. Where it is not, the original description of the species should become the type.

H. Summary and Conclusions

Latinization of species epithets is no longer a major issue within the ICTV. Latinization may turn out to be a useful device for some viruses, but its universal use is no longer envisaged.

Some English vernacular names, some established numbers or letter-number combinations and some place names should form readily acceptable species epithets.

Beyond these, however, there remain two groups of viruses which present serious problems: those with entrenched English vernacular names, and those without such names, but which nevertheless have many distinct viruses within a single genus.

Problems concerning the first group really have more to do with virologists than with the viruses themselves. Three realities need to be faced by members of study groups:

1. Many multiword English vernacular names and some single word names are totally unsuitable as species epithets, either because of their unwieldiness, or because they duplicate information in the generic name, or because they would cause severe confusion in another language.
2. For viruses with such multiword English vernacular names attempts should be made to find a one word contraction or variant of the English vernacular, or a vernacular name in another language, that would be suitable as a species epithet. In some instances, an entirely new name may need to be devised.
3. It should not be insisted that all species names include some aspect that relates immediately to the virus in question. Given the time taken for a new generation of virologists to emerge, say about 30 years, the new names will soon acquire appropriate connotations — at least for important viruses.

The second problem is one to do with the viruses rather than virologists, i.e., where a relatively large number of distinct viruses without vernacular names occur in a single genus. To take as an example the 34 members of the *Mastadenovirus* genus which infect humans (Table 1, Chapter 2). The ICTV has approved the letter-number series h1-h34 as species epithets. This series will suffer from the disadvantages listed under Section VI.B but what are the alternatives? It is really quite difficult to envisage inventing 34 new species epithets (in English, Latin, or Esperanto) for such a series.

As with the problem of delineation of species, the naming problem would be made easier by a ''lumping'' philosophy. If the five subgenera in the human adenovirus series had been taken as the species it should not have been too difficult a task to find five suitable and distinctive one word names for them.

Most of the above remarks really apply at present mainly to viruses infecting vertebrates. Those studying viruses of invertebrates will probably follow the pattern set over the next few years in the vertebrate field.

As noted earlier, the CMI/AAB Descriptions of Plant Viruses delineate about 240 distinct viruses. Most of these could be readily adopted now by the ICTV as plant virus species. However there would be little point in allocating approved species epithets for these. The vernacular names should prove to be adequate until such time as genera of plant viruses are delineated. The present exceptions are those viruses in the two established plant genera *Phytoreovirus* and *Fijivirus*.

Future problems concerning specific epithets are probably most severe among the viruses infecting bacteria. Existing names are chaotic (Table 2) but the temptation will be great to retain these names or to invent new series of letters, numbers, or combinations of these.

VII. THE SPECIES-STRAIN INTERFACE — A FUTURE PROBLEM

Guideline 6 approved by the ICTV in 1981 states, ''ICTV is not concerned with the classification and naming of strains, variants or serotypes. This is the responsibility of specialist groups.'' This guideline reflected a consensus view among members of the Executive Committee of the ICTV that delineation and naming of strains is the responsibility of individual laboratories, groups of workers with related interests, or of other international bodies. The WHO system for delineating and naming influenza A types and variants within types is a good working example.

However for many families, genera, and viruses, there are no such well-established organizations dealing with variants. There are two kinds of situations where guideline 6 breaks down. First, in genera where new species are being delineated and named as with *Mastadenovirus* and *Aviadenovirus*, the definition of species for these genera is the basis for deciding which isolates are species and which merely variants and thus the study group involved must make decisions involving strains. Second, for most families and genera in which species have not yet been formally delineated there are lists of ''other members''. These lists are drawn up by appropriate study groups using criteria of their own making. The study groups decide whether any particular isolates described in the literature should be a new ''other member'' or whether it is merely a variant or synonym for a virus already named in the list. This is usually done using criteria that are not explicitly formulated and published. For some ''other members'' lists the problem has not yet been faced. Many of these lists will be shown later to contain variants, strains, and synonyms as well as distinct viruses.

Thus for many virus families in spite of guideline 6 the ICTV is inevitably involved in ''lumper-splitter'' decisions at the interface between species and strains. These decisions are often made difficult or impossible by the problem discussed in the next section.

VIII. ADEQUATE DESCRIPTIONS OF NEW VIRUSES

Descriptions of new virus isolates are frequently published without detailed data that would allow the isolate to be described as a new virus, a strain of a known virus, or an isolate identical to a previously described virus. The literature becomes cluttered with data that are essentially useless, and which frequently can cause confusion.

This problem is one of concern for all the major host groups of viruses as has been emphasized in Chapters 2 to 6. A reduction in the number of such publications in the future will require the efforts of individual virologists, their laboratories, and the editors of journals publishing virological papers. It is particularly to be hoped that editors of the smaller regional or national journals publishing descriptions of new viruses will tighten their standards with respect to the minimal amount of information that constitutes a satisfactory description of a newly isolated virus.

IX. VIRAL TERMINOLOGY

On several occasions in recent years it has been suggested that the ICTV should involve itself in providing a formal standardization of terms used in virology. Up to 1981, the Executive Committee has considered this to be an inappropriate task for the ICTV. Virology has been a very rapidly developing branch of biology over the past 20 years and this has necessitated the coining of new terms and giving new meanings to old words. When some new structure, enzyme, or process is first discovered several different words may be applied more or less simultaneously by different laboratories. At present, this synonomy and consequent confusion is sorted out by quite haphazard processes. For example, when one or a few leading virologists adopt a particular term for a particular meaning, others tend to follow suit, but the process of standardization may take years.

The problem has been brought home to me forcefully when preparing the Third and Fourth Reports of the ICTV.[33,34] The sections on individual virus families are submitted by study groups or individual virologists. Frequently the same term will be used with a different meaning or different words will be used with the same meaning by different virologists. Thus, under present conditions, the President of the ICTV has the choice of using a variety of synonomous terms (or the same term with different meanings) or of making a decision to standardize on a particular term. I believe that the second course is definitely to be preferred; but that decision requires that a set of definitions be drawn up for the terms as used in the report. Such definitions can be difficult to formulate in a way that properly reflects the current best usage among virologists. For example the following terms relating to viral coats and membranes appear in the Third Report of the ICTV:[33] *coat* (p160); *tegument; the envelope, a bilayer membrane surrounding the tegument* (p165); *envelope of unit membrane structure* (p170); *flexible envelope* (p173 and 198); *lipid as an integral part of the icosahedral shell* (p174); *membrane* (p174); *envelope* (p174, 222, 233); *outer shell; flexible inner coat* (p183); *inner shell* (p184); *protein coats* (p199). This situation has by no means been resolved in the Fourth Report of the ICTV.[34]

As another example, it is difficult at the present time to make clear and brief definitions for the terms RNA replicase, RNA transcriptase, and RNA polymerase that cover all possibilities. However, terms that cannot be defined should have no place in publications such as the ICTV reports.

I believe that the time has come for the ICTV to play a more positive role in the standardization of terminology used in all branches of virology. A step in this direction is being taken by the Standing Subcommittee for Virus Data, with the preparation of a document called "Definitions in Virology", as a requirement for computerized storage of virus data (Chapter 7).

X. AGENTS POSSIBLY RELATED TO VIRUSES

Various agents that have some properties suggesting possible relationships with viruses occur in several groups of organisms, particularly fungi and plants. It will be necessary in the future for the ICTV to keep a watch over developing knowledge about these agents.

In the fungi, unusual forms of nucleic acid are widespread. A great number of fungal species contain dsRNA molecules, often in multiple segments. Sometimes these RNAs are associated with virus-like particles, but such particles have not yet been shown to be infectious. Several species of fungi accumulate unusual forms of covalently closed circular DNA which on the basis of physical properties could be regarded as plasmids.[35] Most of these nucleic acids are not known to cause disease or have any function. At present it is not clear whether they represent cryptic viruses, plasmids, or some other kind of genetic element.

Viroids are agents causing disease in plants. They consist of a single strand of covalently closed circular RNA with a high degree of secondary structure and a molecular weight of $\simeq 120,000$. This RNA is not known to specify any proteins. Various viroids causing different diseases differ in base sequence. The relationship of these agents to viruses has not been established.

However Francki and colleagues[36] have isolated a small icosahedral virus with very interesting properties from a native tobacco in Australia. The virus contains two RNA species — a linear molecule with molecular weight of $\simeq 1.5 \times 10^6$ and a viroid-like circular RNA with $\simeq 1.2 \times 10^5$ mol wt. Both these RNAs are required for infectivity. It is possible that future work on this and related viruses may establish a link between viruses and viroids. If this occurs, viroids will clearly come into the domain of the ICTV.

XI. PROJECTED NUMBERS OF VIRUSES

There is a final question of interest in relation to the future of viral taxonomy. How many new viruses await discovery and description? It is impossible to answer this question with any precision but some indications can be obtained.

One procedure which is perhaps a simplistic one, would be to enumerate the viruses found in nature in well-studied species, e.g., man, the tobacco plant, and the honeybee, and then multiply by the number of host species. For example, there are approximately 10 viruses described from tobacco. There are approximately 200,000 angiosperm species. Therefore, on this basis of calculation there might be about 2×10^6 viruses infecting angiosperms compared with the 224 already classified by the ICTV and the 227 awaiting definite allocation to groups that have been already described. This kind of calculation probably gives a gross over-estimate of the actual number of distinct viruses. The honeybee, the tobacco plant, other well-studied agricultural species and man himself have all been affected ecologically, and therefore virologically, by man's activities. As an example, *Solanum laciniatum* is an indigenous species in the New Zealand flora, and no viruses have been found in individuals in their native habitat. However five well-known viruses were found infecting plantations of this species within 10 years of its being put into commercial use for alkaloid production.

A more conservative estimate, and perhaps a more realistic one as far as viral taxonomy over the next few decades is concerned, is to examine the rate at which new viruses have been described up to the present time. Figure 6 shows that for higher plants new viruses have been described over the past 30 years at the almost constant rate of 4.6 per year. It is reasonable to suppose that over the next two decades about 100 new viruses infecting higher plants will require taxonomic consideration. Similarly, for viruses infecting humans we can predict that on the average, about one new virus per year will be described over the next two decades.

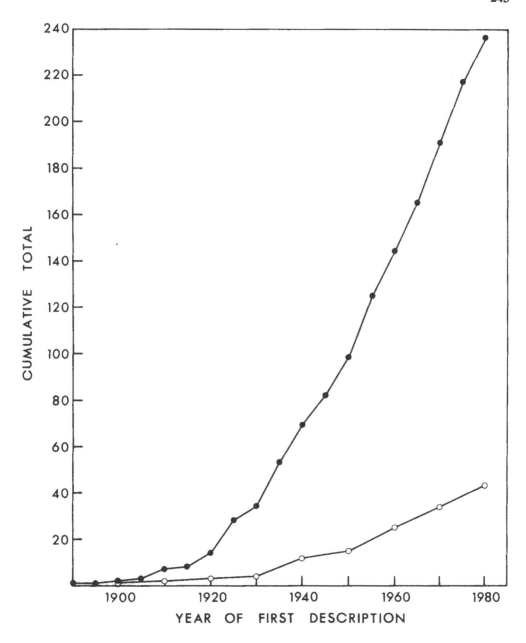

FIGURE 6. ●—● The cumulative total number of plant viruses recorded. First publication dates were taken from the 237 CMI/AAB Descriptions of Plant Viruses. ○—○ The cumulative total number of viruses described as infecting humans. Data kindly provided by Dr. W. R. Dowdle.

XII. CONCLUSIONS

Families — The family taxa set up by the ICTV are soundly based and set for a stable future. Most of these families appear to represent clusters of viruses with a common evolutionary origin. For the viruses infecting vertebrates, most of the families that will be required have probably already been delineated. In other host groups and particularly for viruses infecting invertebrates or fungi, more families will undoubtedly be needed.

Virus species — The major problem confronting the ICTV over the next few years will be the delineation and naming of further sets of virus species, particularly among viruses infecting vertebrates where a generic structure is already well advanced in several families.

Viruses infecting plants — Most of the plant virus groups set up by the ICTV provide a stable framework for the taxonomy of plant viruses. Many of these groups correspond to families, and the ICTV should move in the near future to give these family status. The CMI/AAB Descriptions of Plant Viruses have delineated about 240 distinct viruses, most of which could be readily adopted as plant virus species. The current English vernacular names would suffice for some time into the future, at least until it became apparent that genera were needed in any particular family.

Viruses infecting invertebrates, fungi, lower plants, prokaryotes, and other microorganisms — Much more detailed and extensive studies of individual viruses infecting these groups will be necessary before a really effective taxonomy can be developed.

REFERENCES

1. **Lapage, S. P., Sneath, P. H. A., Lessel, E. F., Skerman, V. B. D., Seeliger, H. P. R., and Clark, W. A., Eds.,** *International Code of Nomenclature of Bacteria,* American Society for Microbiology, Washington, D.C., 1975, 180.
2. **Hollings, M.,** Mycoviruses: viruses that infect fungi, *Adv. Virus Res.,* 22, 1, 1978.
3. **Kazama, F. Y. and Schornstein, K. L.,** Ultrastructure of a fungus herpestype virus, *Virology,* 52, 478, 1973.
4. **McDonald, J. G. and Heath, M. C.,** Rod-shaped and spherical viruslike particles in cowpea rust fungus, *Can. J. Bot.,* 56, 963, 1978.
5. **Barton, R. J. and Hollings, M.,** Purification and some properties of two viruses infecting the cultivated mushroom *Agaricus bisporus, J. Gen. Virol.,* 42, 231, 1979.
6. **Schnepf, E., Soeder, C. J., and Hegewald, E.,** Polyhedral viruslike particles lysing the aquatic phycomycete *Aphelidium* sp., a parasite of the green alga *Scenedesmus armatus, Virology,* 42, 482, 1970.
7. **Lisa, V., Boccardo, G., and Milne, R. G.,** Double-stranded ribonucleic acid from carnation cryptic virus, *Virology,* 115, 410, 1981.
8. **Holmes, F. O.,** *Handbook of Phytopathogenic Viruses,* Burgess Publishing Co., Minneapolis, 1939, 221.
9. **Bawden, F. C.,** *Plant Viruses and Virus Diseases,* 2nd ed., Chronica Botanica, Waltham, Mass., 1943, 294.
10. **van Vloten-Doting, L., Francki, R. I. B., Fulton, R. W., Kaper, J. M., and Lane, L. C.,** Tricornaviridae - a proposed family of plant viruses with tripartite, single-stranded RNA genomes, *Intervirology,* 15, 198, 1981.
11. **Lwoff, A., Horne, R., and Tournier, P.,** A system of viruses, *Cold Spring Harbor Symp. Quant. Biol.,* 27, 51, 1962.
12. **Matthews, R. E. F.,** A classification of virus groups based on the size of particle in relation to genome size, *J. Gen. Virol.,* 27, 135, 1975.
13. **Melnick, J. L.,** Taxonomy of viruses, 1980, *Prog. Med. Virol.,* 26, 214, 1980.
14. **Mayr, E.,** *Principles of Systematic Zoology,* McGraw-Hill, New York, 1969, 428.
15. **Linneaus, C.,** *Critica Botanica,* 1737.
16. **Bessey, C. E.,** The taxonomic aspect of the species question, *Am. Nat.,* 42, 218, 1908.
17. **Levin, D. A.,** The nature of plant species, *Science,* 204, 381, 1979.
18. **Raven, P. H.,** Systematics and plant population biology, *Syst. Bot.,* 1, 284, 1977.
19. **Sylvester-Bradley, P. C., Ed.,** *The Species Concept in Palaeontology,* The Systematics Association Publication Number 2, 1956, 145.
20. **Davis, P. H. and Heywood, V. H.,** *Principles of Angiosperm Taxonomy,* Oliver and Boyd, Edinburgh, 1963, 556.
21. **Stanier, R. Y. and van Niel, C. B.,** The main outlines of bacterial classification, *J. Bacteriol.,* 42, 437, 1941.
22. **Balch, W. E., Fox, G. E., Magrum, L. J., Woese, C. R., and Wolfe, R. S.,** Methanogens: reevaluation of a unique biological group, *Microbiol. Rev.,* 43, 260, 1979.
23. **Riley, M. and Anilionis, A.,** Evolution of the bacterial genome, *Ann. Rev. Microbiol.,* 32, 519, 1978.
24. **Burnet, F. M.,** *Principles of Animal Virology,* Academic Press, New York, 1955, 490.

25. **Wildy, P.,** Herpes: history and classification, in *The Herpesviruses,* Kaplan, A. S., Ed., Academic Press, New York, 1973.
26. **Soeda, E., Maruyama, T., Arrand, J. R., and Griffins, B. E.,** Host-dependent evolution of three papovaviruses, *Nature (London),* 285, 165, 1980.
27. **Matthews, R. E. F.,** *Plant Virology,* 2nd ed., Academic Press, New York, 1981, 897.
28. **Gibbs, A.,** Plant virus classification, *Adv. Virus Res.,* 14, 263, 1969.
29. **Gibbs, A.,** How ancient are the tobamoviruses? *Intervirology,* 14, 101, 1980.
30. **Palukaitis, P. and Symons, R. H.,** Nucleotide sequence homology of thirteen tobamovirus RNAs as determined by hybridisation analysis with complementary DNA, *Virology,* 107, 354, 1980.
31. **Green, M., Mackey, J. K., Wold, W. S. M., and Rigden, P.,** Thirtyone human adenovirus serotypes (Ad1-AD31) from five groups (A-E) based upon DNA genome homologies, *Virology,* 93, 481, 1979.
32. **Wadell, G., Hammarskjold, M.-L., Winberg, G., Varsanyi, T. W., and Sundell, G.,** Genetic variability of adenoviruses, *Ann. N.Y. Acad. Sci.,* 354, 16, 1980.
33. **Matthews, R. E. F.,** Classification and nomenclature of viruses, Fourth Report of the International Committee on Taxonomy of Viruses, Karger, Basel, 1982, 199.
34. **Matthews, R. E. F.,** Classification and nomenclature of viruses: Third Report of the International Committee on Taxonomy of Viruses, Karger, Basel, 1979, 160.
35. **Lemke, P. A., Ed.,** *Viruses and Plasmids in Fungi,* Marcel Dekker, New York, 1979, 653.
36. **Gould, A. R., Francki, R. I. B., and Randles, J. W.,** Studies on encapsidated viroid-like RNA. IV. Requirement for infectivity and specificity of two RNA components from velvet tobacco mottle virus, *Virology,* 110, 420, 1981.

Printed and bound by CPI Group (UK) Ltd, Croydon, CR0 4YY

22/10/2024

01777633-0014